Earthquake Engineering for Nuclear Facilities

Masanori Hamada · Michiya Kuno
Editors

Earthquake Engineering for Nuclear Facilities

Springer

Editors
Masanori Hamada
Waseda University
Shinjuku, Tokyo
Japan

Michiya Kuno
Chubu Electric Power Company
Nagoya
Japan

Original Japanese language edition published by Kajima Institute Publishing Co., Ltd. Genshiryoku Taishin Kougaku Copyright © 2014, Masanori Hamada, Satsuya Soda, Michiya Kuno.

ISBN 978-981-10-2515-0 ISBN 978-981-10-2516-7 (eBook)
DOI 10.1007/978-981-10-2516-7

Library of Congress Control Number: 2016953318

Printed on acid-free paper

This Springer imprint is published by Springer Nature
The registered company is Springer Nature Singapore Pte Ltd.
The registered company address is: 152 Beach Road, #22-06/08 Gateway East, Singapore 189721, Singapore

Preface

In 2011, a gigantic tsunami, triggered by the Great East Japan Earthquake, caused a catastrophe at Fukushima Daiichi Nuclear Power Plant in Japan. The local situation remains critical, and it will take several decades to resolve the consequences of the accident. Many victims of the nuclear disaster are still forced to live away from their homes; they bear various inconveniences, and the accident will have serious impacts on their life and health far into the future. Many social and technical problems must be solved before we can reach closure on the Fukushima Daiichi accident. The accident at Fukushima has undoubtedly damaged people's trust in nuclear technology. To restore the status of nuclear power generation to that achieved before the Fukushima accident, trust in nuclear technology must be rebuilt. To achieve this, it is essential to analyze in depth the causes of the accident and subsequent responses, and to explain these clearly to people worldwide.

Japan is one of the most earthquake- and tsunami-prone countries in the world, and has been pursuing the development and advancement of seismic engineering technologies for nuclear power plants since the 1950s, through huge efforts and investments. Seismic safety engineering of nuclear power plants in Japan has been acknowledged globally and is the pride of Japanese scientists and engineers. However, the accident at Fukushima destroyed people's confidence in nuclear technologies, including that of Japanese experts.

While this book was being written, public opinion remained divided around the question of whether it is plausible to continue using nuclear energy into the future. Not only the public, but also politicians, critics, scientists, and engineers involved in nuclear power are now split between two different attitudes toward this question. At all levels of society in Japan, it is important to take sufficient time to form a calm, scientific judgment.

Even if Japan were to discontinue nuclear power in the future, the country would still need advanced technologies for the reactor decommissioning and radioactive waste disposal. Therefore, Japan must continue to develop these technologies and accumulate knowledge for the future. However, with growing concerns about global warming and unstable petroleum prices, an increasing number of Asian countries are seeking to use nuclear power for energy security and to reduce carbon dioxide emissions.

I believe that the experience of the severe accident at Fukushima will contribute significantly to improving the safety of nuclear power plants against earthquakes and tsunamis around the world, and that the technology and knowledge that Japan has accumulated contributes greatly to the efforts of other countries to improve the safety of their nuclear power plants. The objective of writing this book was to compile, in a systematic manner, information on nuclear power plant seismic safety technology and knowledge, pass it on to future generations, and share this information with people worldwide.

Tokyo, Japan
July 2016

Masanori Hamada

Acknowledgments

I extend my sincere gratitude to the people of International Nuclear Energy Development of Japan, Co., Ltd., and the JAIF International Cooperation Center for their great support for the publication of this book.

I am also indebted and grateful to Dr. Masato Nakajima of the Central Research Institute of Electric Power Industry for his providing useful information on the technical standards for earthquake-resistant design around the world. I express my highest appreciation, as well, to Mr. Tomonori Kitaori of Chubu Electric Power Company, and to my colleagues of Hamada Laboratory, particularly to the secretary, Ms. Naoko Kiyohara, for the outstanding support in arranging text, various figures, and tables.

Contents

Part I Earthquake- and Tsunami-Resistant Design of Nuclear Power Plants

1 Introduction to Earthquake-Resistant Design of Nuclear Power Plants 3
Masanori Hamada, Michiya Kuno and Satsuya Soda

2 Assessment of Standard Seismic Motion 33
Masanori Hamada and Michiya Kuno

3 Stability Assessment of Foundation Ground and Surrounding Slope 51
Masanori Hamada and Michiya Kuno

4 Earthquake-Resistant Design of Building and Structure 65
Masanori Hamada, Michiya Kuno and Satsuya Soda

5 Earthquake-Resistant Design of Equipment and Piping 91
Masanori Hamada and Michiya Kuno

6 Earthquake-Resistant Design of Important Civil Engineering Structures 109
Masanori Hamada and Michiya Kuno

7 Tsunami-Resistant Design 125
Masanori Hamada and Michiya Kuno

8 Seismic Probabilistic Risk Assessment 143
Tatsuya Itoi, Michiya Kuno and Masanori Hamada

9 **International Standards and National Regulation on Seismic Safety Assessment** 171
Tatsuya Itoi, Michiya Kuno and Masanori Hamada

10 **Planning, Construction, Operation, and Decommission of Nuclear Power Plants** 185
Michiya Kuno and Masanori Hamada

11 **Radioactive Waste Treatment and Disposal Technique** 211
Michiya Kuno and Masanori Hamada

12 **Future Technology for the Seismic Safety of Nuclear Power Facilities** 229
Satsuya Soda, Masanori Hamada and Michiya Kuno

Part II Fundamentals of Seismology and Earthquake Engineering

13 **Earthquake, Ground Motion, and Tsunamis** 249
Masanori Hamada and Michiya Kuno

14 **Dynamic Response Analysis** 269
Masanori Hamada

15 **Earthquake-Resistant Design** 287
Satsuya Soda, Michiya Kuno and Masanori Hamada

Part I
Earthquake- and Tsunami-Resistant Design of Nuclear Power Plants

Chapter 1
Introduction to Earthquake-Resistant Design of Nuclear Power Plants

Masanori Hamada, Michiya Kuno and Satsuya Soda

Abstract The aim of the earthquake-resistant design of nuclear power plants is to retain three crucial functions, even in the event of a major earthquake and tsunami: to shut down the reactor (*shut down*), to cool down the reactor under a specified temperature and maintain a stable condition (*cool down*), and to confine so as to prevent radioactive materials from being released into the surrounding environment (*confine*). This chapter explains the mechanism of nuclear power generation and the safety assurance of nuclear power plants, and gives an overview of the earthquake-resistant design aiming to retain the three crucial functions, *shut down*, *cool down*, and *confine*. Furthermore, the damage to the Kashiwazaki-Kariwa Nuclear Power Plant caused by the 2007 Niigata-Ken Chuetsu-Oki earthquake and the catastrophic disaster that affected the Fukushima Daiichi Nuclear Power Plant as a result of the 2011 Great East Japan earthquake are described.

Keywords Mechanism of nuclear power generation · Reactor control · Nuclear fission · Principle of safety assurance · Earthquake-resistant design · Seismic classification · Accidents at the Fukushima Daiichi nuclear power plant

1.1 Mechanism of Nuclear Power Generation

Nuclear power generation technology produces electricity using the thermal energy released by nuclear fission that occurs when U235 contained in nuclear fuel absorbs neutrons. As shown in Fig. 1.1, unstable nuclei called fission products are created when a fission reaction occurs, and fast-moving neutrons are released simultaneously. These fast-moving neutrons are not readily absorbed by U235, and nuclear fission thus does not occur easily. Therefore, fast neutrons are converted to thermal

M. Hamada (✉) · S. Soda
Waseda University, Shinjuku, Tokyo, Japan
e-mail: hamada@waseda.jp

M. Kuno
Chubu Electric Power Company, Nagoya, Japan

M. Hamada and M. Kuno (eds.), *Earthquake Engineering for Nuclear Facilities*,
DOI 10.1007/978-981-10-2516-7_1

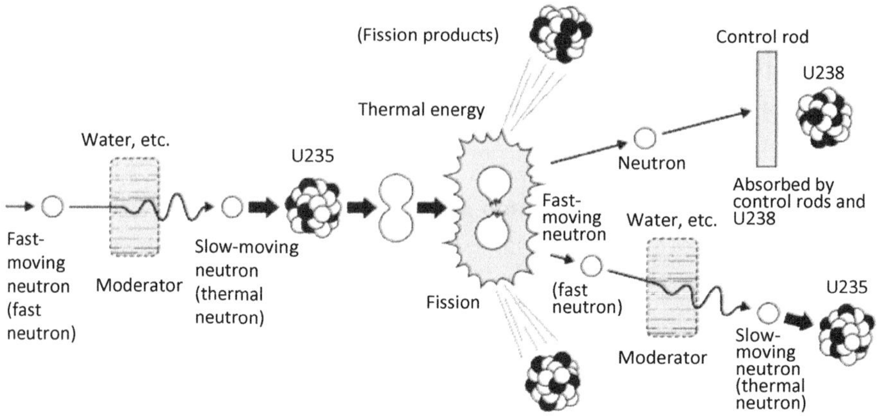

Fig. 1.1 Nuclear fission reaction [1]

neutrons that are readily absorbed by U235 by moderating the fast-moving neutrons with, for example, water (i.e., a moderator) inside the reactor. Some of the thermal neutrons are absorbed by U235, resulting in a fission reaction. In this process, thermal energy is created and fast-moving neutrons are released. Thermal energy is continuously created through the chain reaction of this phenomenon.

The nuclear fuel used for nuclear power generation contains 3–5 % of U235, which is likely to cause a fission reaction, and 95–97 % of U238, which is much less likely to cause a fission reaction. In the nuclear fission chain reaction, some of the thermal neutrons produced are absorbed by the U238 that accounts for the majority of the fuel and by the control rods containing neutron-absorbing material (such as boron and hafnium). The inside of the reactor is controlled in such a manner that the nuclear fission of one uranium atom will induce another uranium atom to undergo fission. This status is called criticality. In this status, thermal energy will not keep on increasing with the repetition of nuclear fission reactions between U235 and thermal neutrons, and the release of thermal energy can be maintained at a certain amount. In contrast, atomic bombs use a type of U235 that is enriched to almost 100 %, thus causing a chain reaction to occur immediately and triggering a massive explosion. The thermal energy released by these nuclear fission reactions is extremely large. The amount of energy produced by the reaction of 1 g of U235 is equivalent to the amount of energy obtained by burning approximately 2,000 L of oil.

The thermal energy produced by nuclear fission reactions is used to evaporate water to form steam, which is then used to rotate a turbine generator and thus produce electricity. The steam is sent to a condenser after rotating the turbine generator, where it is cooled by sea water taken in from the sea, and returned to the reactor as water. As shown in Fig. 1.2, the power generation process consisting of converting water into steam, and sending the steam to rotate the turbine generator,

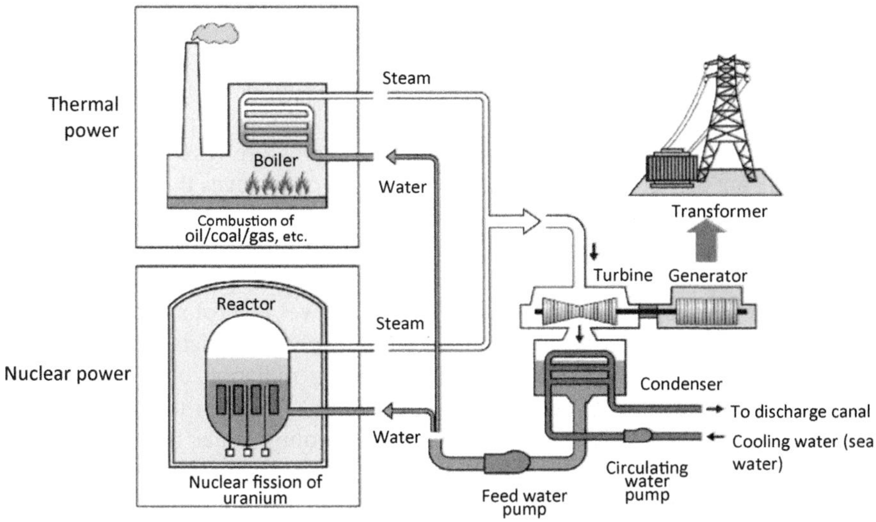

Fig. 1.2 Thermal and nuclear power generation methods [2]

is basically the same as that of thermal power generation; the difference is that the water is also used to cool the reactor in nuclear power generation.

In most nuclear power plants in Japan, the power generation system uses light water (ordinary water) as the neutron moderator and the reactor coolant, and the turbine generators are driven by the force of the water (steam). Reactors that operate using this type of power generation system are called light water reactors. Boiling water reactors (BWRs) and pressurized water reactors (PWRs) are types of light water reactors. In BWRs, turbine generators are directly rotated by the steam produced by the nuclear fission reaction in the reactor, while in PWRs, the water is heated and pressurized by thermal energy produced by the nuclear fission reaction in the reactor. The turbine generators are rotated by the steam produced in the steam generators by high-pressure hot water (for details, refer to Appendix 1.1).

1.2 Safety Assurance of Nuclear Power Plants

Concerning the safety-related functions of nuclear power plants consisting of *shut down*, *cool down*, and *confine*, this section explains the reactor control under normal conditions and when an abnormality is detected, and describes the basic principles for maintaining these three functions.

1.2.1 Reactor Control

Reactor control under normal conditions such as startup, operation, and shutdown mainly consists of controlling the pressure, water level, and output of the reactors. In the case of a BWR, the pressure inside the reactor pressure vessel is controlled by opening/closing the turbine control valves and the bypass valves through the reactor pressure control system. The reactor output is controlled by inserting/withdrawing control rods through the control rod drive system, and by changing the flow rate in the core by adjusting the rotation speed of the recirculation pump. Meanwhile, the reactor water level is secured by the reactor water level control system.

Most nuclear fission products are unstable radioactive materials, which continue to emit heat by repeatedly decaying while releasing radiation until they become stable. It is thus necessary to keep cooling the reactors even after they have stopped operation. The state in which the temperature of cooling water inside the reactor falls below 100 °C and the pressure falls to atmospheric pressure is called cold shutdown. Thus, stable shutdown of a reactor means that the reactor has successfully reached the cold shutdown state.

A nuclear reactor is designed to be shut down in an emergency (scram) (by automatic insertion of control rods into the fuel assembly within a few seconds) in the case that a particular abnormality is detected by sensors of the reactor facility. In the scram of the reactor, the turbine generators also stop generating power. One type of abnormality detected by sensors is large shaking due to an earthquake. Even during the scram of the reactor, in the case of BWRs, for example, offsite power continues to supply electricity, and as a result, normal components such as feed water pumps continue to operate normally. Unless an abnormality in the reactor pressure or water level is detected, or an abnormality in the vacuum level of the condenser in the turbine building is detected, the steam generated in the reactor continues to be sent to the turbine building, where it is cooled by the condenser.

When the power supply to normal components such as feed water pumps is stopped, or abnormalities in the reactor water level or in the vacuum level of the condenser are detected, the isolation valve of the main stream pipe, which sends the steam produced in the reactor to the turbine building, automatically closes. Under such circumstances, the condenser in the turbine building can no longer cool the steam. Therefore, the steam in the reactor pressure vessel is led to the pool water inside the containment system via the safety relief valve and is returned to the form of water. Meanwhile, this pool water is cooled by the emergency cooling system. Thus, the function to ultimately release the reactor heat to sea water is maintained. Additionally, to compensate for the reduction of the water level inside the reactor pressure vessel due to steam production, the water level is maintained by injecting water with the steam-driven pump powered by the steam continuously generated in the reactor. However, if the water level continues falling for some reason, the emergency core cooling system will activate and cool the fuel by injecting water. Figure 1.3 shows the mechanism of the emergency core cooling system.

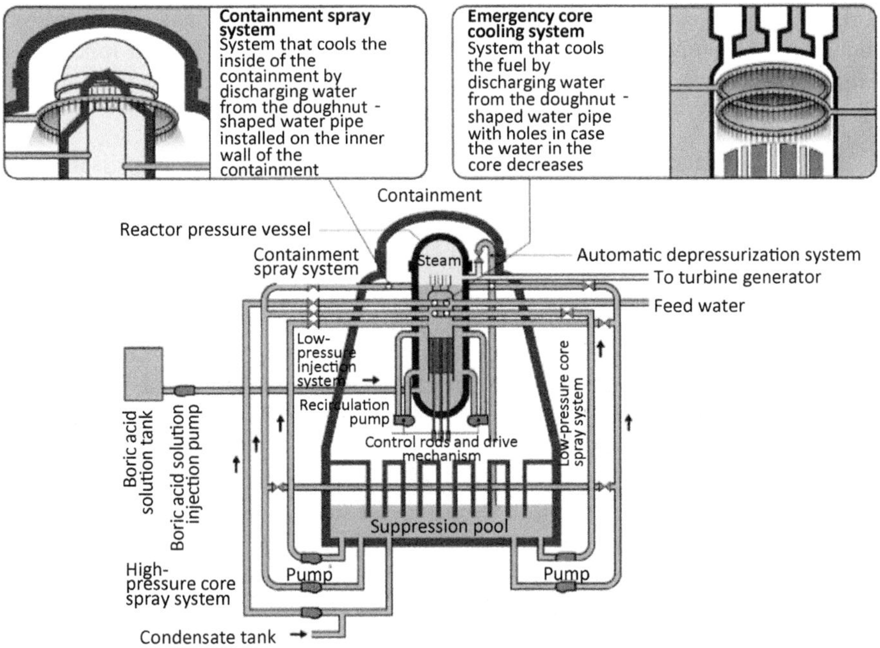

Fig. 1.3 Mechanism of an emergency reactor core cooling system (in the case of a BWR) [2]

When the offsite power supply stops owing to, for example, a blackout, and if the automatic startup feature of the diesel generators functions, a power source for the actuation of the motor-driven pumps that inject cooling water into the reactor and for the control of equipment is secured. Then the function to release heat from the reactor into the sea is maintained. If the emergency diesel generators do not actuate and power exchange from other units is not available, the unit will be in the state of station blackout in which no AC power is available. In this case, the reactor cooling function is secured by shifting the cooling system that uses the above-mentioned steam-driven pumps that do not need electricity as a driving force to inject water stored in the tank and the pool into the reactor, and the DC power source from the battery is used to control the pump's output.

1.2.2 Principles of Safety Assurance

A fundamental principle of safety assurance in nuclear power plants is to prevent radioactive materials from being released to the surrounding environment. To accomplish this goal, nuclear power plants are operated on the principle of defense in depth. Defense in depth means preparing multiple layers of countermeasures by taking second and third safety measures in addition to the first safety measure.

Even if an abnormality occurs, the release of radioactive material to the surrounding environment is prevented by maintaining the safety-related functions to *shut down*, *cool down*, and *confine*. Figure 1.4 shows these safety-related functions.

1. *Shut down*
   In the case that earthquake ground motion greater than a preset value is detected, or if an abnormality occurs in the core cooling system, such as loss of reactor coolant (water), an attempt is made to limit the heat value of the fuel by shutting down the reactor through the prompt insertion of control rods. The control rods are made of materials such as boron and hafnium that efficiently absorb neutrons.
   Additionally, if the control rods fail to actuate, the fission reaction can be stopped by alternate means, including the injection of boric acid–water, which absorbs neutrons.
2. *Cool down*
   Even if the reactor trips and the fission reaction stops, decay heat continues to be generated in the core owing to the decay of fission products. The decay heat immediately after a reactor trip is approximately 7 % of the thermal output before the reactor trip. Failure to properly remove the decay heat will lead to a rise in the fuel temperature and may lead to core damage. It is thus necessary to secure the function that the core cooling system properly removes the decay heat.
   Cooling water is injected by various kinds of pumps, including those of the low-pressure system, high-pressure system, and steam-driven system.

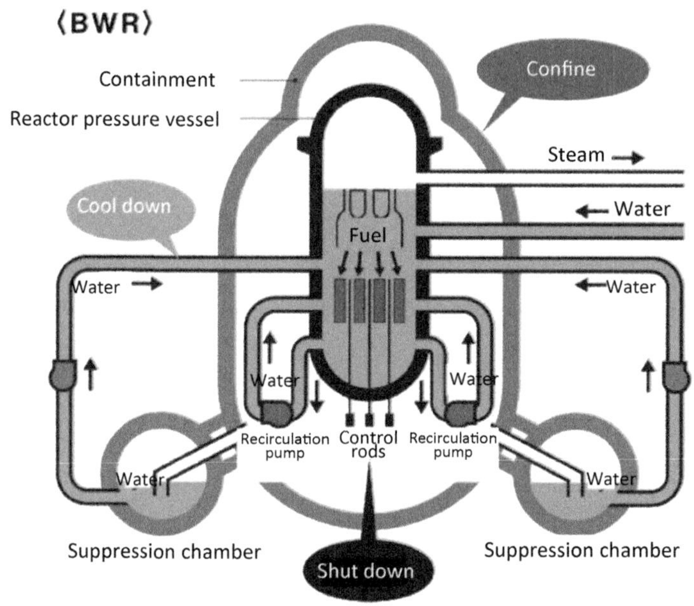

Fig. 1.4 Concept of three safety-related functions

3. *Confine*
   All possible measures must be taken to confine radioactive material and assure the safety of nuclear power plants. The majority of radioactive material is contained in fuel pellets. Part of the radioactive material released from the fuel pellets through diffusion accumulates in the fuel cladding and radioactive material accumulates in the fuel rods. Additionally, even if radioactive material leaks into the coolant in the reactor owing to damage to the fuel cladding, it will remain in the reactor pressure vessel. Moreover, if radioactive material leaks from the reactor pressure vessel, the release into the atmosphere is minimized by the confinement function of the containment and the reactor building and also by the emergency gas treatment system.

1.3 Overview of Earthquake-Resistant Design of Nuclear Power Plants

This section presents the principles and an overview of the earthquake- and tsunami-resistant design of nuclear power plants. First, seismic design forces that are determined depending on the importance of the facilities and structures are explained. An assessment of the stability of the foundation ground of the reactor buildings and structures against seismic forces is then outlined. Furthermore, the design of reactor buildings and other structures, of the components of the equipment and piping systems, and of critical civil engineering structures is explained.

1.3.1 Principles of Earthquake-Resistant Design of Nuclear Power Plants

In the earthquake-resistant design of nuclear power plants, the facilities and structures are designed according to their importance so as to prevent adverse effects on the safety functions to *shut down* and *cool down* the reactor, and to *confine* radioactive materials. The facilities and structures related to the safety functions are classified as the most important category.

The equipment and piping systems directly related to the functions to *shut down* and *cool down*, and the buildings and structures that support these components and *confine* radioactive materials, require a high degree of earthquake resistance to ensure the safety of nuclear power plants. This also applies to critical civil engineering structures that support safety important components and sea water intake facilities that cool the reactor. Furthermore, the safety of the foundation ground on which the buildings and structures are built, and the stability of the surrounding slopes must be assured. In addition, it should be confirmed that the safety-related

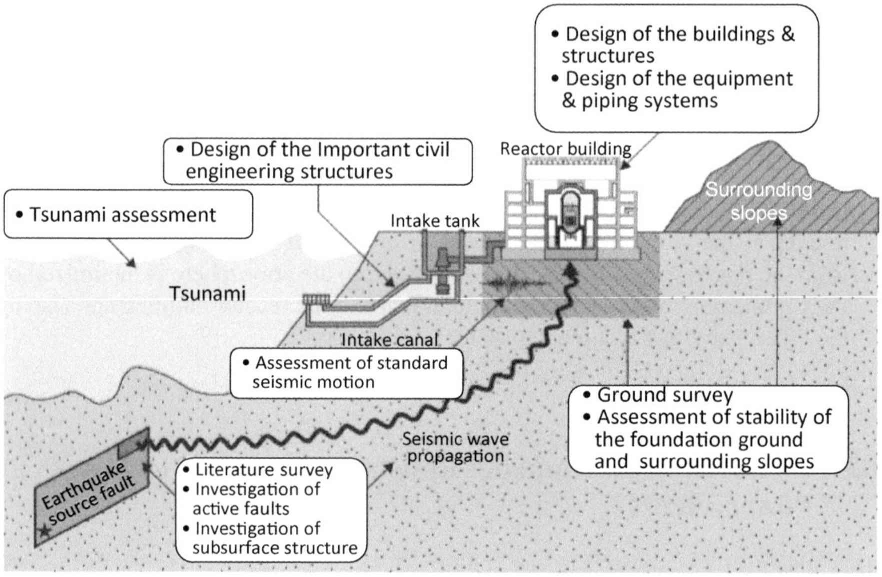

Fig. 1.5 Overview of the earthquake- and tsunami-resistant design of a nuclear power plant

functions of each component would not be lost in the event of a tsunami. Figure 1.5 is an overview of the earthquake- and tsunami-resistant design of a nuclear power plant.

1.3.2 Seismic Classification and Design Force

1.3.2.1 Seismic Classification

Nuclear facilities are classified into three classes—S, B, and C—depending on their importance, to prevent the loss of a safety function in the event of an earthquake and to prevent radiation effects on the public. The facilities with higher importance are designed by considering greater seismic forces to secure a high degree of earthquake resistance.

Table 1.1 presents the classification of facilities for the earthquake-resistant design, where the facilities related to *shut down*, *cool down*, and *confine* are classified as Class S.

Figure 1.6 shows the classification of facilities for the earthquake-resistant design of a BWR. Table 1.2 gives specific examples of facilities and structures falling in each seismic class.

Table 1.1 Classification of facilities for earthquake-resistant design

Seismic class	Classification of facilities
S	• Facilities required to shut down the reactor and cool down the core in earthquake-related events • Facilities that contain radioactive materials • Facilities directly associated with the facilities, the loss of function of which may cause the release of radioactive material to the outside • Facilities that alleviate the effects, if the loss of function of the above-stated facilities may result in an accident, and also those required to reduce the radiation effects on the public • Facilities required to support these important safety functions • Facilities required to prevent the loss of safety functions due to tsunami
B	Among the facilities with safety functions, those whose loss of function has smaller effects than Class S facilities
C	Facilities that require a safety level equivalent to general industrial facilities or public facilities other than those categorized as Class S and Class B

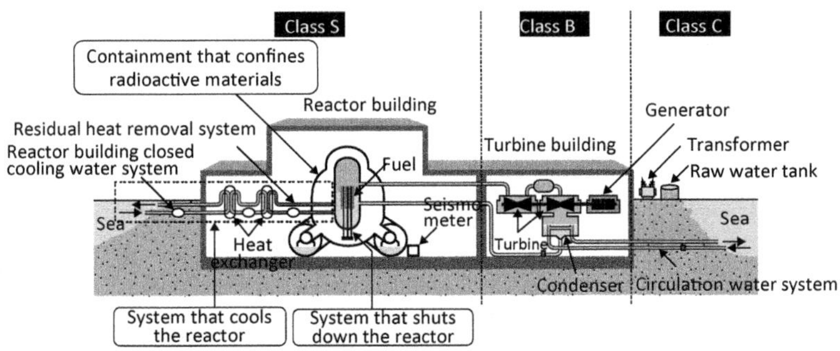

Fig. 1.6 Example of the classification of facilities for the earthquake-resistant design of a BWR [3]

Class S facilities include those related to the insertion of control rods to shut down the reactor, cooling systems such as the emergency core cooling system, the containment to confine radioactive materials, and the spent fuel storage system.

Class B facilities are those whose loss of functions has smaller effect on the safety-related functions than class S facilities.

Class C facilities include transformers and fire extinguishing systems that are not equipped with safety functions and do not contain radioactive materials; their seismic design is equivalent to that of general industrial facilities or public buildings.

The safety of a nuclear power plant is assured only when the overall integrity of the main systems directly related to the functions of the reactor facility, complementary auxiliary systems, and the structures that support them (direct support structures and indirect support structures) is secured. It is therefore necessary to conduct earthquake-resistant design according to the relationships within whole systems.

Table 1.2 Examples of the classification of facilities and structures (for a BWR)

Seismic class	Classification of facilities and structures at BWR
S	• Reactor building (secondary containment facility) • Containment • Reactor pressure vessel • Control rods and the drive system • Cooling systems such as the emergency core cooling system • Vessels, pipes, and pumps that belong to the reactor coolant pressure boundary • Spent fuel storage system, etc.
B	• Waste treatment facility • Fuel pool cleaning system • Turbine system components (turbine generators and their support systems), etc.
C	• Fresh fuel storage system • Main generator and transformers • Fire extinguishing system • Circulation water system, etc.

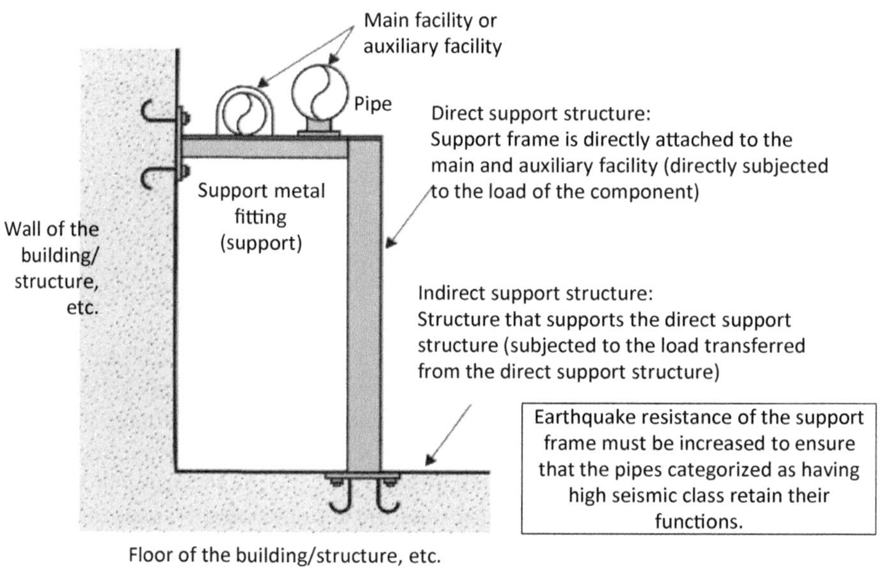

Fig. 1.7 Pipes and support frame [4]

Much of the equipment and piping is installed in buildings and civil engineering structures. If the equipment and piping systems are classified in a higher seismic class, the earthquake resistance of the frame that supports the equipment and piping systems must be designed with consideration of higher seismic forces. Figure 1.7 shows an example of the pipes and support structures.

Additionally, it must be verified that the failure of facilities of a lower seismic class will not affect the facilities of a higher seismic class.

1.3.2.2 Seismic Forces for Design According to Seismic Classification

Two types of seismic forces should be taken into consideration as shown in Table 1.3: the dynamic force, and static force obtained by weighting the seismic forces required for ordinary buildings according to the seismic classification. The dynamic force is considered for Class S facilities.

The static seismic force shall be 3.0, 1.5, and 1.0 times the seismic force required for ordinary buildings, for Class S, B, and C facilities, respectively. The dynamic forces generated by the standard seismic motion S_s and the seismic motion for elastic design S_d should be considered for the earthquake-resistant design of Class S facilities and structures.

The standard seismic motion S_s is determined according to the geological and ground conditions, and seismic activity at the plant site.

The seismic motion for elastic design S_d is derived from the standard seismic motion S_s multiplied by the coefficient α, which has been determined in the range of 0.5–0.75 for existing power plants at the discretion of each licensee. In the elastic design of structures, the deformations of the structures are proportional to the external forces. Generally, there is a good margin between the elastic limit and the ultimate strength of structures. Structures designed within the elastic range have margins before they collapse. Therefore, the seismic motion for elastic design S_d is defined for the design of reactor facilities, and the Class S facilities are designed to be approximately in the elastic range in the event of the ground motion S_d. This ensures safety against the standard seismic motion S_s, and it can be confirmed with high accuracy that the safety functions are retained.

Table 1.3 Seismic force in design to be considered for each seismic class

Seismic class	Static force	Dynamic force
S	3.0 times the seismic force required for ordinary buildings	Time-dependent inertia force generated by the standard seismic motion S_s Time-dependent inertia force generated by the seismic motion for elastic design S_d
B	1.5 times the seismic force required for ordinary buildings	–[a]
C	1.0 times the seismic force required for ordinary buildings	–

[a]For equipment that may cause resonant vibration with earthquake motion, one-half of the time-dependent inertia force caused by the seismic motion for elastic design S_d is considered

1.3.3 Flow of Earthquake-Resistant Design of Nuclear Power Plants

This section explains the outline and flow of earthquake-resistant design consisting of the assessment of the standard seismic motion S_S, the stability assessment of the foundation ground and the surrounding slopes, and the earthquake-resistant design of the buildings and structures, the equipment and piping systems and important civil engineering structures. The section also explains the design for resistance against a tsunami. Figure 1.8 shows the flow of the earthquake-resistant design of nuclear power plants ranging from fundamental surveys to safety verification.

1.3.3.1 Surveys

Surveys on past earthquakes and active faults in the areas surrounding the plant site, geological and soil conditions, and tsunami traces are conducted to evaluate the seismic motion and tsunami height.

1.3.3.2 Assessment of the Standard Seismic Motion S_s (Chap. 2)

The standard seismic motion S_s is assessed according to surveys of past earthquakes and the activities of active faults in the vicinity of the plant. Furthermore, the

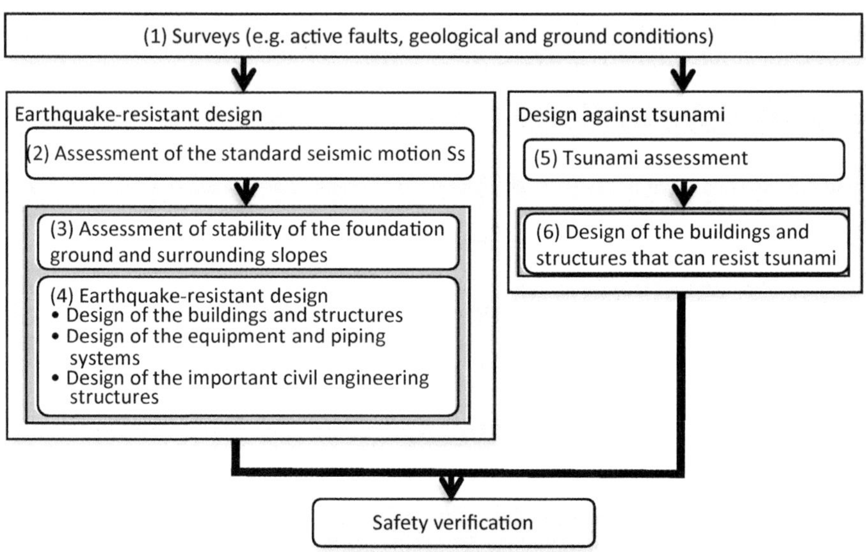

Fig. 1.8 Flow of the earthquake- and tsunami-resistant design of nuclear power plants

characteristics of the seismic wave propagation from the seismic source to the site are investigated according to the survey of geological conditions. The survey on the soil condition of the site is also important in terms of assessing the amplification of the seismic motion. There are two types of assessment of seismic motion: that based on the response spectrum and that based on the fault model. The standard seismic motion S_s is evaluated in the horizontal and vertical directions on the virtually free surface of the bedrock.

1.3.3.3 Assessment of Stability of the Foundation Ground and Surrounding Slopes (Chap. 3)

The stability of the foundation ground and surrounding slopes should be examined to prevent the serious adverse effects on the buildings and structures containing the safety-related equipments. The process of the safety assessment is as follows.

First, the strength and stiffness of the foundation ground and surrounding slopes, and the distribution of weak strata are investigated by conducting borings, elastic wave explorations, and rock tests. The foundation ground and the surrounding slopes are modeled according to the survey results. Next, the forces acting on the ground during earthquakes are dynamically calculated against the standard seismic motion S_s. Then, the stability of the foundation ground and surrounding slopes is assessed by comparing the forces acting on the foundation ground and slopes with their strength.

1.3.3.4 Earthquake-Resistant Design (Chaps. 4, 5 and 6)

The buildings and structures, the equipments and piping, and the important civil engineering structures are designed against the static force according to the seismic class.

In the design of the safety-related buildings and structures, the dimensions of structural members are assumed, and the models of buildings and structures are prepared. The dynamic force generated by the standard seismic motion S_s are calculated by dynamic response analysis. The design force is determined as the greater of the dynamic force and the static force according to the seismic class. The stress of the structural members is obtained by combining the stationary load, operation load, and seismic force.

The operability of equipment such as pumps and valves required to function during earthquakes is examined in shaking table tests.

1.3.3.5 Tsunami Assessment (Chap. 7)

It must be confirmed that a tsunami will not affect the safety-related functions of nuclear power plants. The wave source of a tsunami that is expected to have significant effects on the site is determined according to various surveys, and the tsunami height and flood depth are assessed by the numerical simulation of tsunami propagation and run-up on land.

1.3.3.6 Design of the Buildings and Structures that Can Resist Tsunamis (Chap. 7)

The buildings and structures are designed by taking into consideration the inundation height and wave forces of tsunami, calculated from the tsunami assessments.

1.4 Accidents at Nuclear Power Plants During Past Earthquakes and Subsequent Responses

This section describes accidents of the Kashiwazaki-Kariwa Nuclear Power Plant by the 2007 Niigata-Ken Chuetsu-Oki earthquake, and the Fukushima Daiichi Nuclear Power Plant and the Onagawa Nuclear Power Plant by the 2011 Great East Japan earthquake, and the post-earthquake emergency responses.

1.4.1 Status of the Kashiwazaki-Kariwa Nuclear Power Plant at the Time of the 2007 Niigata-Ken Chuetsu-Oki Earthquake

A moment magnitude (M_W) 6.8 earthquake struck off the coast of Chuetsu in Niigata Prefecture, Japan, on July 16, 2007. In the cities of Nagaoka and Kashiwazaki, and the village of Kariwa in Niigata Prefecture and also in the town of Iizuna in Nagano Prefecture, the seismic intensity measured 6 + on the Japan Meteorological Agency scale (Modified Mercalli Intensity, MMI: IX). Of the seven reactors in the Kashiwazaki-Kariwa Nuclear Power Plant, the four reactors operating at the time of the earthquake (Units 2, 3, 4, and 7) all automatically shut down safely, resulting in a cold shutdown status. The safety functions, namely to *shut down*, *cool down*, and *confine*, were secured in all reactors from Unit 1 to 7. Major damage due to the earthquake included a fire in the Unit 3 onsite transformer caused by a leak of insulation oil from a duct breached by uneven settlement of the ground, and the overflow of water in the spent fuel pool due to sloshing. However, none of the damage affected the safety-related functions.

The observed earthquake motion records show that the maximum acceleration considerably exceeded the design level. At one instant, the measured acceleration was more than twice the design acceleration. Additionally, the maximum response accelerations measured at Units 1–7 were approximately three to six times the average ground motion evaluated using the distance attenuation formula. The analysis of the observed records indicates that the large ground motions, stronger than expected were caused by the seismic source faults, and the three-dimensional geological structure under the plant site amplified the ground motion.

This earthquake provided new insights on earthquake source properties and ground amplification characteristics. In addition, insights on the safety margins of the seismic design of nuclear power plants were gained because the safety-related structures and facilities suffered no problems even though the actual ground motion largely exceeded the design ground motion. One of the factors behind these margins is considered to be the static seismic force required for the safety-related facilities and structures. For example, in the elastic design of the reactor building, the seismic force is weighted to be three times the seismic force required for ordinary buildings. The seismic force acting on the plant components during the Niigata-Ken Chuetsu-Oki earthquake is estimated to have been almost the same level as this. Therefore, each structural member of the reactor building was likely to be within its elastic range and was therefore not damaged (for details, refer to Appendix 1.2).

1.4.2 Accidents at the Fukushima Daiichi Nuclear Power Plant and the Onagawa Nuclear Power Plant at the Time of the 2011 Great East Japan Earthquake

A moment magnitude (M_w) 9.0 earthquake struck off the coast of northeast Japan on March 11, 2011. In the city of Kurihara in Miyagi Prefecture, the seismic intensity measured 7 on the Japan Meteorological Agency scale (MMI: X–). Additionally, in 37 cities, towns, and villages in the four prefectures of Miyagi, Fukushima, Ibaraki, and Tochigi, the seismic intensity measured 6 + (MMX: IX). The source region extended off the coast of Iwate to Ibaraki Prefectures, stretching approximately 500 km from north to south, and approximately 200 km from east to west. The maximum slippage of the plate boundary was estimated to be more than 50 m. This earthquake triggered a gigantic tsunami along the Pacific coast covering regions from Tohoku to Kanto, and caused devastating damage. The damage to the Fukushima Daiichi Nuclear Power Plant and the Onagawa Nuclear Power Plant imparted by this tsunami is introduced as follows.

1.4.2.1 Accident at the Fukushima Daiichi Nuclear Power Plant

When the earthquake occurred, Units 4–6 were in a shutdown state owing to periodic inspection and Units 1–3, which were operating, automatically tripped. As all six supply lines from the offsite power sources were cut off by, for example, damage to breakers and the collapse of transmission line towers, the emergency diesel generators of each unit started. However, the tsunami struck the site tens of minutes after the earthquake, flooding the sea water pumps used for cooling, emergency diesel generators, and switch boards in Units 1–5. Except for one emergency diesel generator in Unit 6, which was not flooded, all emergency diesel generators stopped operation and the plant lost all AC power sources.

As the emergency core cooling function using the AC power source was lost in Units 1–3, attempts were made to actuate the core cooling function of the reactor isolation cooling system using DC power from batteries. However, the core cooling stopped when the batteries ran out, and the cooling was thus switched to the water injection of fresh water or sea water from the fire extinguishing system line. As the pressure inside the reactor increased, this alternate water injection became impossible. The nuclear fuel in the core of each unit started to become exposed because water could not be injected into the reactor pressure vessel for a certain period of time, resulting in core meltdown.

Furthermore, a massive amount of hydrogen was generated by the chemical reaction between the zirconium contained in the fuel rod cladding that was exposed to the atmosphere and steam, and filled the containment. In addition, the loss of the cooling system (without an ultimate heat sink) due to damage to the sea water equipment caused the containment to fail. As a result, explosions occurred in Units 1 and 3 when the leaked hydrogen ignited, entirely destroying the upper floors of the reactor buildings.

The ground motion observed at the reactor buildings was reported to be approximately the same as the standard earthquake ground motion S_s determined by the *Guidelines for the Seismic Design Review of Power Reactor Facilities* in 2006. Although the damage to the structures and facilities due to the ground motion could not be clearly confirmed because of the tsunami, according to a statement made by Tokyo Electric Power Co., Inc., "it is considered that major facilities maintained safety-related functions during and immediately after the earthquake."

The maximum tsunami height assessed at the Fukushima Daiichi Nuclear Power Plant before the earthquake was O.P. + 6.1 m (O.P.: reference level used for the construction of Onahama Port, calculated by deducting 0.727 m from the average sea level of Tokyo Bay T.P.). However, the height of the tsunami that actually struck the plant was reported to be O.P. + 13.1 m. The inundation height onsite was O.P. + 15.5 m at the sites of Units 1–4, and O.P. + 14.5 m on the side of Units 5 and 6. Most of the power sources and functions related to cooling were lost owing to flooding caused by the tsunami (for details, refer to Appendix 1.3).

1.4.2.2 Status of the Onagawa Nuclear Power Plant

When the 2011 Great East Japan earthquake struck, Units 1 and 3 were in operation and Unit 2 was in the process of starting up the reactor at the Onagawa Nuclear Power Plant. All reactors automatically tripped owing to the acceleration exceeding the operational limit. Among the five supply lines from offsite power sources, Matsushima Line No. 2 continued to function. Thus, Units 2 and 3 continued to receive offsite power, and onsite power was secured. In Unit 1, although offsite power could not be received as the startup transformer tripped in response to the earthquake, power was secured by the actuation of the emergency diesel generator.

In Unit 2, the reactor was subcritical and the reactor water temperature was below 100 °C before the earthquake struck. Therefore, the reactor reached the cold shutdown status by shifting the reactor mode switch to the shutdown position. Units 1 and 3 were successfully brought to cold shutdown within approximately 12–13 h.

The ground motion measured at the reactor buildings during the earthquake was estimated to be at approximately the same level as that considered in the design. The confirmed damage due to the earthquake motion included a fire at the high-pressure power panel in the turbine building of Unit 1, and the opening of the blowout panel in the reactor building of Unit 3. However, the safety-related components in the reactor facilities were confirmed to have successfully functioned during and after the earthquake.

The maximum height of the tsunami confirmed at the Onagawa Nuclear Power Plant was O.P. + approximately 13 m, and did not exceed the ground surface elevation of the Onagawa plant site (O.P. is approximately 13.8 m; the ground level subsided by approximately 1 m owing to the tectonic deformation of the earthquake, which is taken into consideration in the calculation of this value). Although the reactor building was partly flooded by the influx of water flowing through the underground trenches and thus the functions of parts of the cooling system equipment were affected, the cooling function was successfully maintained by other cooling systems. In addition, although an oil tank collapsed, it did not affect safety (for details, refer to Appendix 1.4).

1.4.3 Responses After the Earthquakes

The earthquake ground motions observed at the Kashiwazaki-Kariwa Nuclear Power Plant during the Niigata-Ken Chuetsu-Oki earthquake largely exceeded the design ground motion. Furthermore, the fire was caused by uneven settlement of the ground. In response to these phenomena and the accident, initiatives are being implemented at multiple nuclear power plants, involving reevaluation of source properties and ground amplification characteristics. This is done by investigating the geological structures and analyzing observation records of earthquake ground motions, reevaluating and improving the reliability of Class B and C facilities such as transformers, and increasing the reliability of emergency facilities, including the emergency response center.

The lessons learned from the Fukushima Daiichi Nuclear Power Plant accident included that safety functions were lost owing to the station blackout and the loss of the ultimate heat sink as previously stated, and also that the subsequent evolution of a severe accident could not be prevented. In light of this accident, the design basis for the prevention of severe accidents was reinforced, and new standards that could withstand a major accident or terrorist attack were established. In June 2013, *Regulations Regarding the Standards for the Location, Structure and Components*

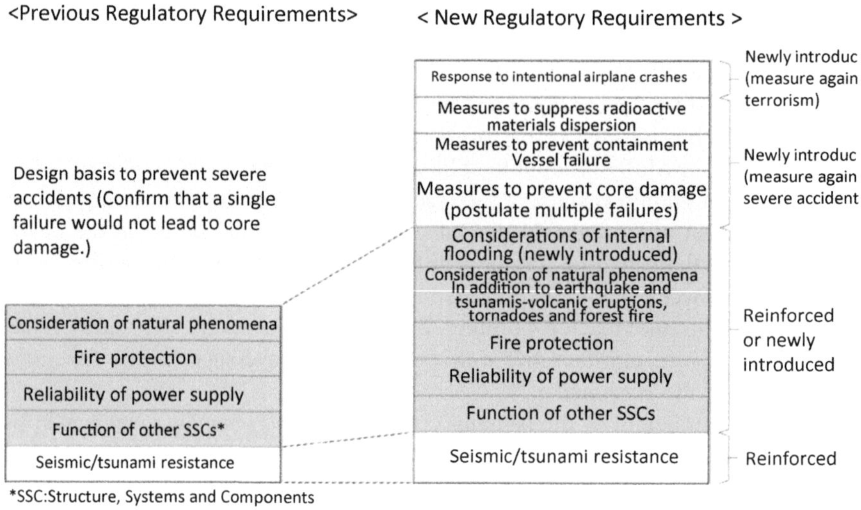

Fig. 1.9 Comparison between previous and new standards [5]

of Commercial Power Reactors and Their Adjacent Facilities (hereafter referred to as *New Regulatory Standards*) were established to reflect these lessons learned as shown in Fig. 1.9, and they came into force in July 2013.

The major items reflecting the lessons learned are as follows.

1. Reinforcement of countermeasures against tsunamis
2. Reinforcement of countermeasures against earthquakes

   - Clarification of the standards of displacement and deformation of the ground in addition to the vibration due to an earthquake
   - Specification of the standards for defining active faults
   - Evaluation of the basic ground motion based on a three-dimensional survey of the geological structure

3. Reinforcement of countermeasures against other natural phenomena (e.g., volcanic eruptions and tornadoes)
4. Countermeasures against events other than natural phenomena (e.g., airplane accidents)
5. Measures that prevent core damage and containment failure, and measures that limit the diffusion of radioactive material.

Appendix 1.1: Boiling Water Reactor and Pressurized Water Reactor

1. BWR
   In a BWR, electricity is generated by turbine generators driven by the steam produced by boiling the reactor water (light water). First, primary coolant (light water) is brought into the reactor pressure vessel. The water boils among the fuel rods, evaporates, and produces steam. The steam is then directly sent to the turbine generators and used to rotate the generators. After rotating the generators, the steam is sent to the condenser, where it is cooled and returned to water, which is sent to the reactor pressure vessel again as primary coolant. The water flowing into the reactor pressure vessel is circulated by the recirculation pumps, and the reactor output is controlled by increasing/decreasing the recirculating flow rate and by inserting/withdrawing control rods.
   A characteristic of BWRs is that radiation control is necessary also in the turbine building because the steam that is sent to the turbine generators contains radioactive materials. Figure 1.10 shows the mechanism of a BWR.
   A more recent type of reactor is the advanced boiling water reactor (ABWR). The ABWR offers the advantage of having a lower center of gravity and higher flexibility in the layout of equipment through the use of a reinforced concrete containment vessel instead of the steel containment of conventional BWRs. Additionally, with the use of internal pumps, the reactor building is more compact than the conventional type.
2. PWR
   In a PWR, electricity is generated by sending heated and pressurized reactor water (light water) to the steam generators. In the steam generators, the water from a different system is transformed to steam. The steam then rotates the turbine generators to produce electricity. The primary coolant system in which the water circulates in the reactor, and the secondary coolant system in which

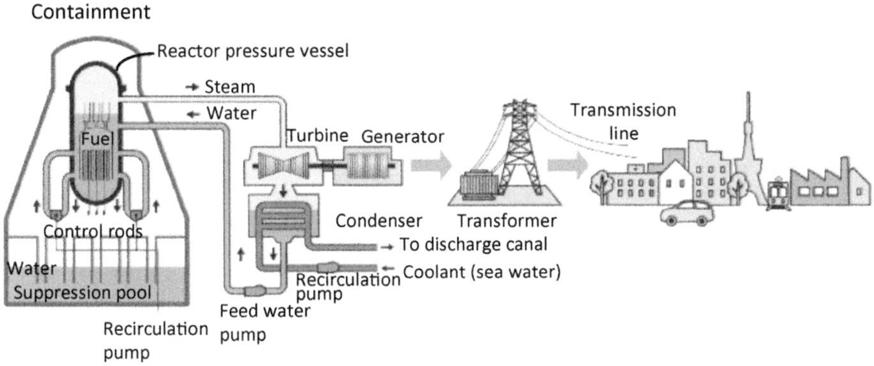

Fig. 1.10 Mechanism of a BWR [6]

the steam produced in the steam generators is sent to the turbine generators, are separated by the steam generators.

In the primary coolant system, light water sent to the reactor vessel by the heat generated by the fission reaction of nuclear fuel becomes hot among the fuel rods and is sent to the steam generators. The water sent to the steam generators is circulated by coolant pumps. The reactor output is controlled by inserting/withdrawing control rods and adjusting the concentration of boric acid dissolved in the coolant.

In the secondary coolant system, the water on the turbine generator side is transformed to steam by the high-pressure hot water of the primary coolant system that is sent to the steam generators. The steam is then sent to the turbine generators to rotate the turbines. After rotating the generators, the steam is cooled by sea water in the condenser and returns to water, and then is sent back to the steam generators.

Characteristics of PWRs include that the radiation control area is smaller than that of BWRs because the steam containing radioactive material does not flow into the turbine, and that heat exchange from the primary coolant to the secondary coolant is necessary in the reactor building. Moreover, a high pressure of 15 MPa is applied to the primary coolant system by the pressurizer, and the temperature reaches up to approximately 320 °C. PWRs are characterized by the high pressure and high temperature of the primary coolant; the pressure of the primary coolant in the BWR reactor pressure vessel is approximately 7 MPa, and the temperature is approximately 280 °C. Figure 1.11 shows the mechanism of the PWR.

PWR technology is being developed to raise the performance of the reactor core system and thus increase the output and duration of cycle operation. There are several plants with advanced PWR systems under construction around the world.

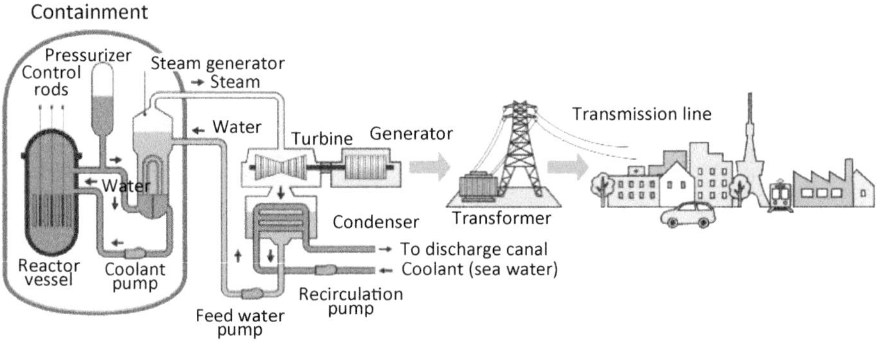

Fig. 1.11 Mechanism of a PWR [6]

Appendix 1.2: Accident at the Kashiwazaki-Kariwa Nuclear Power Plant by the 2007 Niigata-Ken Chuetsu-Oki Earthquake

At 10:13 on July 16, 2007, a moment magnitude (M_w) 6.8 earthquake centered at a depth of 17 km off the coast of Chuetsu in Niigata Prefecture struck. In the cities of Nagaoka, and Kashiwazaki, and the village of Kariwa in Niigata Prefecture and in the town of Iizuna in Nagano Prefecture, the seismic intensity measured upper 6 on the Japan Meteorological Agency scale (MMI, Modified Mercalli Intensity: IX). In the cities of Joetsu, and Ojiya, and the town of Izumozaki in Niigata Prefecture, seismic intensity of lower 6 (MMI: VIII) was measured. In addition to these cities and towns, shaking was measured in much of the western and northern regions of the main island of Japan, but also extending to the Tohoku, Kinki, and Chugoku regions. A tsunami with a height of 32 cm to 1 m was recorded at Kashiwazaki.

Among the seven reactors of the Kashiwazaki-Kariwa Nuclear Power Plant, the reactors under operation or in the process of startup at the time of the earthquake (Units 2, 3, 4, and 7) all automatically shut down safely, and reached a cold shutdown status after cooling and depressurization. Confinement of radioactive material in fuel was maintained in all seven reactors, including the reactors of other units (i.e., Units 1, 5, and 6). Thus, the most important functions required in a time of emergency to ensure reactor safety—*shut down*, *cool down*, and *confine*—were secured.

The acceleration measured at the plant during the earthquake largely exceeded the maximum acceleration assumed in the design phase. For example, a maximum acceleration of 680 gal was recorded in the fifth basement of the Unit 1 reactor building (on the foundation rock), whereas the maximum acceleration assumed in the design phase was 273 gal. The measurements at Units 1–4 were approximately six times the average ground motion evaluated using the epicentral distance attenuation formula, and the measurements at Units 5–7 were approximately three times the design ground motion. However, no damage due to the shaking of the earthquake was found.

The analysis of the records indicates the following causes of the large ground motion.

1. The seismic source caused ground motion 1.5 times that of an earthquake of the same magnitude (owing to the effect of the source properties).
2. The ground motion was amplified by the effect of the irregularity of deep geological structures (approximately 4–6 km deep).
3. There was a difference in the amplification of the ground motion at the site owing to the presence of an old fold structure in the shallow ground (approximately 2–4 km deep).

Major damage caused by the earthquake included a fire at the Unit 3 onsite transformer (Fig. 1.12). The fire resulted from insulation oil leaking from the transformer and being ignited by an arc induced by a short circuit. Although the

Fig. 1.12 View of the transformer fire [7]

Fig. 1.13 View of flooding from the spent fuel pool [7]

transformer itself was supported by foundation piles, the cable duct was supported by a flat foundation. As a result, it was reported that only the foundation of the duct subsided because of earthquake ground motion, and this damaged the bushing (which takes in electric cables from the outside to the equipment such as transformers, and insulates and supports the cables from the equipment and walls), and caused an arc, which ignited the leaked insulation oil.

Additionally, owing to the long-period component of earthquake motion, there was sloshing vibration in Units 1–7, and the spent fuel pool water flooded the units (Fig. 1.13). In Unit 6, the flood water flowed into the discharge water tank (through cable holes, for example), which does not contain radioactive material, and was released into the sea. The amount of radiation release and the radiation dose were approximately 9×10^4 Bq and approximately 2×10^{-9} mSv, respectively.

Additional reported damage was a trace amount of radioactivity being released from the ventilation duct of Unit 7, the failure of a joint of the reactor building ceiling crane of Unit 6, the opening of a blowout panel of the reactor building of Unit 3, and the misalignment of the main ventilation ducts in Units 1–5 (i.e., air conditioning ducts for sending the air exhausted from reactor buildings to air vents). It was also reported that damage to the administrative building and the resulting deformation of the door to the emergency response center prevented personnel from entering the room, and that failure of the facilities inside the room made it difficult to establish the response organization immediately after the earthquake.

Appendix 1.3: Accidents at the Fukushima Daiichi Nuclear Power Plant by the 2011 Great East Japan Earthquake

When the 2011 Great East Japan earthquake struck, Units 1–3 of the Fukushima Daiichi Nuclear Power Plant were operating at the rated output, while Units 4–6 were undergoing periodic inspection. Units 1–3 automatically tripped in response to the earthquake ground motion. At that time, a total of six lines were connected to offsite power sources from the plant, but they all stopped supplying electricity because of, for example, damage to breakers and the collapse of transmission line towers due to the earthquake ground motion. Therefore, the emergency generators of each unit started operation. However, when the tsunami struck the site several dozen minutes later, the sea water pumps for cooling, emergency diesel generators, and switchboards were flooded, causing all emergency diesel generators except that in Unit 6 to stop, and all units except Unit 6 suffered a blackout. In Unit 6, one emergency diesel generator (air-cooled) and the switchboard were not flooded and continued to operate. Additionally, the equipment of the residual heat removal system that releases residual reactor heat to the sea stopped functioning because the sea water pumps for cooling were flooded by the tsunami.

As the emergency core cooling function powered by AC was lost in Units 1–3, attempts were made to actuate the core cooling function using DC power sources. Specifically, the isolation condenser of Unit 1, the reactor core isolation cooling system of Units 2 and 3, and the high-pressure coolant injection system of Unit 3 were actuated. However, the core cooling functions of these DC power sources later stopped when their batteries ran out. Thus, core cooling was switched to the alternate means of injecting fresh water or sea water from the fire extinguishing system line, using fire extinguishing pumps.

As the pressure inside each reactor was very high and water could not be injected into each reactor pressure vessel for some time, the nuclear fuel in the core of each unit started to be exposed, resulting in core meltdown; some of the molten fuel accumulated in the lower part of the reactor pressure vessel. Meanwhile, a huge amount of hydrogen was generated by the chemical reaction between the zirconium contained in the fuel cladding and the steam. The damaged fuel cladding caused the radioactive materials inside the fuel rods to be released to the outside of the reactor pressure vessels. Moreover, in the process of depressurizing the reactor pressure vessels, the hydrogen and radioactive material were released to the containment.

The water injected into the reactor pressure vessel was turned into steam by the heat generated by the nuclear fuel and the pressure inside the reactor pressure vessel continued to rise without core cooling. Therefore, the steam was released to the containment through safety valves. Although the containment can be ultimately cooled down by the cooling system using sea water as a cooling source under normal conditions, this was not possible owing to the damage and collapse of the sea water system components, and the units lost their ultimate heat sink status. Consequently, the temperature and pressure inside the containment gradually rose. In Units 1–3, several attempts were made to release the gas inside the containment from the vapor phase (i.e., the part above the water surface) of the suppression chamber (a water-cooled facility that restricts the pressure rise in the containment caused by, for example, steam pressure) into the atmosphere through air vents (i.e., containment wet-well vents).

At Units 1 and 3, explosions thought to be caused by the hydrogen that leaked from the reactor containment occurred in the upper part of the reactor building after the containment wet-well vent. As a result, the top floor of each reactor building was blown away, causing significant amounts of radioactive materials to be released into the atmosphere. Subsequent to the explosion in the building of Unit 3, there was also an explosion, considered to be caused by hydrogen too, in the reactor building of Unit 4, where all fuel in the core had been conveyed to the spent fuel pool for periodic inspection, and the top of the reactor building was destroyed. At the same time, it was reported that a failure occurred in a location presumed to be around the suppression chamber of the reactor building of Unit 2 (Fig. 1.14).

Attempts were also made to inject water into the spent fuel pools in Units 1–4 in conjunction with the continuous efforts to recover the power sources and to inject water into the reactor pressure vessels. The water level of the spent fuel pool of each unit kept falling owing to the evaporation of water by the heat generated from the spent fuel as the function to cool the pool water had been lost through the loss of power sources. Therefore, water was dropped into the spent fuel pools by helicopters and water cannons by the self-defense forces, fire brigade, and police. After all these attempts, concrete pump trucks were used to inject fresh water from a nearby reservoir to cool the pools.

The maximum acceleration of 550 gal that was observed on the foundation ground (i.e., the bottom of the fourth floor of the reactor building of Unit 2)

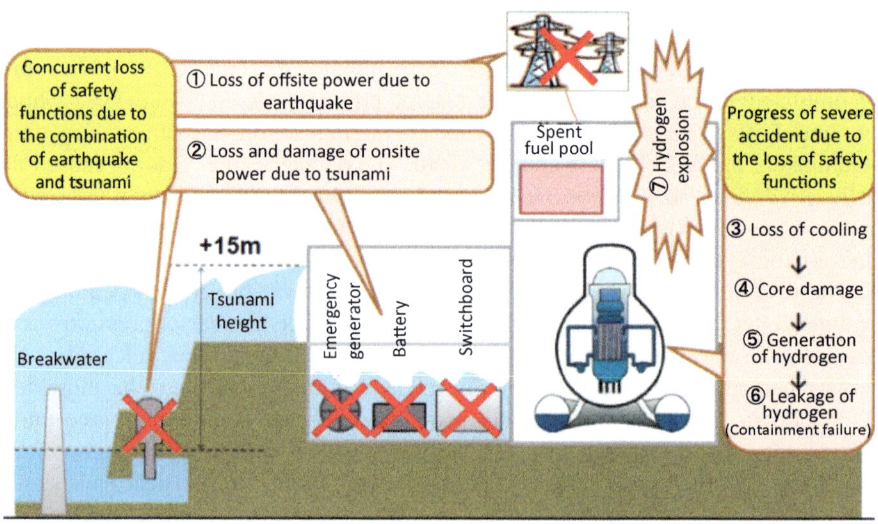

Fig. 1.14 Status of the accident at the Fukushima Daiichi Nuclear Power Plant [5]

exceeded the basic earthquake ground motion for the original design S_s, but most other measured accelerations were below the design acceleration.

The damage to facilities caused by the earthquake ground motion could not be clearly confirmed owing to the tsunami that struck approximately 40 min after the earthquake. Tokyo Electric Power Co., Inc. stated that "it is considered that major facilities maintained their safety-related functions during and immediately after the earthquake," on the basis of instrument measurements showing the plant status and alarm issuance records, dynamic response analysis of the reactor buildings and major facilities using observed acceleration on the basement of the buildings, and onsite visual inspections. In this regard, all the reports on the accident made by various organizations concluded that the direct cause of the accident was the blackout caused by the tsunami, and only the Independent Investigation Commission of the National Diet questioned limiting the accident cause to the tsunami, stating "it cannot be assumed that important safety facilities were not damaged by the earthquake ground motion."

Although the tsunami height for the original design was O.P. + 6.1 m, the actual tsunami height was O.P. + 13.1 m. The inundation height at the site was reported to be O.P. + 15.5 m around Units 1–4, and O.P. + 14.5 m around Units 5 and 6 (Fig. 1.15).

Damage by the tsunami included the loss of function of the emergency diesel generators and the switchboards due to flooding, and the loss of function of the residual heat removal system in the reactor and the cooling system due to the flooding of the sea water pumps. Thus, the power sources and most of the functions

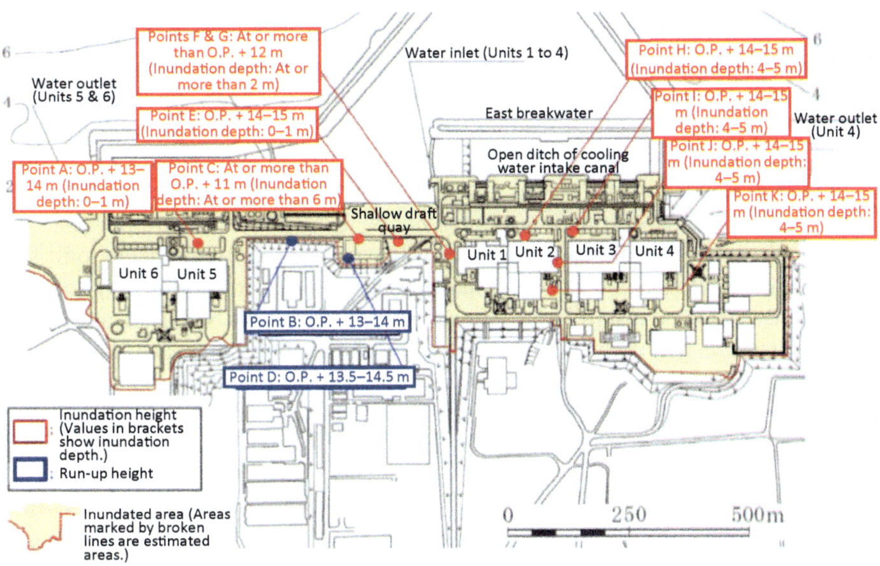

Fig. 1.15 Status of inundation due to the tsunami at the Fukushima Daiichi Nuclear Power Plant [8]

Inundated power panel on the first floor of the Unit 1 turbine building

The north side of the fourth floor of the waste treatment building

Fig. 1.16 Status of damage due to tsunami at the Fukushima Daiichi Nuclear Power Plant [7]

related to cooling were lost (Fig. 1.16). In addition, innumerable drifting objects, including a destroyed and drifting heavy oil tank and vehicles, filled the plant site and impeded emergency responses and activities.

Appendix 1.4: Status of the Onagawa Nuclear Power Plant by the 2011 Great East Japan Earthquake

When the Great East Japan earthquake struck, Units 1 and 3 were in operation and the reactor of Unit 2 was in the process of starting up at the Onagawa Nuclear Power Plant. The reactors of all units were automatically tripped by the limit signal from measured accelerations. Among the five supply lines from the offsite electric power sources, Matsushima Line No. 2 continued to function. Thus, Units 2 and 3 continued to receive offsite power and onsite electric power was secured.

In Unit 1, although offsite power could not be received as the startup transformer tripped in response to the earthquake, power was secured by the actuation of the emergency diesel generator. Additionally, the reactor core cooling system was successfully actuated using a DC power source and could be used to cool the reactor. Pressure was controlled by the main steam relief valve, and after depressurization, water was injected into the reactor by the control rod drive water pressure system. The suppression chamber and the reactor were cooled by the residual heat removal system, and approximately 12 h after the earthquake, the unit was brought to the cold shutdown status.

In Unit 2, the reactor was subcritical and the reactor water temperature was below 100°C before the earthquake struck. Therefore, the reactor was brought to the cold shutdown status by shifting the reactor mode switch to the *shut down* position. The cooling water system (B) of the reactor building and the cooling water system of the high-pressure core spray lost their functionality because the sea water flowing from the sea water intake canal lifted the lid of the level gauge box installed in the sea water pump room. The sea water then flowed into the underground trenches through the pipe penetrations after passing through the cable trays and finally poured into the reactor building. However, as another reactor cooling water system (A) remained intact, the reactor cooling function of the residual heat removal system (A) was secured.

In Unit 3, the level gauge of the sea water pump was damaged by the spilling waters of the tsunami, resulting in the tripping of the circulation pump. Furthermore, as the cooling sea water system of the turbine building lost its functionality owing to the influx of sea water, the reactor feed water pump was stopped and the reactor core isolation cooling system was started manually to cool the reactor. In addition, the pressure was controlled by the main steam relief valve, and the water was injected into the core after depressurizing the reactor with the makeup water condensate system. The suppression chamber and the reactor were cooled by the residual heat removal system, and approximately 13 h after the earthquake, the unit was brought to the cold shutdown status.

Additionally, although the cooling system for the spent fuel pool automatically tripped in response to the earthquake motion, it was restarted after confirming that there was no abnormality in the system, and no significant increase in the temperature of the fuel pool was observed.

Despite the confirmation of the fire at the high-pressure power panel in the turbine building of Unit 1, the failure of the glass windows in the fuel handling machine room in Unit 2, the opening of the blowout panel in the turbine building of Unit 3, and the collapse of the ceiling panel of the Unit 1 main control room in response to the earthquake, it was confirmed that important safety-related functions was properly maintained.

The height of the tsunami at the plant was O.P. + approximately 13 m at the maximum and did not exceed the ground height of the site (O.P. + approximately 13.8 m). (The ground level subsided by approximately 1 m owing to tectonic deformation, which is taken into consideration in the calculation of this value.) From the beginning of the design and construction of Unit 1, countermeasures against a tsunami were repeatedly discussed at survey meetings with external experts. As a result, the height of the plant site was set to be 14.8 m, reflecting the survey result that "the height of the plant site shall be the primary countermeasure against tsunami and O.P. + 15 m is considered to be a sufficient height." This is one of the reasons why damage to the plant due to the tsunami was limited.

Although the tsunami caused some trouble for cooling water systems of the reactor and the turbine buildings, and automatic tripping of the circulation pump due to damage of the sea water pump level detector, the cooling function was successfully maintained. Additionally, although a heavy oil tank collapsed, it did not affect the safety functions (Figs. 1.17, 1.18 and 1.19).

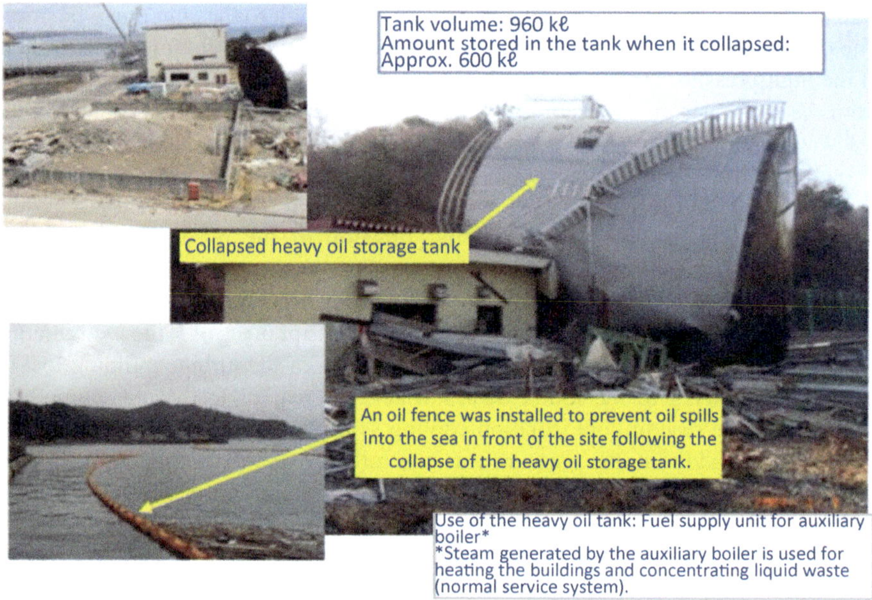

Fig. 1.17 Collapse of the heavy oil tank at Onagawa Unit 1 [9]

Fig. 1.18 Burnt-out of the high-pressure power panel at Onagawa Unit 1 [9]

Fig. 1.19 Influx of sea water into the reactor building at Onagawa Unit 2 [9]

References

1. "Genshiryoku-hatsuden Pocket Book (Nuclear Pocket Book) (2010)", Chubu Electric Power Company, 2010
2. "Graphical Flip-chart of Nuclear & Energy Related Topics 2012", The Federation of Electric Power Companies of Japan, 2012
3. Chubu Electric Power Company website
4. "Technical Guidelines for Seismic Design of Nuclear Power Plants (JEAG4601-1984)", The Japan Electric Association, 1984
5. Nuclear Regulation Authority of Japan website
6. "Graphical Flip-chart of Nuclear & Energy Related Topics 2011", The Federation of Electric Power Companies of Japan, 2011
7. Tokyo Electric Power Company website
8. "Fukushima Daiichi Genshiryoku Hatsudensho oyobi Fukushima Daini Genshiryoku Hatsudensho ni okeru Heisei 23 nen Tohoku-Chihou Taiheiyou Oki Jishin ni yori hasseishita tsunami no Chousakekka wo fumaeta Taiou ni tsuite (Shiji) (Actions based on lessons learned from the survey results of tsunami in the 2011 Tohoku Earthquake which hit the Fukushima Daiichi and the Fukushima Daini Nuclear Power Plants) (Directive)", former Nuclear and Industrial Safety Agency of Japan, April 13, 2011
9. "Tohoku Chihou Taiheiyou Oki Jishin ni yoru Onagawa Genshiryoku Hatsudensho no Joukyou ni tsuite (The situation of the Onagawa Nuclear Power Plant in the 2011 Tohoku Earthquake)", Tohoku Electric Power Company website, April 26, 2011

Further Readings

10. "Niigata-ken Chuetsu Oki Jishin ni-yoru Eikyou ni kansuru Genshiryoku Anzen Iinkai no Taiou (NSC's Opinion and Future Actions about the Niigata Chuetsu Earthquake's Effects)", Nuclear Safety Commission of Japan, July 30, 2007
11. "Niigata-ken Chuetsu Oki Jishin wo fumaeta Genshiryoku Hatsudensho nado no Taishinanzenseihyouka ni Han-ei subeki Jikou ni-tsuite (Matters that should be reflected on Seismic Safety Assessments for Nuclear Power Plants, etc. in consideration of the Niigata

Chuetsu Earthquake) (interim report)", Nuclear and Industrial Safety Agency of Japan, December 27, 2007
12. "Technical Code for Seismic Design of Nuclear Power Plants (JEAC4601-2008)", The Japan Electric Association, 2008
13. "Niigata-ken Chuetsu Oki Jishin wo fumaeta genshiryoku hatsudensho nado no taishinanzenseihyouka ni han-eisubeki jikou ni-tsuite (Matters that should be reflected on Seismic Safety Assessments for Nuclear Power Plants, etc. in consideration of the Niigata Chuetsu Earthquake)", Nuclear and Industrial Safety Agency of Japan, September 4, 2008
14. "Niigata-ken Chuetsu Oki Jishin wo uketa Kashiwazaki-Kariwa Genshiryoku Hatsudensho ni kakaru Genshiryoku-Anzen-Hoan-In no Taiou (NISA's Actions Taken after the Kashiwazaki-Kariwa Nuclear Power Plant hit by the Niigata Chuetsu Earthquake) (the third interim report)", Nuclear and Industrial Safety Agency of Japan, April 8, 2010
15. "Report of Japanese Government to the IAEA Ministerial Conference on Nuclear Safety – The Accident at TEPCO's Fukushima Nuclear Power Stations", Nuclear Emergency Response Headquarters, Government of Japan, June and September, 2011
16. "Fukushima Nuclear Accidents Investigation Report", Tokyo Electric Power Company, June 20, 2012
17. "Final Report of Investigation Committee on the Accident at the Fukushima Nuclear Power Stations of Tokyo Electric Power Company", Investigation Committee on the Accident at the Fukushima Nuclear Power Stations of Tokyo Electric Power Company, Cabinet Secretary, July 23, 2012
18. "Fukushima Daiichi Nuclear Power Plant Accident and Four Accident Investigation Commissions", National Diet Library of Japan, August 20, 2012

Chapter 2
Assessment of Standard Seismic Motion

Masanori Hamada and Michiya Kuno

Abstract Earthquake surveys in areas that may affect plant sites and assessment of ground motion are indispensable for earthquake-resistant design to prevent the loss of safety-related functions of nuclear power plants and to avoid the effects of radiation on the public in a large affected region. The standard seismic motion S_s is assessed based on the source properties of the active faults, the propagation characteristics of the seismic waves from the source to the plant site, the amplification characteristics of the ground motion through the subsurface soil, and the geological and soil conditions at the site. Furthermore, ground motion in cases where fault surveys cannot clearly identify active faults in advance should be also assessed. The seismic motion for elastic design S_d, is determined from the standard seismic motion S_s.

Keywords Standard seismic motion · Seismic motion for elastic design · Survey on seismic source · Virtual free surface of bedrock · Empirical Green's function method · Direct method by fault model

2.1 Overview of Assessment of Standard Seismic Motion

This section presents the standard seismic motion S_s assessment process, identification of the seismic source parameters that may affect the site, and the method of assessing ground motion.

Figure 2.1 is a flowchart for the assessment of standard seismic motion S_s. The standard seismic motion S_s is assessed by first identifying the seismic sources.

M. Hamada (✉)
Waseda University, Shinjuku, Tokyo, Japan
e-mail: hamada@waseda.jp

M. Kuno
Chubu Electric Power Company, Nagoya, Japan

M. Hamada and M. Kuno (eds.), *Earthquake Engineering for Nuclear Facilities*,
DOI 10.1007/978-981-10-2516-7_2

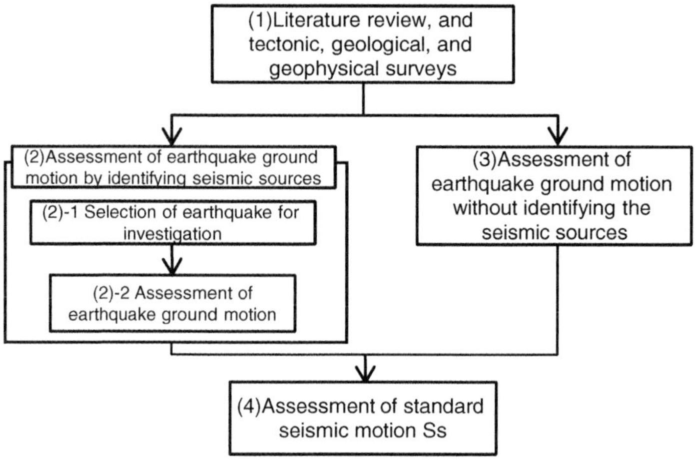

Fig. 2.1 Assessment of standard seismic motion S_s

(1) Literature review, and tectonic, geological and geophysical surveys (Sect. 2.2).
Literatures on past earthquakes are reviewed, and tectonic, geological, and geophysical conditions are surveyed to identify earthquakes that may significantly affect the site.

(2) Assessment of earthquake ground motion by identifying seismic sources (Sect. 2.3).
Seismic sources of earthquakes that may have significant impacts on the site are identified, and the standard seismic motion S_s is assessed.

  (2)-1 Selection of earthquakes for investigation.
  The earthquakes for the investigation are selected from inland crustal earthquakes, interplate earthquakes, and intra-oceanic plate earthquakes.

  (2)-2 Assessment of earthquake ground motion.
  The standard seismic motion S_s is assessed in the form of response spectra and time histories based on source models of the selected earthquakes. Various uncertainties about seismic source parameters for the assessment are taken into consideration.

(3) Assessment of earthquake ground motion without identifying the seismic sources (Sect. 2.4).
In the case when the seismic sources cannot be identified by the active fault survey, ground motion records of past inland crustal earthquakes are analyzed, and the standard seismic motion S_s is assessed based on these analyses.

(4) Assessment of standard seismic motion S_s (Sect. 2.5).
The standard seismic motion S_s in horizontal and vertical directions is assessed by identifying as well as not identifying the earthquake sources.

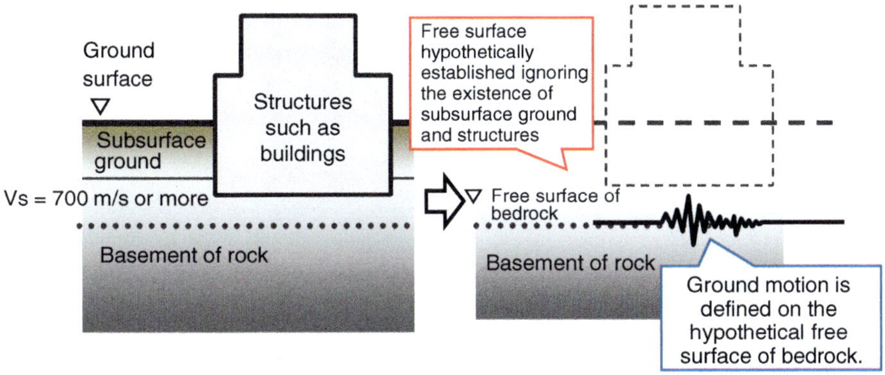

Fig. 2.2 Concept of the virtual free surface of bedrock

The standard seismic motion S_s is assessed on the virtual free surface of the bedrock, where it is assumed that no subsurface ground and structures exist on the ground surface (Fig. 2.2). The hypothetical free surface of the bedrock is defined as being an extensive, flat surface. The S wave velocity of bedrock is required to be more than 700 m/s. Figure 2.2 shows the concept of the hypothetical free surface of the bedrock.

2.2 Literature Review, Tectonic, Geological, and Geophysical Surveys

To model the seismic sources, past earthquakes, and active faults [faults active after the Pleistocene period of the late Quaternary (approximately 120,000–130,000 years ago) and faults that may become active in the future] in the area surrounding the site must be investigated.

A review of the existing literatures, including ancient documents and seismic observation records, investigations on tectonic geomorphology based on topographical and geological conditions, and geological and geophysical surveys, will be conducted. The literature review will provide an understanding of the geology and geological structure of a wide area, while the tectonic topography will be identified by the geomorphological investigation. Additionally, the geomorphological survey will identify the detailed characteristics and activeness of faults. The geophysical investigation will identify the locations and extent of faults and subsurface structures, including the seismic wave velocity distribution. The effects of the three-dimensional subsurface structure of the site and surrounding area on the characteristics of seismic wave propagation and ground motion amplification must be studied based on the above-mentioned series of investigations.

Additionally, assessments of the foundation ground bearing capacity and slope stability are carried out. Furthermore, a tsunami assessment survey is important for oceanic interplate earthquakes. These surveys are described in Chaps. 3 and 7.

2.2.1 *Literature Review*

Based on the existing literatures, the magnitudes of past earthquakes and the damage they caused, and the size and activity of faults in the areas surrounding the site will be surveyed. Figure 2.3 is one of examples of the existing literatures on active faults in Japan. Based on the ground motion recorded during past earthquakes, geodetic records using GPS, tectonic geographical surveys, the characteristics of faults that are highly probable to cause earthquakes, and the mechanism of the occurrence of earthquakes will be identified (refer to Part II, Chap. 13).

2.2.2 *Tectonic Geomorphological Survey*

The purpose of the tectonic geomorphological survey is to assess faults that may become active in the future (refer to Appendix 2.1) by observing the development process, the origin, and activeness of the topography, and focusing on the characteristic landscapes created by the displacement of active faults. Landscapes that may have been formed by fault activity are identified from aerial photographs and aerial laser surveys. For oceanic areas, seabed topography maps based on seawater depth measurements are used. When slippages caused by fault activity come close to the ground surface, they frequently leave continuously slipped landscapes on mountain ridges, valleys, and cliffs. However, these landscapes may also be eroded and covered by sedimentation. Therefore, the active faults must be assessed comprehensively with reference to the results of other surveys.

Fig. 2.3 Fault map of Japan

Fig. 2.4 Example of a landscape formed by displacement of an active fault [1]

Figure 2.4 is an example of a landscape created by the displacement of an active fault. In this figure, valleys and ridges are displaced in the same direction across the fault, and triangular cliffs are formed on the mountain ridges, which are cut by the fault. These characteristics indicate the presence of right-lateral faults.

2.2.3 Geological Survey

Geological surveys will be conducted to identify topographical structures in an extensive area around the site based on the results of the literature review, and the tectonic and geomorphological investigations, and also to clarify the subsurface structure of the areas surrounding the faults.

Figure 2.5 shows geological surveys such as outcrop observation, trench surveys, and borings. A geological survey is conducted to discover faults and deformation in geological formations, and to find slippage and disorders in geological structures positioned across faults. Based on the results, the period during which the faults were active, the magnitude and direction of displacement, and the history of faulting are determined. Geological deformation is measured by trench surveys and boring surveys are used to study deep geological structures.

2.2.4 Geophysical Survey

Geophysical surveys are conducted to identify faults that are difficult to find by ground surface investigations, the extent of faults, gradient, and folding of geological formations, and the distribution of seismic wave velocity of the surface ground. Particularly in oceanic areas where it is impossible to directly confirm the deformation of geological structures by aerial photos and the outcrop observation, subsurface structures, and the locations and extent of faults can be identified by geophysical surveys.

Figure 2.6a shows seismic wave exploration (measurement of *P* and *S* wave's velocity) in a bored hole, and Fig. 2.6b shows elastic wave exploration to measure the gradient and folding of faults and geological formations, and the distribution of

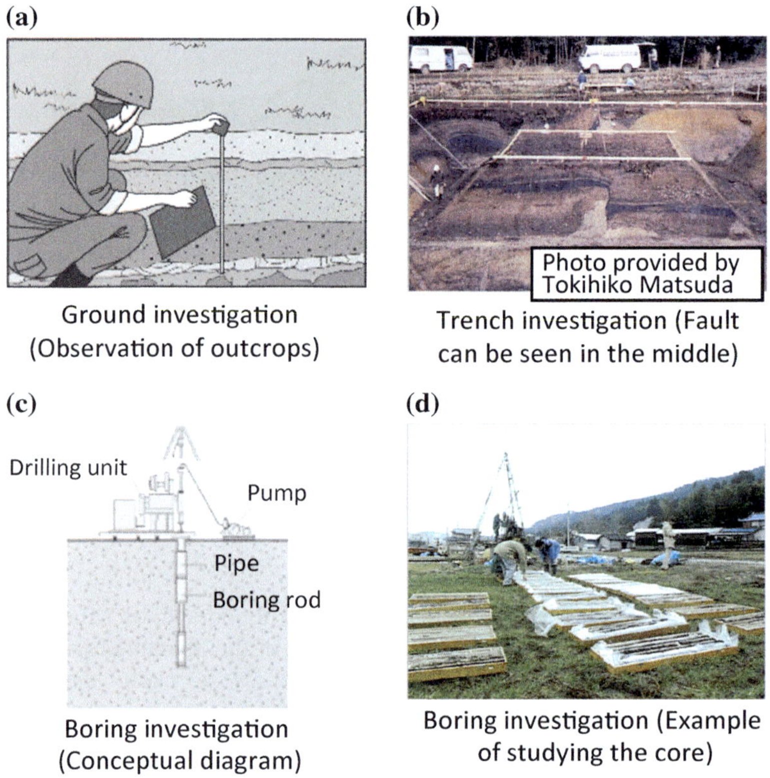

Fig. 2.5 Geological investigations. **a** Outcrop observation, **b** trench survey, and **c**, **d** boring survey [2]

seismic wave velocity in the ground by observing the propagation of artificially produced elastic waves. Microtremor observations are made by a vertical array of seismometers to determine seismic wave velocity structure. Earthquake observation records can provide information on the dynamic response characteristics of the subsurface ground. Moreover, the gravity prospecting method is useful to determine soil and rock density.

2.2.5 Analysis of Earthquake Observation Records

Since the 1995 Kobe earthquake, earthquake ground motion including micro-earthquake-caused motion have been measured on land and the seabed surrounding Japan by the Japan Meteorological Agency (JMA), the National Research Institute for Earth Science and Disaster Prevention, the Japan Agency for Marine-Earth Science and Technology, and several universities. The source

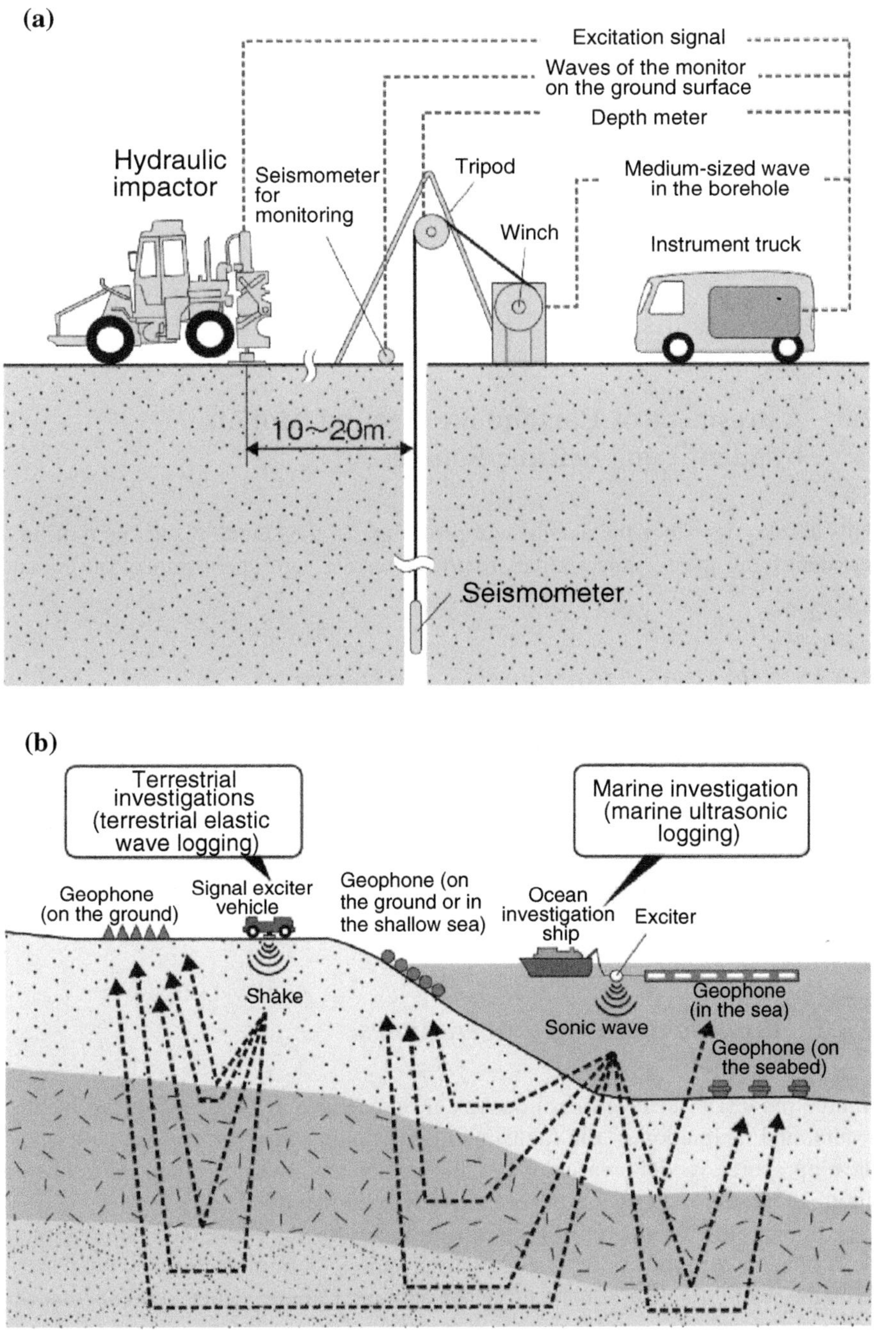

Fig. 2.6 Geophysical surveys, **a** physical logging (PS logging), **b** elastic wave logging

mechanism and propagation characteristics of seismic waves have been studied by analyzing these records. The seismic intensity distribution of earthquakes reported by JMA can be used to create the seismic source model. The subsurface structure model is developed to assess ground motion at the site, with reference to the distribution of seismic wave velocities and the damping characteristics of the surface ground.

Earthquake ground motions have been observed by seismometers installed in buildings and ground at nuclear power plants. The earthquake ground motions observed is used to verify the assumptions and the ground motion assessment process.

2.3 Assessment of Earthquake Ground Motion by Identifying Seismic Sources

This section describes the assessment procedure of the standard seismic motion S_s, by identifying the seismic sources that may have a significant impact on the site.

2.3.1 Selection of Earthquakes for Design

Earthquakes are selected for the earthquake-resistant design according to the earthquake type (inland crustal, interplate, and intra-oceanic plate earthquakes) based on the geological and geophysical survey results. The standard seismic motion S_s is assessed as time history record and/or response spectra based on the seismic source data for the selected earthquakes (refer to Part II, Chap. 14, Sect. 14. 1 for the response spectrum).

2.3.2 Considerations of Uncertainties

Source models of selected earthquakes are developed by identifying the location, extent, and inclination of the faults, and by assuming the fault rupture patterns. To develop source models, variations of the parameters must be taken into consideration, because not all of the uncertainties will be identified.

Examples of considerations for uncertainties in an earthquake source model development are shown in Fig. 2.7. The survey committee of the Nankai Sea Trough mega-thrust earthquake affiliated with the Cabinet Office developed several

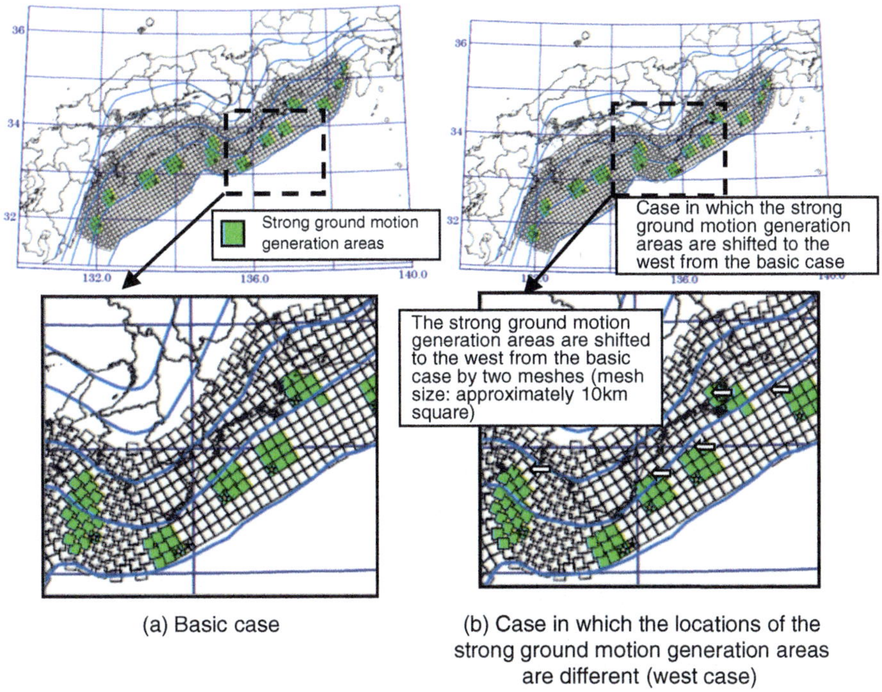

Fig. 2.7 Various seismic source models that consider uncertainties [3]

models with different areas for the generation of earthquake ground motion. The reason for the adoption of the different source areas is the fact that the earthquake ground motion may change compared with that of past earthquakes along the Nankai Trough.

2.3.3 Assessment of Standard Seismic Motion S_s by Source Models

The standard seismic motion S_s is assessed on the virtual free surface of the bedrock by the two methods as shown in Fig. 2.8. In the first method, the response spectra are evaluated from the magnitude of the earthquake source and the hypocentral distance. The time histories of the seismic motion are obtained to fit them with the spectra. In the second method, the time histories of the ground motion are directly developed based on the earthquake source models and the wave propagation characteristics from the source to the site.

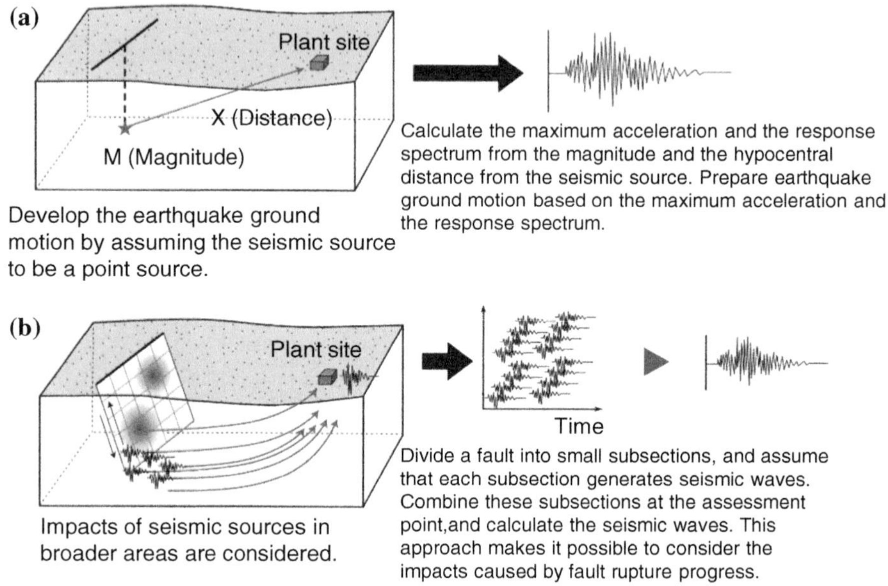

Fig. 2.8 Two methods of assessing of basic earthquake ground motion, **a** ground motion assessment based on the response spectrum, **b** ground motion assessment by the fault model

2.3.3.1 Method by Response Spectra

In this method, the response spectra of the standard seismic motion S_s on the virtual free surface of the bedrock are assessed according to a distance attenuation formula. Generally, there are two types of distance attenuation formulas. The first one provides a maximum value such as acceleration and velocity, and the second method provides the response spectra. The second method is used to assess basic earthquake ground motion S_s.

The distance attenuation formulae show the relationship between the intensity of ground motion, e.g., maximum accelerations and velocities, and the distance from the seismic source to the site. The formulae are developed by statistical analysis of the observed seismic motions during past earthquakes. Figure 2.9 compares the maximum accelerations observed during the 1995 Kobe earthquake and a proposed attenuation formula.

Figure 2.10 shows the process of assessing seismic motion based on the response spectrum. First, the response spectrum of the seismic motion on the virtual free surface of the bedrock is calculated from the magnitude of the earthquake and the equivalent hypocentral distance. Here, the equivalent hypocentral distance is the distance between a point representing the whole fault plane and the site. The response spectrum for the earthquake-resistant design is then developed as the envelope response spectra. Next, the time histories of the earthquake ground motion on the virtual free surface of the bedrock are produced to fit them with these

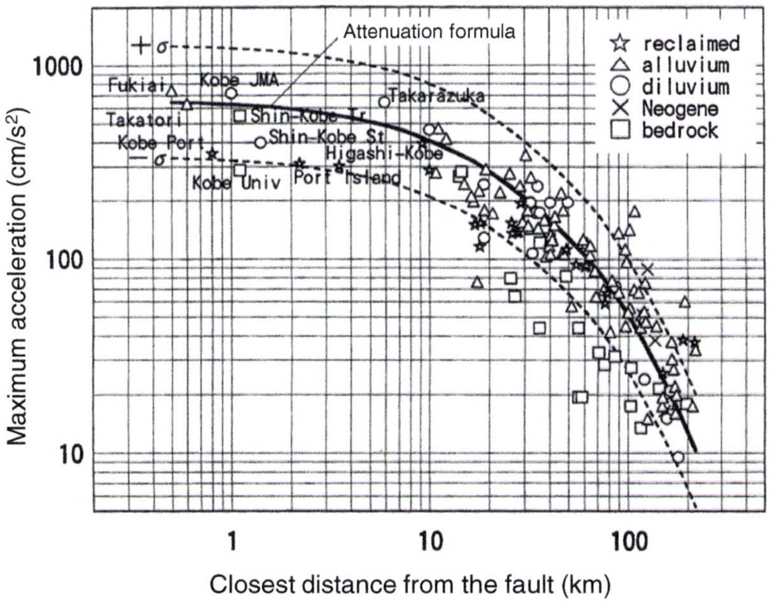

Fig. 2.9 Comparison of the maximum accelerations observed during the 1995 Kobe earthquake and those obtained from the attenuation formulae [4]

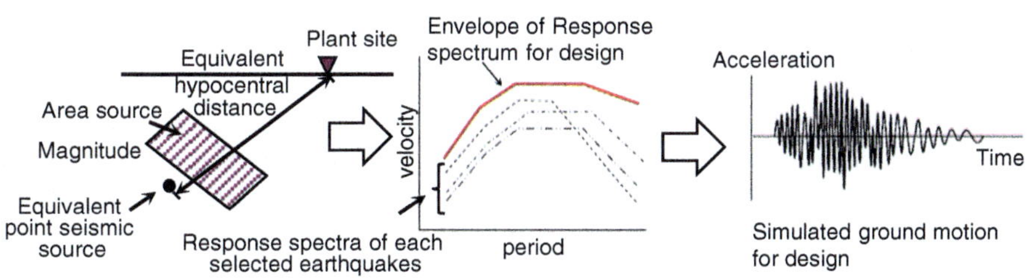

Fig. 2.10 Assessment of response spectra and time histories of earthquake ground motion based on the magnitude of the earthquake and the equivalent hypocentral distance

response spectra. The details of ground motion assessment based on response spectra are presented in the Technical Guidelines for the Seismic Design of Nuclear Power Plants issued by the Japan Electric Association (JEAG 4601-2008) (refer to Appendix 2.2).

2.3.3.2 Ground Motion Assessed Directly by a Fault Model

The source faults are divided into multiple small segments as shown in Fig. 2.11. It is assumed that small earthquakes are caused by each fault segment according to the propagation of the fault rupture. The effect of the propagation of the fault ruptures

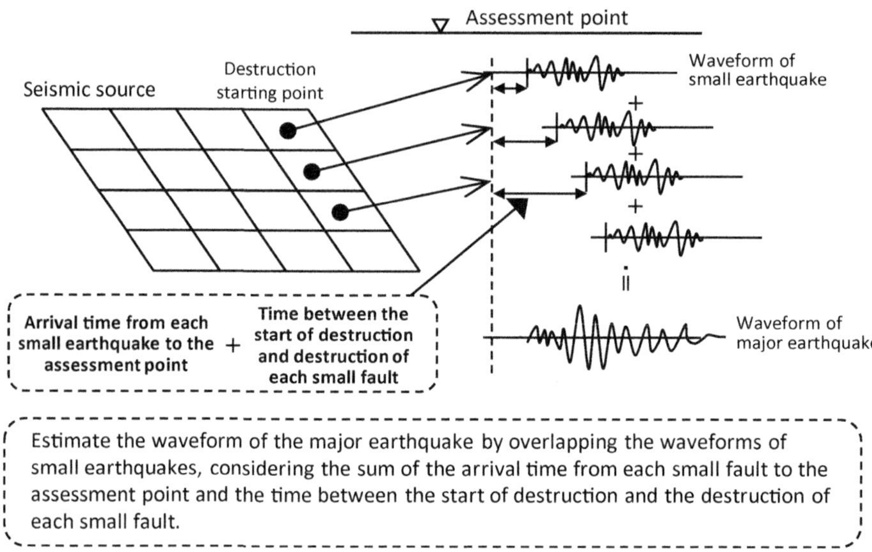

Fig. 2.11 Assessment of earthquake ground motion by fault models

and of the areas of the ruptured zone can be taken into consideration for the assessment of the earthquake ground motion. The standard seismic motion S_s is the sum of the ground motions caused by each of its segments.

The semiempirical methods (empirical Green's function method, statistical Green's function method) and theoretical methods are used to assess earthquake ground motion by the fault model. There is also an approach called the hybrid method, which combines the assessment of short period components of earthquake ground motion by semiempirical methods, and the assessment of long period components by theoretical methods.

As shown in Fig. 2.12, response spectra are calculated from the time histories of the earthquake ground motion assessed by the fault model and are compared with the response spectra assessed by the response spectra method mentioned in the previous section. If the response spectra by the fault model exceed the ones by the response spectrum method, the time history of the earthquake ground motion by the fault model is adopted as the standard seismic motion S_s for the earthquake-resistant design.

2.4 Assessment of Earthquake Ground Motion Without Identifying the Seismic Source

It is generally difficult to identify all of the inland crustal earthquakes that may occur in the area surrounding the site in advance, even by detailed geological surveys and precise active fault investigations. Therefore, in addition to the

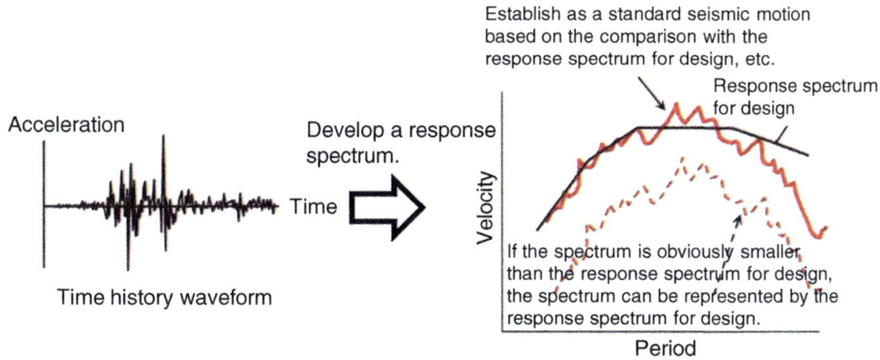

Fig. 2.12 Comparison of response spectrum of earthquake ground motion for the seismic design

earthquake ground motion based on the identified active faults, the earthquake ground motion will be estimated without identifying the seismic sources. To assess this type of earthquake ground motion, ground motions observed in the neighborhood of the epicenter of past inland crustal earthquakes, for which seismic sources and active faults had not be identified in advance, are collected. Based on the observed earthquake ground motions, the response spectra for the design are then derived by considering the amplification characteristics of the subsurface ground at the site, and the time histories of the earthquake ground motion to fit with the response spectra are developed.

2.5 Assessment of the Standard Seismic Motion S_s

The standard seismic motion S_s on the hypothetical free surface of the bedrock is determined based on the seismic motion assessed by both identifying and not identifying the seismic source.

Examples of the tripartite response spectra and the time histories of the standard seismic motion S_s are shown in Figs. 2.13 and 2.14, respectively. The tripartite response spectra (refer to Part II, Chap. 14) are a type of graph in which the response spectrum for each displacement, velocity, and acceleration is presented in one figure. The vertical axis is the velocity response, and the horizontal axis is the periods of seismic motion, while the 45° axis increasing towards the left and the 45° axis increasing towards the right are the acceleration and displacement responses, respectively.

The standard seismic motion S_s is used to verify the safety-related functions of the plant. The seismic motion for the elastic design S_d are developed from the standard seismic motion.

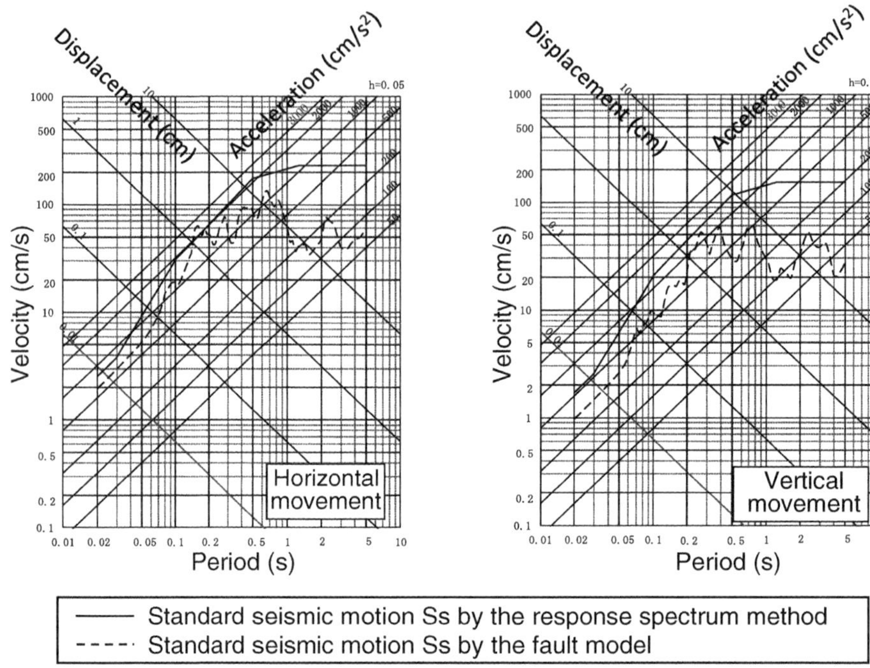

Fig. 2.13 Tripartite response spectra of the standard seismic motion S_s

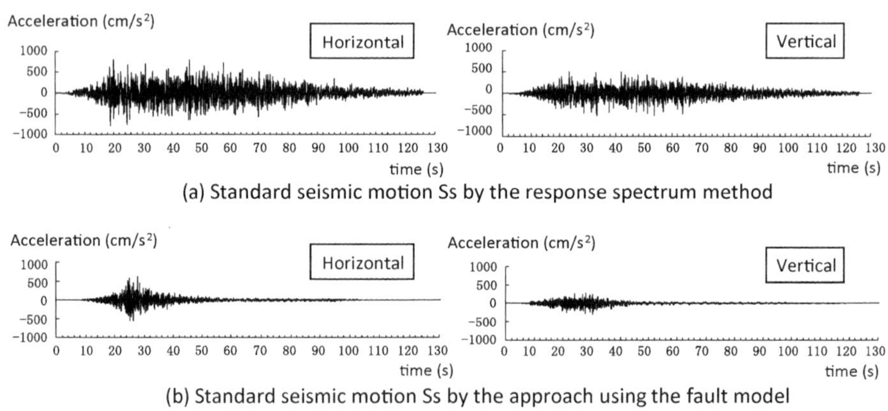

Fig. 2.14 Standard seismic motion S_s (time histories)

Appendix 2.1: Active Fault for Earthquake-Resistant Design

Besides the active faults that may cause earthquakes, faults that may cause permanent displacement, such as slippage in the plant's foundation during earthquakes should be taken into consideration for earthquake-resistant design. Additionally, faults that cannot be proven to have been inactive since the Late Pleistocene period (approximately 120,000–130,000 years ago) should be considered for the design. If the activeness of faults cannot be clearly determined, it must be assessed by carrying out comprehensive studies on the landscape, geological conditions and structures, and stress condition of the crust after the Middle Pleistocene (approximately 400,000 years ago) (Fig. 2.15).

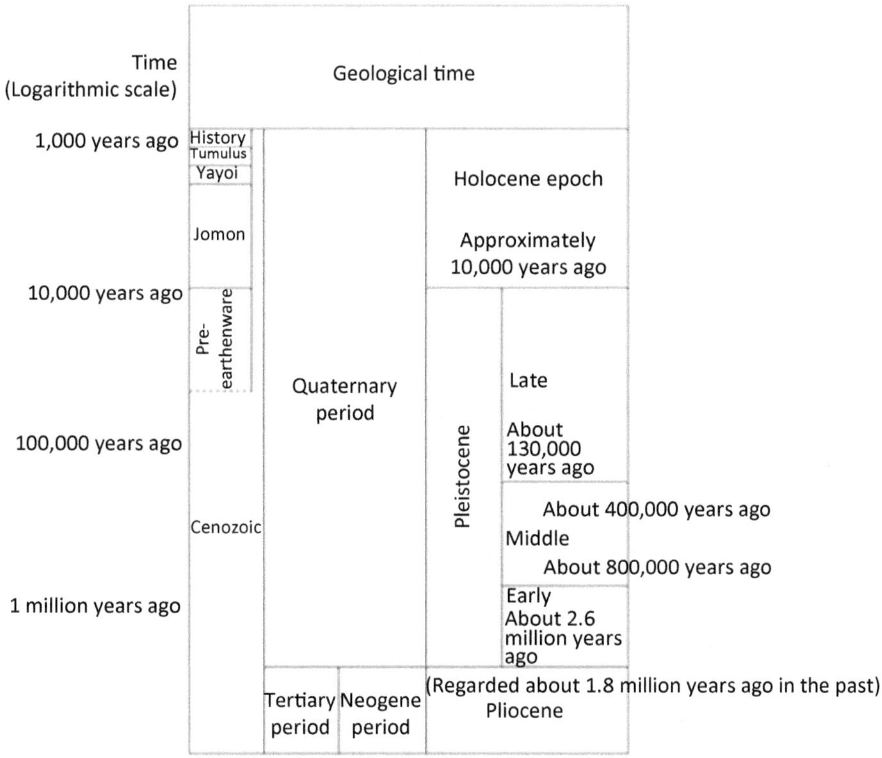

Fig. 2.15 Geological age [5–7]

Appendix 2.2: Assessment of Standard Seismic Motion S_s by Response Spectrum Method: Guidelines for the Seismic Design of Nuclear Power Plants (JEAG 4601-2008)

This method was developed for deriving response spectra on the bedrock based on statistical analysis of ground motions observed during past earthquakes (214 horizontal components and 107 vertical components) on bedrock in the Kanto and Tohoku regions, Japan.

JMA (Japan Meteorological Agency) magnitudes, hypocentral depths, and epicentral distances of the observed seismic motions were >5.5, <60, and <200 km, respectively. The applicability of this method was verified by comparing the seismic motion records observed at 23 locations in Japan, as well as 14 overseas locations with those calculated by this method. The parameters for the seismic motion assessment include earthquake magnitude M_J (JMA), equivalent hypocenter distance X_{eq}, and the S wave velocities of the bedrock.

As Fig. 2.16 shows, the equivalent hypocentral distance is defined by taking into account the effects of the area of the fault plane, and the distribution of the strong ground motion generation areas. The equivalent hypocentral distance is the distance at which the seismic wave energy released from each segment of the seismic source fault becomes equivalent to the energy released from a single specific point.

Response spectra in the horizontal and vertical directions on the virtual free surface of the bedrock are calculated by considering the effects of the site amplification of the ground. The effects of seismic wave amplification characteristics at the site are evaluated by analyzing existing earthquake ground motion records.

To develop the time histories of the earthquake ground motion, the envelope curves E (T), as shown in Fig. 2.17, are applied. The periods for the changing points of the curves, T_b and T_c, can be defined by the magnitude (M_J) and the equivalent hypocentral distance X_{eq}. Figure 2.18 shows an example of the time histories and the response spectrum of the earthquake ground motion development.

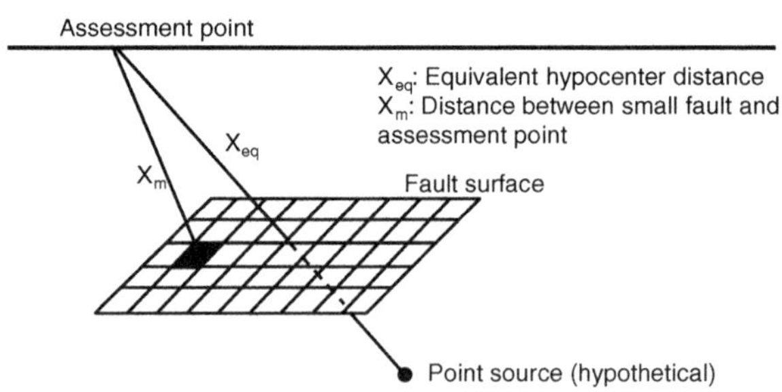

Fig. 2.16 Concept of equivalent hypocenter distance [8]

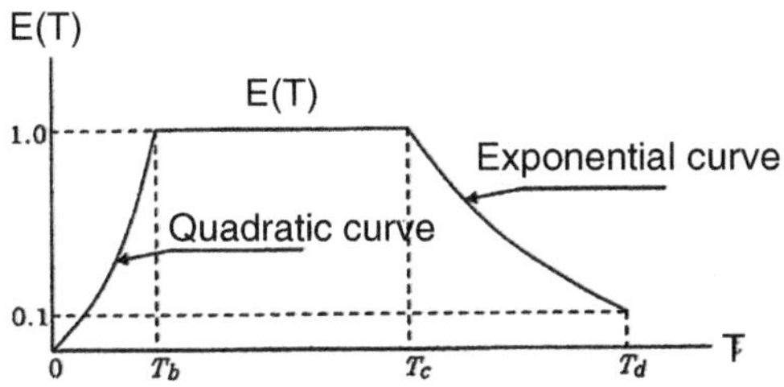

$$E(T) = (T/T_b)^2 \qquad (0 < T \le T_b)$$

$$E(T) = 1 \qquad (T_b < T \le T_c)$$

$$E(T) = e^{\frac{\ln(0.1)}{T_d - T_c}(T - T_c)} \qquad (T_c < T \le T_d)$$

$$T_b = 10^{0.5M-2.93}$$

$$T_c - T_b = 10^{0.3M-1.0}$$

$$T_d - T_c = 10^{0.17M+0.54\log X_{eq}-0.6}$$

Fig. 2.17 Temporal changes in the amplification envelope [8]

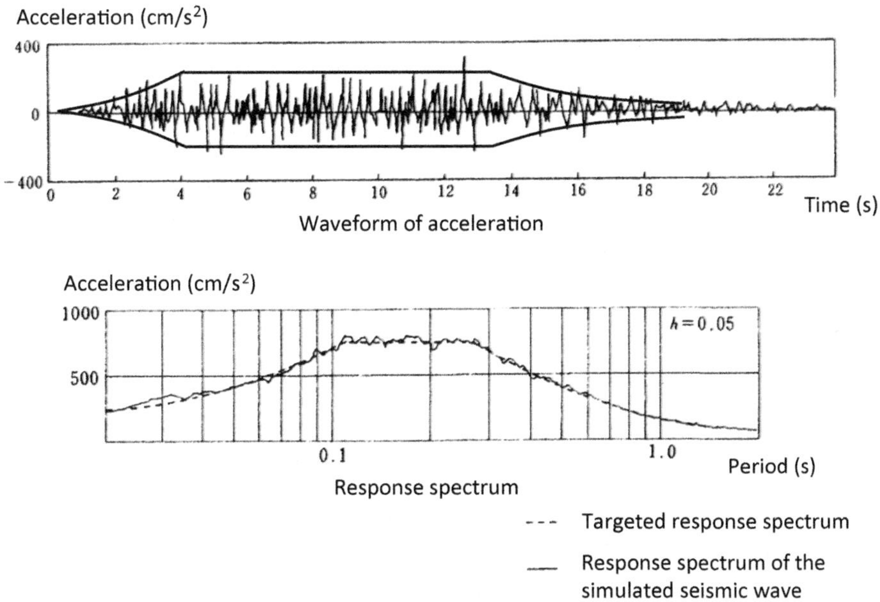

Fig. 2.18 Example of earthquake ground motion development [8]

References

1. "*Nihon no Jishin Bosai, Katsudanso* (Japan's Seismic Disaster Prevention; Active faults)", Ministry of Education of Japan, Culture, Sports, Science and Technology, 2004
2. The Headquarters for Earthquake Research Promotion, Japan website
3. "*Nankai Torafu no Kyodai Jishin Moderu Kentoukai* (The investigation committee of the Nankai Trough mega-thrust earthquake)", Cabinet Office, Government of Japan, 2012
4. "Characteristics of Observed Peak Amplitude for Strong Ground Motion from the 1995 Hyogoken Nanbu (Kobe) Earthquake", Yoshimitsu Fukushima, Kojiro Irikura, Tomiichi

Uetake, and Hisashi Matsumoto, Bulletin of the Seismological Society of America, 90, 3, pp. 545–565, June 2000
5. "*Nihon no Chikei* (Geographical Features of Japan)", Sohei Kaizuka, Iwanami Paperbacks, 1977
6. "*Nihon-rettou no Tanjou* (Birth of the Japanese Archipelago)", Asahiko Taira, Iwanami Paperbacks, 1990
7. Japan Association for Quaternary Research website
8. "Technical Guidelines for Seismic Design of Nuclear Power Plants (JEAG4601-2008)", The Japan Electric Association, 2008

Further Readings

9. "Shinpen, Nihon no Katsudansou (New Edition of Active Faults in Japan); Active Fault Research", University of Tokyo Press, 1991
10. "Kijun-jihindou oyobi Tiashin-sekkei Houshin ni kakaru Shinsa Gaido (Review Guide for Design Basis Ground Motion and Seismic Design Policy)", Nuclear Regulation Authority, 2013

Chapter 3
Stability Assessment of Foundation Ground and Surrounding Slope

Masanori Hamada and Michiya Kuno

Abstract In designing earthquake-resistant nuclear power plants, it is essential to verify the stability of the foundation ground upon which the reactor building and other important buildings/facilities will be built; assessing the stability of surrounding slopes is also important. If the foundation ground fails as a result of slippage, important safety-related functions in the nuclear power plant may be seriously affected. Therefore, it is necessary to verify that the safety-related functions of the plant would not be seriously affected by either foundation ground failure or slope collapse by the standard seismic motion, S_s. This chapter introduces the methods used to determine the strength and stiffness of the foundation ground and bedrock for assessing the stability of foundation ground and surrounding slope, and explains how the test data was used in the stability analysis.

Keywords Foundation ground · Surrounding slope · Rock test · Shear test · Flat plate loading test · Static analysis · Dynamic analysis · Finite element method · Slippage

3.1 Stability Assessment

The stability assessment for the foundation ground and surrounding slope is done by numerical analysis, taking into consideration the physical properties of the foundation ground and slope by site and laboratory tests. The stability assessment flowchart is shown in Fig. 3.1.

M. Hamada (✉)
Waseda University, Shinjuku, Tokyo, Japan
e-mail: hamada@waseda.jp

M. Kuno
Chubu Electric Power Company, Nagoya, Japan

M. Hamada and M. Kuno (eds.), *Earthquake Engineering for Nuclear Facilities*,
DOI 10.1007/978-981-10-2516-7_3

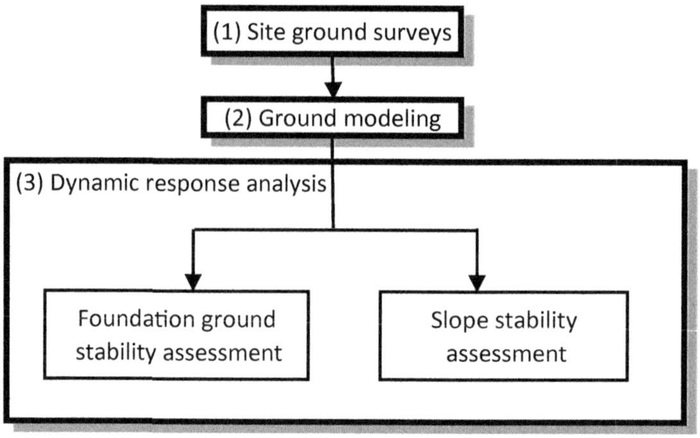

Fig. 3.1 Stability assessment flow

Ground surveys are performed to identify geological features and structures and determine the physical properties and static/dynamic mechanical characteristics of the foundation ground and surrounding slope (Sect. 3.2).

Based on the site ground survey, the grounds and the slope are grouped into different zones according to rock type and physical properties. Geological and rock classification maps are prepared, and a ground model is constructed for the analysis.

Basically, the grounds and slope stability during earthquakes is examined by focusing on slippage in the foundation ground and slope and on settlement of the reactor buildings by dynamic response analysis. Input ground motion for the analysis is also derived from the standard seismic motion S_s (Sects. 3.3 and 3.4).

3.2 Site Ground Survey

Geological surveys (boring explorations, etc.) identify geomorphic features and geological structures, and rock/bedrock testing determines their strength and stiffness.

Rock/bedrock testing are classified into two types: in situ testing performed on site, and laboratory testing of core samples from boreholes and/or specimens from exploratory adits. In principle, hard rock structures are addressed mainly by bedrock testing because the strength and deformation characteristics are largely determined by irregular surfaces at joints, while soft rock structures are addressed mainly by rock testing because its properties are hardly affected by joints and cracks. Geological surveys have already been discussed in Chap. 2, Sect. 2.2. The following section mainly describes in situ and laboratory testing for strength and stiffness, which are important for stability assessment.

3.2.1 *Bedrock Testing*

3.2.1.1 Shear Test of Bedrock

To determine bedrock strength and stiffness of the foundation ground, an in situ test is performed in an exploratory adit. There are two types of rock shear test: the block shear test in which a concrete block is cast over the target bedrock and a force is applied via the concrete block to shear the bedrock below it. The shear test in which the target bedrock is carved in situ to form a projection and a direct force is applied to shear it. The details of the block shear test are described below.

As shown in Fig. 3.2, the test apparatus consists of loading equipment (hydraulic jacks, counterforce bearing concrete block) and measuring instruments (load cells, etc.). A concrete block is cast over the target bedrock (specimen). In the test, the specimen is subjected to a certain vertical load (vertical counterforce) by hydraulic jacks, the inclined load from the lateral hydraulic jack is then gradually increased until a shear fracture occurs on the bedrock surface under the block. This shear strength is measured for several specimens under various vertical loads, and the relationship between the shear strength and vertical load is plotted as shown in Fig. 3.3. From these results, the shear strength characteristics (coherence c, and friction angle of shearing resistance φ) are derived using Coulomb's failure criterion.

3.2.1.2 Flat Plate Loading Test

To determine the stiffness and bearing capacity of the bedrock for assessing the deformation caused by the weight of the buildings and structures, and inertia forces during earthquakes, a plate loading test is performed at the site.

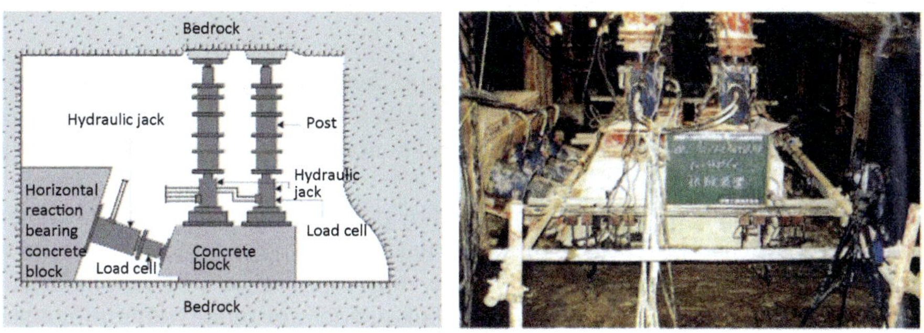

Fig. 3.2 Block shear testing of bedrock [1]

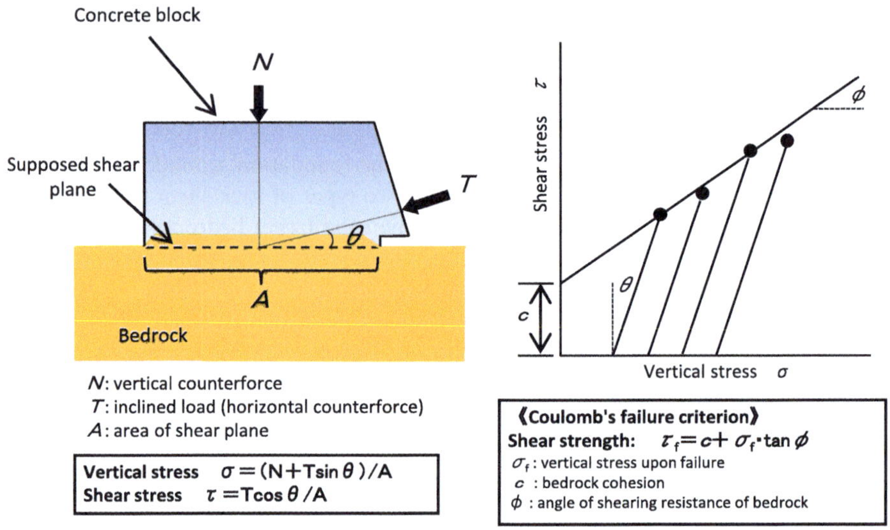

Fig. 3.3 Strength characteristics by the bedrock shear test

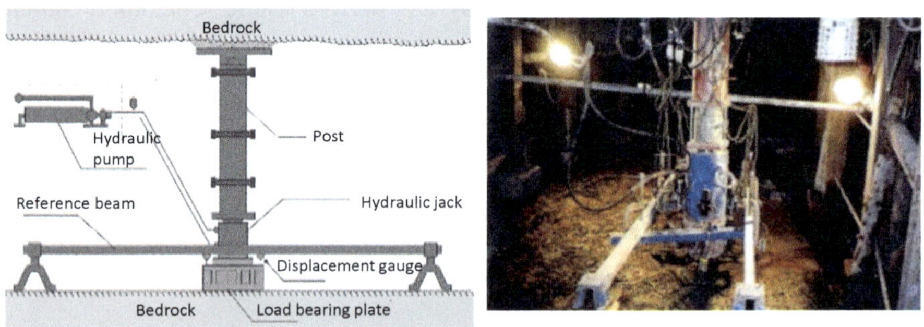

Fig. 3.4 Flat plate loading test of bedrock [1]

As shown in Fig. 3.4, the test apparatus consists of a load bearing plate, hydraulic jack, post, reference beam, and displacement gauges. With a load applied vertically by jacks, the displacement of the reference beam is measured to obtain the bedrock deformation characteristics.

As shown in Fig. 3.5, the test is preceded by preliminary loading. The maximum load is then reached in repeated cycles of loading and unloading. The maximum loading level is usually set to one to three times the design load caused by the standard seismic motion S_s. With soft bedrock, the loss of rigidity after prolonged loading (creep) needs to be considered: after the completion of the loading and unloading cycles, the maximum load is kept to obtain the creep characteristics.

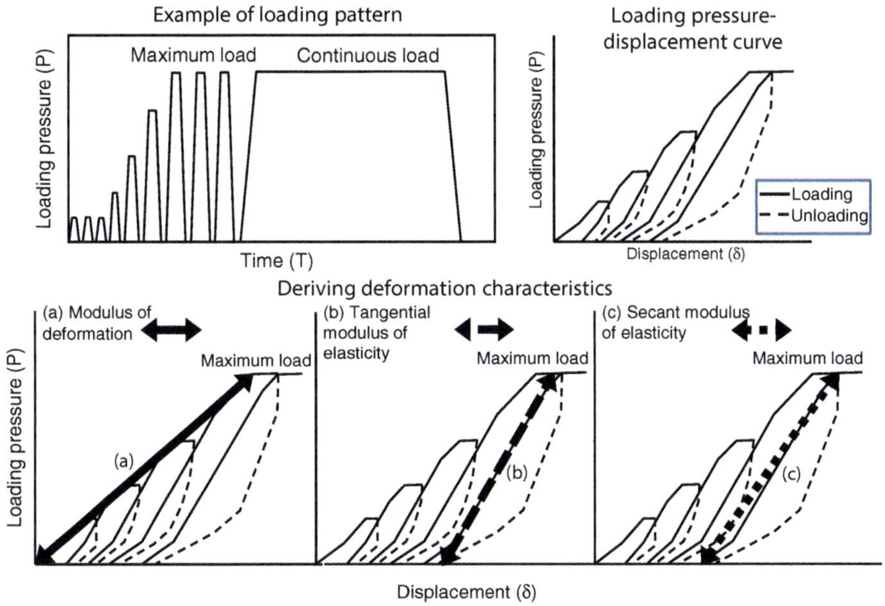

Fig. 3.5 Load-displacement relationship in plate loading test

From the applied load and the resulting displacement, a load displacement curve is derived to determine the bedrock stiffness. As shown in Fig. 3.5, the bedrock deformation characteristics are defined in terms of (a) the modulus of deformation, (b) the tangential modulus of elasticity, and (c) the secant modulus of elasticity. These are referred selectively depending on the purpose of the bedrock stability assessment.

(a) Modulus of deformation:
Represents the deformation characteristics considering the effect of the softening of the bedrock, and is used for the elastic analysis where the softening is taken into consideration. The modulus of deformation is derived from the envelope of the repetitive loading cycles as shown in Fig. 3.5a.

(b) Tangential modulus of elasticity:
Represents the deformation characteristics of bedrock when it behaves elastically, and is used as the modulus for linear analysis and initial elastic modulus in nonlinear analysis. The modulus of tangential elasticity is derived from the inclination of the linear part of the unloading curve from the maximum loading as shown in Fig. 3.5b.

(c) Secant modulus of elasticity:
Represents the deformation characteristics when the bedrock behaves nonlinearly, and is used when analyzing deformation, including the effects of fissures. The secant modulus of elasticity is derived from the inclination of the line connecting the start point of the cycle loading and the maximum load. The

load at which the displacement begins to increase quickly is defined as the ultimate bearing capacity of the bedrock.

The modulus of deformation D, and the modulus of elasticity E can be calculated as follows:

$$D \text{ or } E = 1/2 \cdot \pi a\left(1 - v^2\right) \cdot \Delta P/\Delta\delta$$

here

- v Poisson's ratio of bedrock
- a load bearing plate radius
- ΔP load increment in the section chosen for modulus estimation
- $\Delta\delta$ displacement increment in the section chosen for modulus estimation.

3.2.2 Rock Test

A rock test is conducted in the laboratory using specimens obtained from boreholes and exploratory adits. Rock tests fall into two categories: a test to determine physical properties such as specific gravity, and a mechanical test to determine strength and stiffness.

The results of the rock test are compared with the results of in situ bedrock tests and used to estimate the spatial distribution of bedrock having different properties. They are also used to prepare an analytical model for foundation ground and surrounding slopes, described in Sect. 3.3.

Table 3.1 lists mechanical test methods and their suitability for different types of rock. Tests are implemented after choosing an appropriate test method in consideration of the type of rock to be tested. When a relatively weak zone exists, such as either a fault or a fractured layer, it is particularly important to appropriately assess the mechanical characteristics of such layers.

3.3 Foundation Ground Stability Assessment

This section describes the method for foundation ground stability assessment, analytical model, ground stability analysis and dynamic analysis.

3.3.1 Analytical Model

A ground model for assessing the stability of the foundation ground is prepared based on geological survey results, e.g., boring explorations, and those from

Table 3.1 Bedrock/rock test methods and assessment of strength and deformation characteristics

	Static			Dynamic		
	Deformation characteristics		Strength characteristics	Deformation characteristics		Strength characteristics
	Modulus of static elasticity E	Static poisson's ratio	Ground constant C, φ	Modulus of dynamic shearing elasticity G	Damping constant h	Cohesion C Angle of internal friction φ
Soft rock	Triaxial compression test		Triaxial compression test Bedrock shear test	Seismic wave velocity logging Cyclic triaxial compression test		Utilization of static test results
Hard rock	Flat plate loading test	Uniaxial compression test	Bedrock shear test	Seismic wave velocity logging	Conventional value (2–3 %)	
Faults and weak layers	Triaxial compression test			Cyclic triaxial compression test		

rock/bedrock tests. Hard rocks are classified into several zones according to the state of weathering and distribution of joints. Soft rocks are also classified considering the type, composition, and degree of consolidation. Because the presence of weak layers, such as faults, fractured zones, and seams and cracks, affect the stability assessment, their origins, continuity, and friability are taken into consideration for the foundation ground analytical model.

3.3.2 Ground Stability Analysis

The methods used in assessing the foundation ground stability include the conventional method, static analysis, and dynamic analysis. The conventional method is simple and easy, but it is not very accurate or reliable. Both static and dynamic analyses of foundation ground require a more detailed definition of the physical rock properties. The dynamic analysis method requires earthquake bedrock motion as the input to the ground model. In the past, the conventional method was used in the primary design phase, while dynamic analysis was used in the detailed design phase. In recent years, however, dynamic analysis has been used more extensively.

(1) Conventional Method
The conventional method is a simple means for assessing the foundation ground stability. Assuming that static inertia forces act on the slip planes through the weak layers distributed either beneath the foundation or in the ground, a method such as Bishop's or Janbu's method is used to assess the stability against the static force, and the weight of the buildings and the ground. Additionally, ground settlement is assessed using the secant modulus of elasticity from both the plate loading and creep tests.

(2) Static Analysis
The foundation ground stability beneath the buildings and slippage are examined by the finite element method (FEM), whereby static force is taken into consideration. The finite element model of the foundation ground is developed based on information about the geomorphic features, geological structure, and the physical rock properties. The assessment is performed using the slip plane method; ground stability is assessed by comparing the stress along the slip planes and the shear strength of the slip plane. In static analysis, the linear relationship between the stress and the strain is applied. However, nonlinear analysis is performed when either the ground deformation characteristics show any significant nonlinearity or when a significant increase in plasticity as a result of stress concentration is expected.

(3) Dynamic Analysis
The dynamic behavior of the foundation ground is analyzed by the finite element method (FEM) using the time histories of seismic motion. Combining the normal-state stress derived from static analysis with the dynamic response stress caused by simultaneous horizontal/vertical excitation, the dynamic

response of the foundation ground is computed. The assessment of foundation ground stability is performed using the slip plane method.

Dynamic analysis methods include direct integration in the time domain, modal analysis, and integration in the frequency domain. Usually, however, the method of integrating in the frequency domain is used with equivalent linearization (for details of the direct integration and modal analysis methods see Part II, Chap. 14, Sect. 14.2)

3.3.3 Assessment by Dynamic Analysis

In the assessment of foundation ground stability by dynamic analysis, the following three items are examined: slippage of the buildings in the horizontal direction and inclination of the building foundations, and vertical bearing capacity of the foundation ground as shown in Fig. 3.6. Dynamic analysis by the finite elemental method is usually carried out by the equivalent linear method, where the linear analysis is repeated by changing the stress–strain relationship and damping constant depending on the strain level of the foundation ground. Figure 3.7 shows the process of foundation ground stability assessment by dynamic analysis. Figure 3.8 shows an example of the FEM foundation ground and reactor building models (for the basic concept of FEM, see Part II, Appendix 14.1).

As shown in Fig. 3.7, the foundation ground stability assessment process, the normal-state stresses generated by the weight of the building/structure and the ground are calculated first. The earthquake ground motion for input at the bottom level of the FEM model is then defined from the standard seismic motion S_s on the virtual free surface of the bedrock and the dynamic response of the foundation

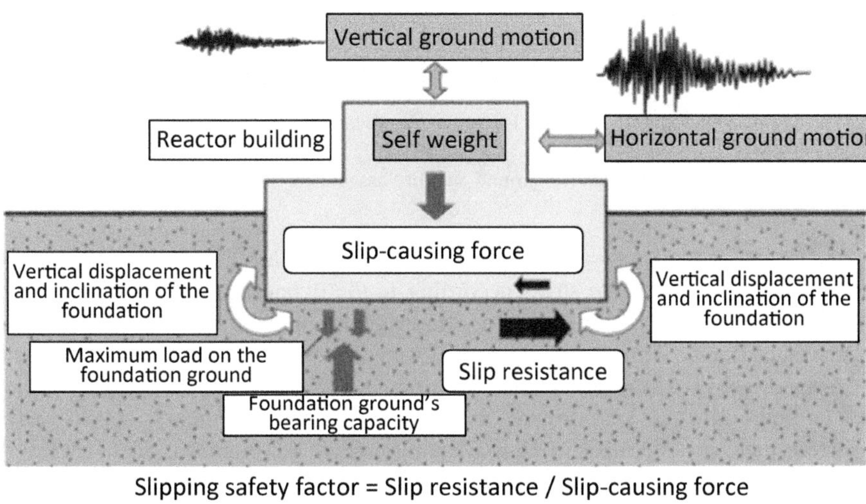

Fig. 3.6 Reactor building foundation ground stability assessment [2]

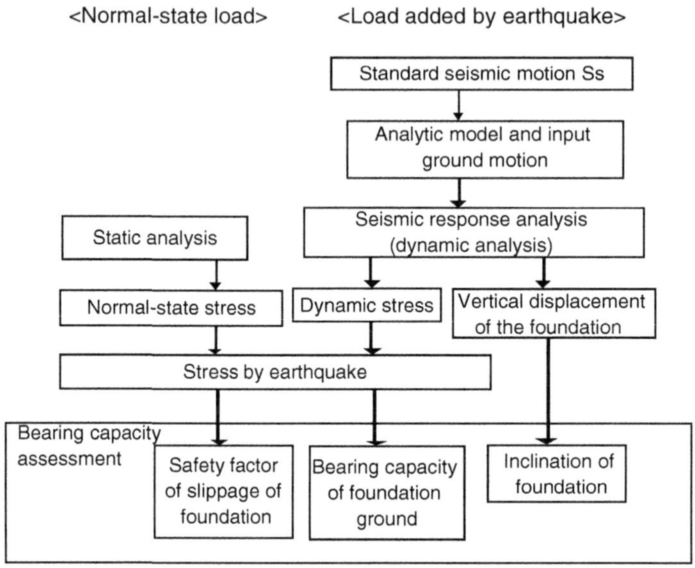

Fig. 3.7 Foundation ground stability assessment by dynamic analysis

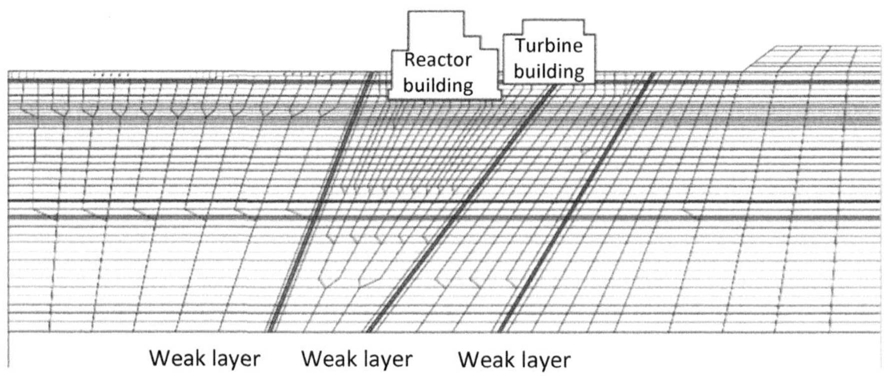

Fig. 3.8 FEM model for foundation ground stability assessment

ground is analyzed. Total stress by an earthquake is determined by combining the normal-state stress with the stress according to the dynamic analysis.

As shown in Fig. 3.9, the ground motion input at the bottom of the FEM model is derived as the incident wave propagating vertically upward in the bedrock, which is obtained from the standard seismic motion S_s defined on the virtual free surface of the bedrock. The one-dimensional wave propagation theory (Multi wave reflection theory) is applied for this calculation (Part II, Chap. 13, Sect. 13.3).

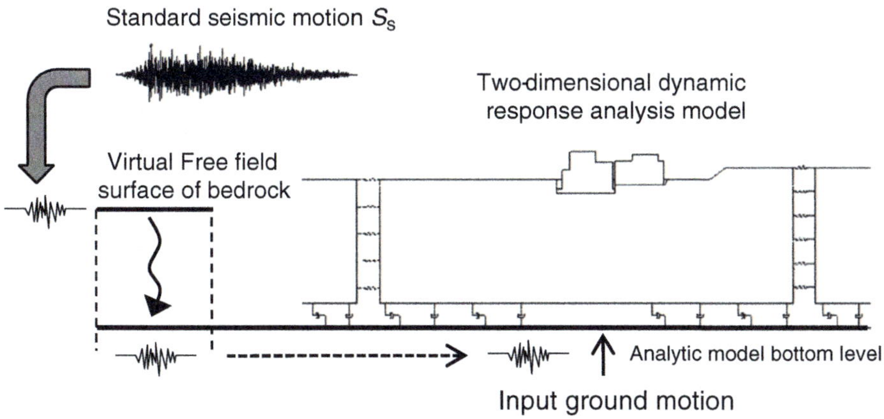

Fig. 3.9 Deriving the input ground motion at the bottom of the analytical model

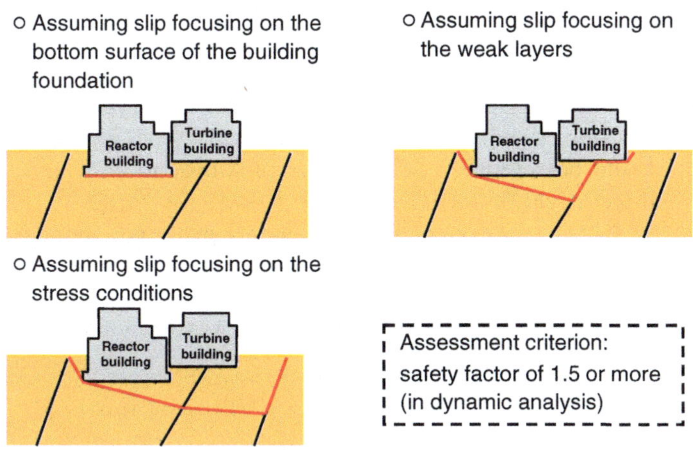

Fig. 3.10 Assumption of slippage planes

Slippage along the bottom of the buildings and on assumed slippage planes, which include weak layers and fractured zones, will be examined based on the shear stress on the planes obtained by the dynamic analysis, as shown in Fig. 3.10.

The slippage safety factor is obtained by dividing the sum of shear strength on the slip plane by the sum of the stresses caused by the weight of the ground and buildings, and dynamic forces. Several probable slippage planes are assumed for the foundation slippage safety assessment. In the dynamic analysis, the slippage safety factor should be >1.5 against the basic earthquake ground motion S_s.

The foundation ground bearing capacity will be examined in both the normal state and during earthquakes. The bearing capacity in the normal state is assessed by comparing the yield stress from the plate loading tests and the compressive stress as a result of the building's weight. The bearing capacity during earthquakes is assessed by comparing the ultimate strength from the plate loading tests and the stress during earthquakes.

The inclination of the foundation as a result of the buildings' weight and the seismic inertia force is examined to ensure that the structural integrity of the buildings and facilities will not be lost because of uneven ground settlement. The inclination of the building's concrete foundation slab is obtained by the dynamic analysis. The relative displacement in the vertical direction between the two ends of the foundation slab is estimated, and the inclination is calculated by dividing it by the width of the foundation. The predicted inclination degree must be smaller than the allowable inclination derived from the specifications of equipment and facilities.

3.4 Slope Stability Assessment

Rock slopes are created on the plant grounds as a result of site grading. If slope collapse poses any potential risk to the safety-related functions of the reactor facilities, a dynamic analysis using the standard seismic motion S_s should be performed to assess slope stability by the slippage plane method. The slopes requiring assessment should be selected based on their proximity to critical facilities and buildings, and the height of the slope.

The slope stability assessment process by dynamic analysis is the same as that of the buildings' foundation ground stability assessment shown in Fig. 3.7. The normal-state stress by the weight of the ground determined by the static analysis is combined with the stresses by the dynamic analysis. As shown in Fig. 3.11, the slip

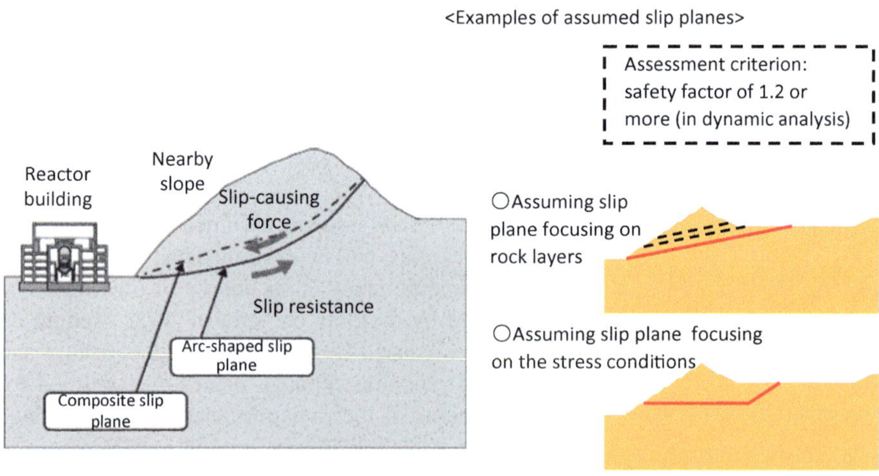

Fig. 3.11 Slope stability assessment on the reactor building grounds [2]

planes addressed by the stability assessment not only include the arc-shaped slippage planes typically used in slope stability assessment, but also those assumed when considering geological layer inclination, weak layers, and potential slip planes. The slip safety factor is the ratio between the shear strength and shear force acting along the slip plane. The slippage safety factor against standard seismic motion S_s should be >1.2.

References

1. Chubu Electric Power website
2. "*Genshiryoku Hatsudensho no Taishin Anzensei* (Seismic Safety of Nuclear Power Plants)", Japan Nuclear Energy Safety Organization, 2007

Further Readings

3. "*Ganbanjou no Oogata Kouzoubutsu Kiso* (Rock Foundation of Large-Scale Structures)", Japan Society of Civil Engineers, 1999
4. "Method of Classification of Geomaterials for Engineering Purposes (JGS0051)", The Japanese Geotechnical Society, 2000
5. "Technical Guidelines for Seismic Design of Nuclear Power Plants (JEAG4601-2008)", The Japan Electric Association, 2008

Chapter 4
Earthquake-Resistant Design of Building and Structure

Masanori Hamada, Michiya Kuno and Satsuya Soda

Abstract A nuclear power plant comprises various buildings, such as the reactor building, turbine building, and exhaust stack. Such buildings and structures must continue to fulfill their functional requirements in the event of earthquakes to ensure that the nuclear power plant remains safe. This chapter discusses the earthquake-resistant design of buildings and structures at nuclear power plants. First, the flow of the earthquake-resistant design based on dynamic response analysis is explained. Focusing on the reactor building, the eternal loads considered in the earthquake-resistant design and their combinations with the stationary load are explained. In addition, static and dynamic design approaches are described. Dynamic response models of the buildings and the foundation ground for earthquake-resistant design, where the effects of the soil–structure interaction are taken into consideration, are explained. Furthermore, a method of verifying the earthquake resistance of facilities employing shaking table tests is described.

Keywords Reactor building · Earthquake-resistant design · Dynamic response analysis · Performance requirement · Load combination · Allowable stress · Standard seismic motion S_s · Ultimate strength · Shaking table test

4.1 Buildings and Structures for Earthquake-Resistant Design

Among buildings and structures that must be designed to withstand earthquakes, the reactor building and other structures important for safety are explained. The functional requirements of buildings and structures in terms of their earthquake-resistant design are introduced.

M. Hamada (✉) · S. Soda
Waseda University, Shinjuku, Tokyo, Japan
e-mail: hamada@waseda.jp

M. Kuno
Chubu Electric Power Company, Nagoya, Japan

M. Hamada and M. Kuno (eds.), *Earthquake Engineering for Nuclear Facilities*,
DOI 10.1007/978-981-10-2516-7_4

4.1.1 Buildings and Structures of Nuclear Power Plant

The buildings and structures of a nuclear power plant are outlined as follows, taking a boiling-water reactor (BWR) plant as an example. The major buildings at a BWR plant include the reactor building, which contains the reactor, the control building, which contains facilities such as the main control room for the centralized operation control and monitoring of the nuclear power plant, the turbine building, which contains turbine generators, and the radioactive waste disposal building used to contain and dispose radioactive waste. Figure 4.1 shows the buildings of an advanced boiling-water reactor (ABWR) plant.

4.1.2 Structure of Reactor Building

The reactor buildings of an ABWR plant are constructed using reinforced concrete with steel frame construction in parts. The building, which mostly has a box wall structure with the partial use of frame construction, is composed of walls, floor slabs, and the foundation slab. The walls, designed to withstand seismic loads, are as thick as approximately 2 m at lower story levels; the foundation slab is as thick as 5–6 m. A large open space is provided above the operation floor (where refueling is performed); the overhead crane of the reactor building is attached to and travels across the ceiling. Around this large open space, a steel–concrete hybrid frame that

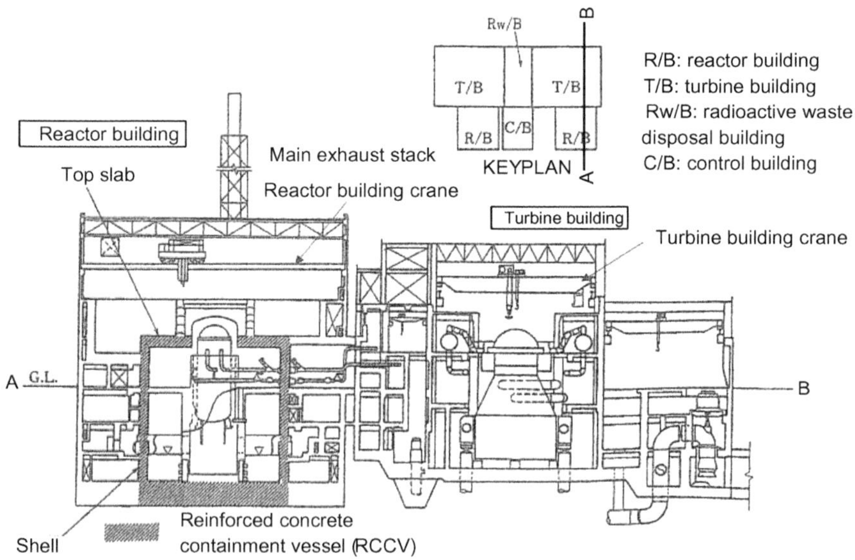

Fig. 4.1 Buildings of an ABWR plant of 1.35 million kWe class (single-box type) [1]

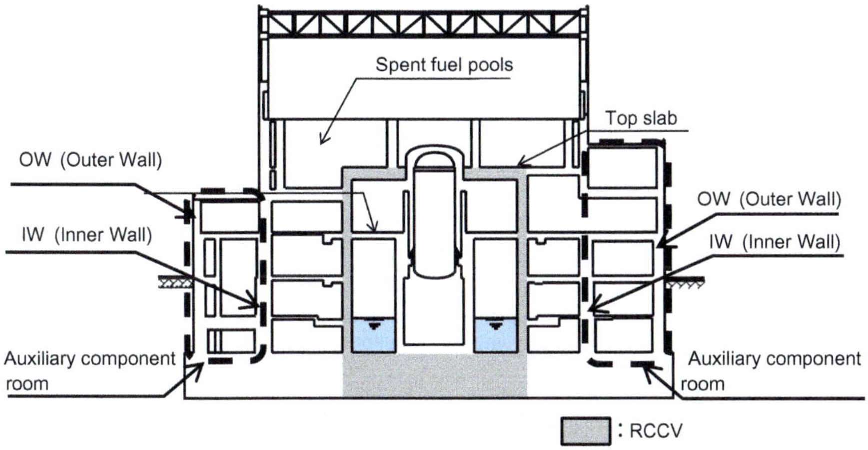

Fig. 4.2 Double-box-type reactor building (double-box type)

supports the crane and ceiling is constructed. To provide this large open space, the roof is supported by a steel frame truss.

The reinforced-concrete containment vessel (RCCV) is composed of reinforced-concrete walls, designed to withstand the pressure inside the vessel, and steel plate liners attached to the walls on the interior side for air tightness. The RCCV is connected with the reactor building by floor slabs at each story level. The bottom of the RCCV is embedded in the foundation slab of the reactor building. There are two types of reactor building: the single-box type and the double-box type. The single-box-type reactor building has a single seismic wall around the RCCV as shown in Fig. 4.1. The double-box-type reactor building, as shown in Fig. 4.2, has double seismic walls comprising the outer box wall (OW) and the inner box wall (IW) with a space for auxiliary components between them.

4.1.3 Performance Requirements of Building

Taking the case of an ABWR plant, the following explains the seismic classification of nuclear facilities and the functional requirements of buildings and structures.

The most important function required of reactor buildings and other important structures is the containment of radioactive substances. It is thus necessary to maintain subatmospheric pressure, to prevent the leakage of radioactive materials, and to prevent consequent impacts. Another important requirement is to support equipment/piping having important safety functions. Figure 4.3 shows the functional requirements for buildings of nuclear power plants. Table 4.1 outlines a design policy that ensures that the functional requirements are satisfied.

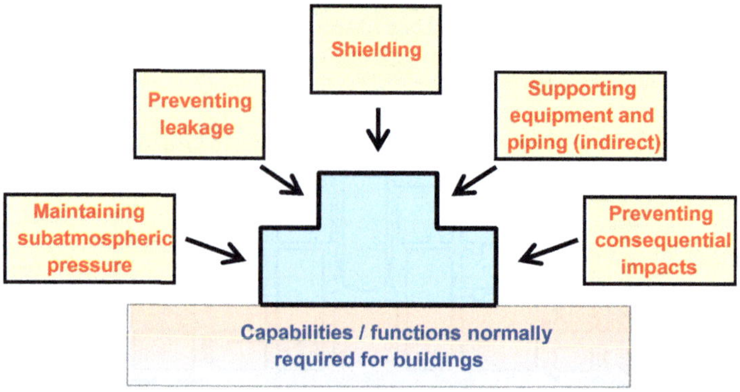

Fig. 4.3 Functional requirements for buildings and structures of nuclear power plants

Table 4.1 Design policy that ensures functional requirements [2, 3]

Functional requirement	Design policy
Maintaining subatmospheric pressure	The pressure of some spaces inside the reactor building should be kept below subatmospheric level to prevent radioactive substances leaking to the external environment in the event of an accident involving radioactive release.
Preventing leakage	Airtightness should be ensured by defense-in-depth provided by the reactor containment vessel, the secondary containment facilities, etc. Water-tightness should be ensured by means of steel liners attached to concrete structures.
Shielding	Reinforced concrete construction should be chosen in principle, for the Shielding.
Supporting equipment/piping (indirect)	Wherever a structural component (such as walls and floors) of a building or structure supports facilities that rank high in the seismic classification, the building/structures should be constructed as retaining its support function against the seismic class of the given facility.
Preventing consequential impacts	Ensure that facilities ranked in high seismic classification should not be affected by consequential impacts from the failure of walls, floors.

Figures 4.4 and 4.5 show the relationship between the functional requirements and the aseismic classification of buildings and structures in the case of an ABWR plant.

As shown by the example in Fig. 4.4, all facilities that are expected to directly relate to a safety function, such as the RCCV for the containment capability (leakage prevention), should be designed depending on their seismic classification. All buildings and structures that indirectly support equipment and piping in high seismic categories need to be designed to ensure that the function of the supported facility is maintained. Furthermore, the safety functions of facilities having high

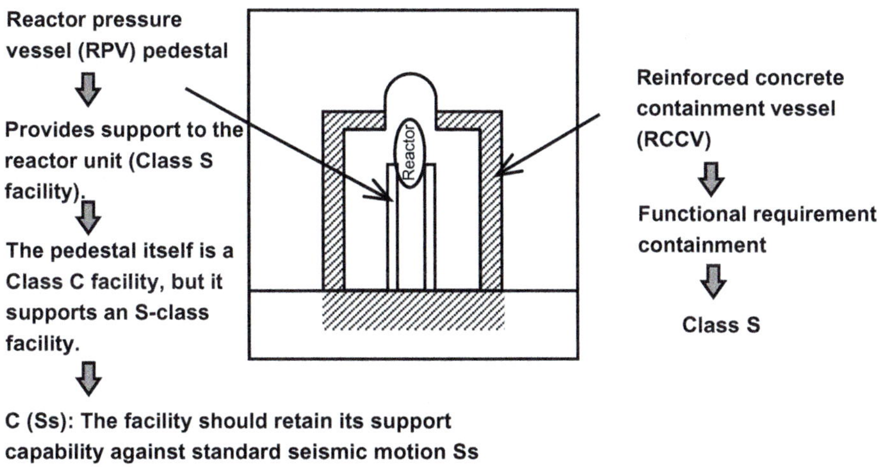

Fig. 4.4 Example 1 of the functional requirements and aseismic classification in the case of an ABWR

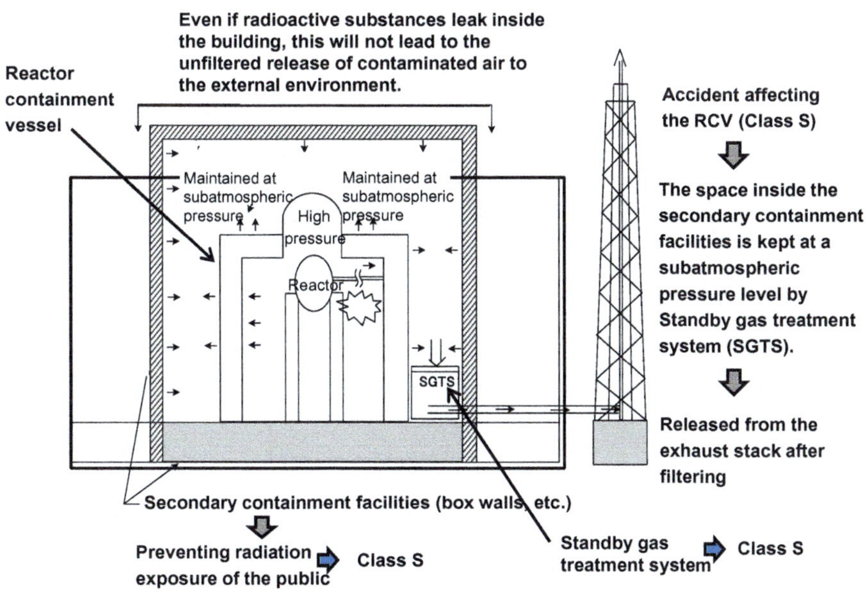

Fig. 4.5 Example 2 of the functional requirements and the seismic classification in the case of the ABWR

seismic classification will not be affected by consequential impacts from the failure of facilities having lower seismic classification.

As shown by the example in Fig. 4.5, the consequence of a leakage of radioactive substances from the reactor containment vessel is mitigated by the airtight design that maintains subatmospheric pressure inside the secondary containment facilities through the action of the standby gas treatment system (i.e., a subatmospheric pressure maintaining capability). This prevents the external release of radioactive substances. To ensure that an external release is prevented, a pair of airtight doors with an interlock that prevents simultaneous opening is provided at the gate of the boundary of the secondary containment facilities.

4.2 Flow of Earthquake-Resistant Design

Figure 4.6 shows the flow of earthquake-resistant design including dynamic response analysis and the seismic force determination of buildings and structures.

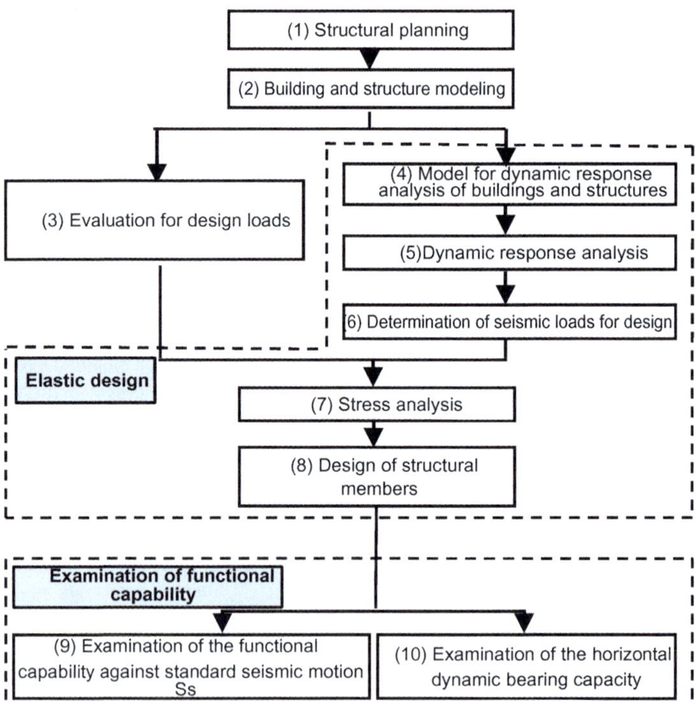

Fig. 4.6 Flow of the earthquake-resistant design of buildings and structures

(1) Structural planning

The basic structural design is planned according to the seismic classification of buildings and structures and the functional requirements of the installed facilities.

(2) Building and structure modeling

Buildings and structures are modeled as multiple-masses–springs-damper systems for dynamic response analysis (Part II, Chap. 14, Sect. 14.2) for the calculation of bending and shearing deformation in the horizontal and vertical directions.

(3) Evaluation for design loads

The earthquake-resistant design of buildings and structures considers multiple external forces: the dead loads due to the weight of buildings and structures, equipment, and piping, live loads due to the operation of equipment, snow and wind loads, and water and earth pressure. In addition, other loads, such as the load provided by the crane on the operation floor, are considered. For the design of the RCCV and the temperature loads, the dynamic water pressure and reactor-specific loads anticipated during abnormality and during test operations should be considered (Sect. 4.3).

(4) Model for dynamic response analysis of buildings and structures

The dynamic interaction of forces between the ground and structure is taken into consideration (Sect. 4.4).

(5) Dynamic response analysis

The dynamic response of each floor of the buildings is analyzed employing a ground–structure interaction model. The results of the analyzed response of each floor are used for the examination of the earthquake-resistance of equipment and piping installed on the floor (Sect. 4.4).

(6) Determination of seismic loads for design

The seismic loads used for design are determined according to the dynamic forces obtained in dynamic response analysis and the static loads prescribed according to the seismic classification (Sect. 4.4).

(7) Stress analysis

Stresses of each member of buildings and structures such as walls, columns and beams, generated by seismic loads, dead loads, and other loads, are calculated using a structural model (Sect. 4.5).

(8) Design of structural members

The size and strength of each structural member are set such that the stress generated by design loads is less than the allowable stresses (Sect. 4.5).

(9) Examination of the functional capability against the standard seismic motion S_s

The functions of buildings and structures in which equipment and facilities of S class are installed should be examined using dynamic loads induced by the standard seismic motion S_s. In the case of a building with seismic walls of reinforced concrete, dynamic analysis employing the standard seismic motion S_s as the input motion is performed while taking into consideration the nonlinear characteristics of the walls. There should be an adequate safety margin for the ultimate deformation (i.e., the ultimate deformation that corresponds to the ultimate strength (Sect. 4.6).

(10) Examination of the horizontal dynamic bearing capacity

Buildings and structures are assessed for their horizontal bearing capability to confirm there is a sufficient safety margin for the ultimate strength (Sect. 4.6).

4.3 Loads and Their Combination for Earthquake-Resistant Design

The earthquake-resistant design of buildings and structures requires the consideration of stationary loads, operation loads, seismic loads, and reactor-specific loads. The following explains the loads used for the design of buildings and structures.

4.3.1 Loads for Design

1. Stationary loads

Stationary loads are loads that act constantly on buildings and structures; they include fixed loads, borne loads, and earth pressure and water pressure. Fixed loads are vertical loads produced by the self-weight of building components, such as the floors, walls, and beams. Borne loads are vertical loads produced by the weight of the equipment and piping supported by the building and/or by the weight of furniture, fixtures, and equipment inside the building. Earth pressure and water pressure act on the walls of underground parts of structures.

2. Operation loads

Operation loads are loads that act on buildings and structures as a result of plant operation; they include the loads produced by the operation of equipment and piping (as a result of their thermal expansion, for example) and the pressure and temperature loads arising from operation.

3. Seismic loads

Seismic loads are loads that act on buildings and structures during earthquakes; they include inertia forces transmitted from equipment and piping, the dynamic earth pressure during earthquakes, and the dynamic pressure of the pool water. The seismic loads is considered as dynamic and/or static forces for the design.

4. Reactor-specific loads

Reactor-specific loads are loads that appear during an abnormality or during the testing of reactor facilities. They include pressure and temperature loads, as well as high-temperature and high-pressure jet stream loads that are produced by equipment and piping.

5. Other loads

Other loads include the snow and wind load and the load produced by cranes in the reactor building overhead.

4.3.2 Load Combinations and Allowable Limits

When designing each structural member of buildings and structures, appropriate combinations of stationary loads, operation loads, and seismic loads are considered to ensure that the resulting stress and deformation do not exceed allowable limits.

The designing of Class S facilities requires the consideration of both static and dynamic loads in the horizontal direction. The dynamic loads are assessed according to dynamic response analyses using the standard seismic motion S_s and the seismic motion for elastic design S_d.

As for the static seismic loads, a shear coefficient three times that for normal buildings (3 Ci) must be considered according to a provision in the Building Standards Act in Japan. The coefficient Ci is determined as a product of the regional coefficient depending on the seismic activity, coefficients related with the dynamic characteristics of the structures and the ground, and the coefficient of the distribution over height (Ai) (Part II, Chap. 15, Sect. 15.4). Vertical seismic loads are accounted for by the vertical seismic coefficient Cv (≥ 0.3).

Stationary and operation loads are combined with the seismic loads generated by the seismic motion for the elastic design S_d. Each structural member of buildings and structures is designed to ensure that the stress resulting from the combination of loads will not exceed the allowable stress. Furthermore, buildings and structures are required to have sufficient margins of strength and ductility against the combination of the stationary loads, operation loads, and seismic loads generated by the standard seismic motion S_s (Sect. 4.5).

Specifically, it is necessary to verify that the buildings and structures as a whole have sufficient deformability to withstand the standard seismic motion S_s (i.e., ensure that the shear strain of seismic walls and the ductility factor of steel frame structures remain within the allowable ranges).

In the case of Class B and C facilities, stationary and operation loads are combined with static seismic forces (1.5 Ci for Class B facilities, 1.0 Ci for Class C

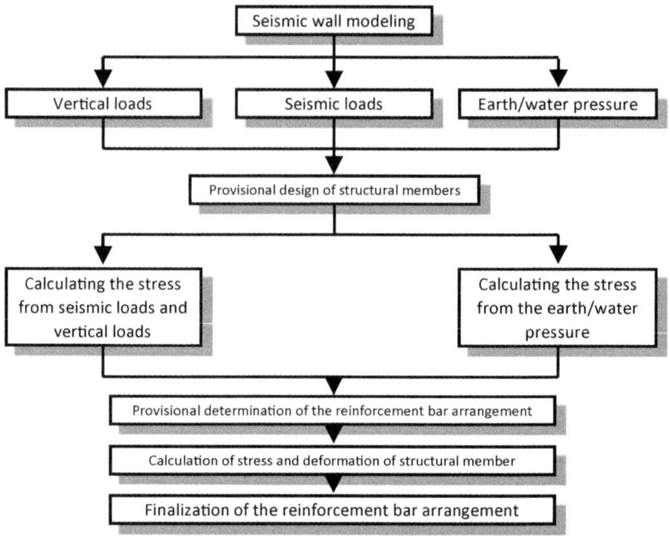

Load combinations			Assessment
Vertical loads	Seismic loads	Earth/water pressure	
Fixed loads Peration loads	-	Static earth/water pressures Dynamic earth pressure during earthquake	Allowable stress for long-term load
	Seismic load from Sd		Allowable stress for temporary load
	Seismic load from Ss		Examination of the functional requirements' Maximum shear strain

Fig. 4.7 Design flow of a seismic wall and examples of load combination

facilities), and each structural member is determined to ensure that the stress resulting from the combination of loads will not exceed the allowable stress specified by standards and regulatory criteria.

Figure 4.7 shows the design flow for seismic walls of a reactor building with a load combination. Using an analytical model of the wall, stresses are calculated for combined forces including seismic loads and earth and water pressure. The volume and arrangement of reinforced steel bars are determined so that the stresses of the concrete and steel bar remain below allowable stresses.

4.4 Dynamic Response Analysis

A dynamic response analysis is required for buildings and structures in which facilities of a high-rank seismic class are installed. The dynamic response of buildings and structures against the input earthquake ground motion is obtained by time history analysis, or modal analysis using the response spectrum. The following is an overview of the dynamic response analysis conducted to determine the seismic loads.

4.4.1 Flow of Dynamic Response Analysis

Figure 4.8 shows the flow of the dynamic response analysis, taking into account the interaction of the dynamic forces acting between the buildings and structures, and the ground. For the analysis of the interaction of the dynamic forces, a sway-rocking model (SR model) of the ground, where the ground is modeled by springs, or a finite element method (FEM) model of the ground, where the ground is treated as a continuous body is applied. Buildings and structures are modeled by bending and shearing beams with concentrated masses.

For the dynamic response analysis of the reactor building in the vertical direction, the columns are modeled as bar elements and the horizontal members of the roof are modeled by bending and shearing beams with concentrated masses.

The elastoplastic relationship between shear stress τ and shear strain γ and that between the bending moment M and curvature of the beam model φ of the seismic wall are taken into consideration. Here, the $\tau - \gamma$ and $M - \varphi$ relationship are obtained from the results of loading tests. Figure 4.9 shows $\tau{-}\gamma$ and $M{-}\varphi$ relationships.

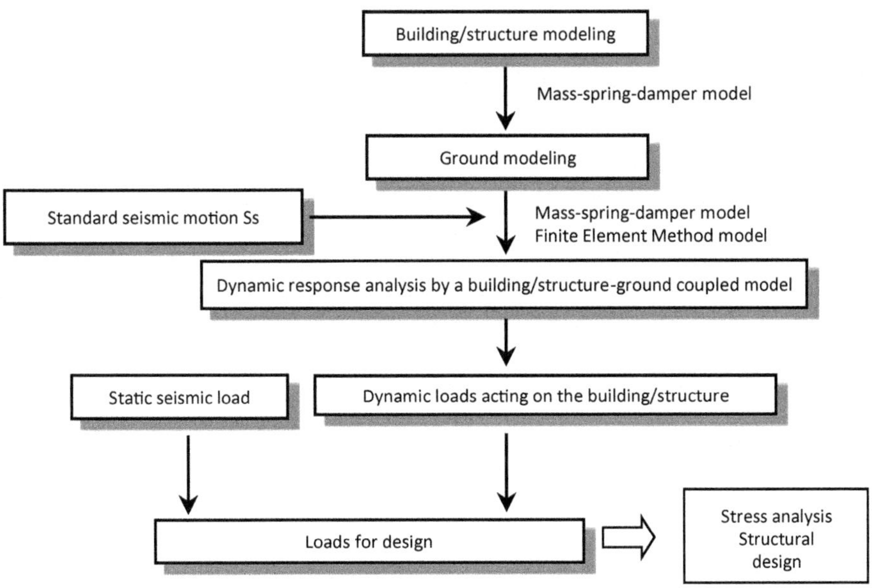

Fig. 4.8 Flow of dynamic response analysis

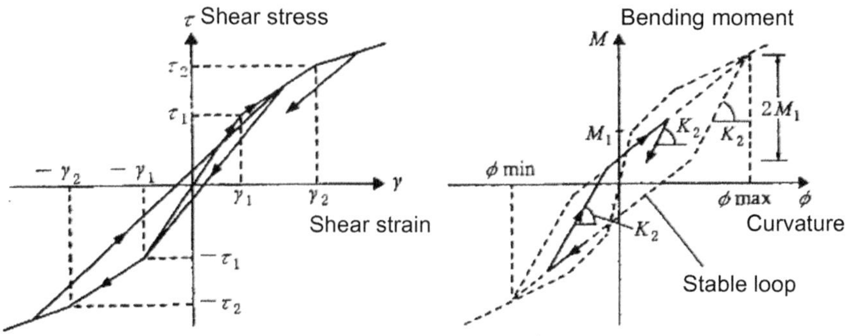

Fig. 4.9 Trilinear expression of the shear stress–shear strain relationship of seismic walls and the bending moment–curvature relationship of beam members [1]

Both the standard seismic motion S_s and the seismic motion for elastic design Sd are defined on a virtually free surface of the bedrock. Therefore, the input ground motion for the model of dynamic response analysis is derived from the seismic motions S_s and S_d on the virtual free surface to the basement of the analytical model according to one-dimensional shear-wave propagation theory.

4.4.2 *Dynamic Interactions Between Building/Structure and Ground*

Buildings and structures of nuclear power plants are heavy and rigid. The analysis of their dynamic response should therefore consider dynamic interactions of the buildings and structures with the ground.

The dynamic interaction refers to the mutual interaction of dynamic forces between buildings/structures and ground. From the viewpoint of seismic wave propagation, upward incident waves propagate into the buildings and structures from the ground, while downward waves reflect from buildings and structures and return to the ground. Figure 4.10 illustrates the concept of dynamic interactions between the buildings/structures and the ground.

Fig. 4.10 Concept of dynamic interaction between buildings/structures and ground

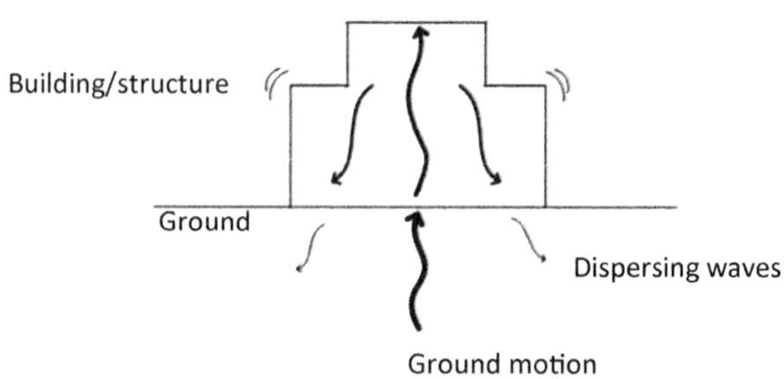

4.4.3 Analytical Model of Building and Structure

A model that consists of multiple masses, springs, and dampers is normally used for the dynamic analysis of buildings/structures, where the masses of the buildings/structures are concentrated at multiple mass points that are connected by springs that represent the stiffness of beams, columns, and walls (Part II, Chap. 14, Sect. 14.2).

Because the floor slabs of buildings are highly rigid, it is often assumed that the floor slab can be modeled as a rigid body, and the dynamic displacement in the horizontal direction is uniform over the whole of the floor slab. However, in the case that the rigidity in the in-plane direction is not enough, floor slabs are also modeled by multiple masses and springs. Since floors have greater mass than other structural members, it is assumed that masses of other structural elements such as walls, beams, and columns are concentrated to floor slabs. A multi-story building is modeled by multiple masses, springs, and dampers with one degree of freedom in the horizontal direction, where the number of mass points is equal to the number of stories of the building. The reactor building, however, is represented by a mass model with multiple degrees of freedom, where the number of degrees of freedom depends on the type of building structure and the layout of seismic walls. The seismic walls have connections with floors that are assumed to be rigid bodies. Figure 4.11 shows an example of a reactor building model at an ABWR plant.

With seismic walls (i.e., OWs, IWs, and RCCV walls), the extent to which they bear the mass of each building floor is defined. The masses in these divided zones appear as mass points in the multi-axes, multi-masses model. Figure 4.11 also

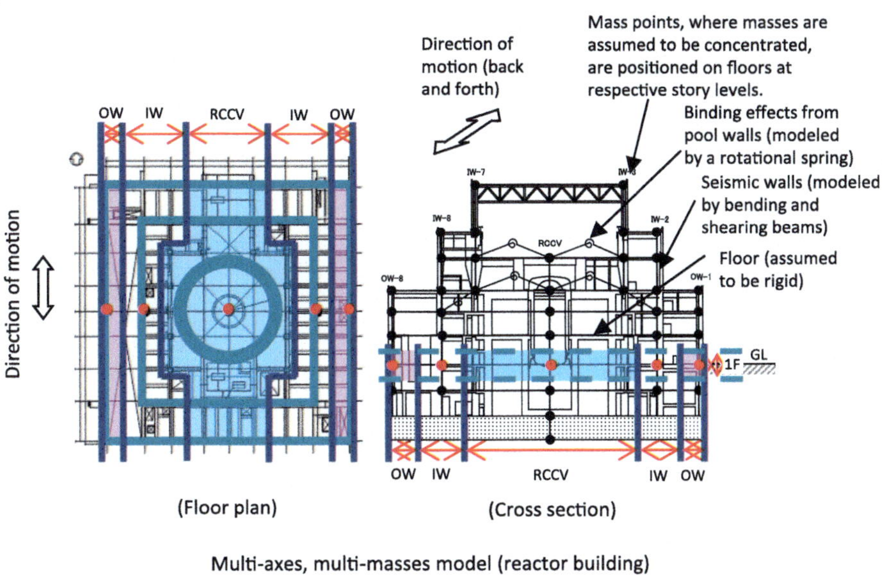

Fig. 4.11 Example of building model

shows the connections between seismic walls and floors. Around the spent fuel pools that form a large opening (discontinuity) in the floor, the walls at both ends restrict the rotational deformation of the RCCV. This restriction is modeled by a rotational spring.

4.4.4 Analytical Model of Ground

The dynamic response of the foundation ground of the reactor building is generally analyzed using the sway-rocking (SR) model. The ground is modeled as springs in horizontal and vertical directions as well as in rotation at the bottom of the foundation slabs. The coefficients of these ground springs are evaluated employing elastic wave theory under an assumption of semi-infinite and uniform ground under the building's foundation. In the analysis in the vertical direction, the ground is also modeled by a spring in the vertical direction.

Figure 4.12 illustrates the SR model and Fig. 4.13 illustrates the evaluation of the coefficients of the ground spring. The effects of the embedding of the foundation are taken into consideration by the embedded SR models.

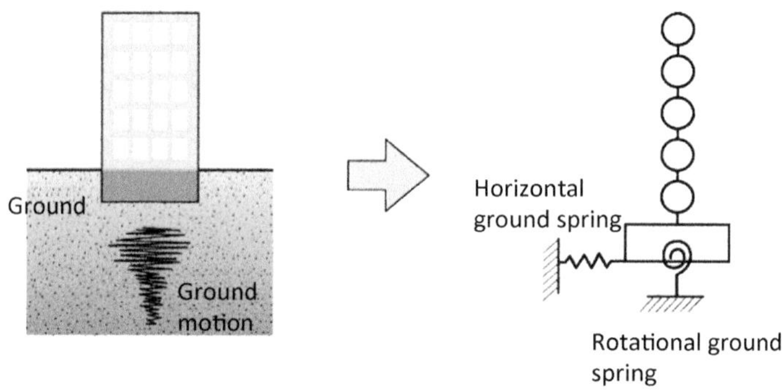

Fig. 4.12 Sway-rocking model

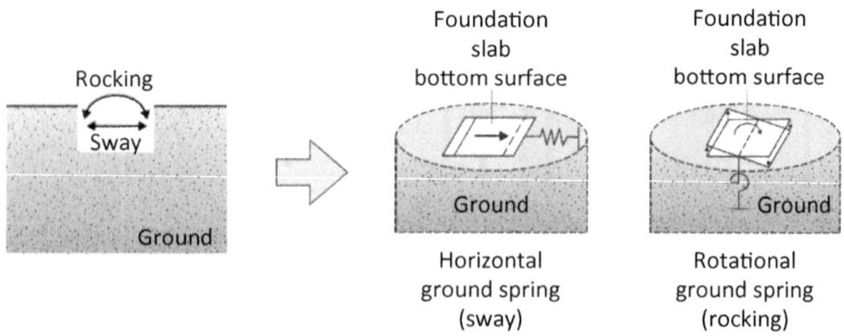

Fig. 4.13 Ground springs

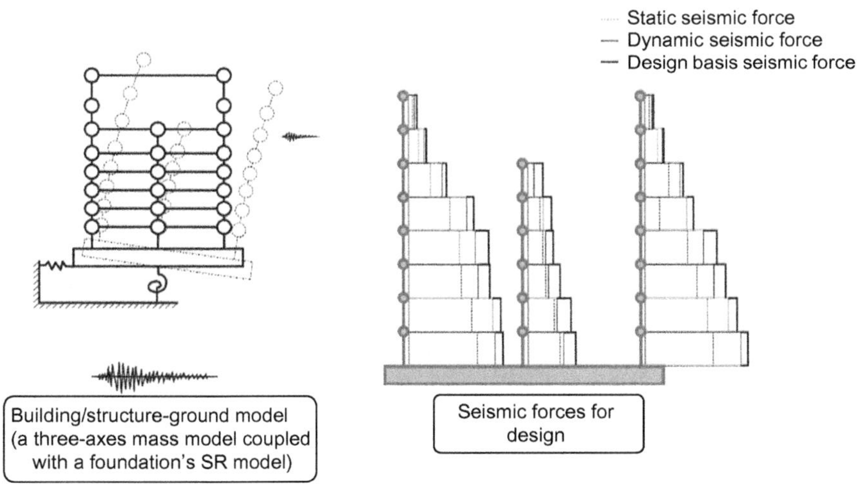

Fig. 4.14 Seismic forces for the design of buildings/structures

4.4.5 Seismic Loads for Design of Building and Structure

Static and dynamic seismic loads are determined for the design of structural members such as beams and columns. As shown in Fig. 4.14, static and dynamic seismic loads are obtained on each floor of the buildings and structures through dynamic response analysis using a building/structure–ground interaction model.

4.5 Design of Structural Members

Figure 4.15 shows three kinds of relationships between the external force and the deformation angle of buildings and structures. Type (a) has high strength and stiffness, but low ductility, while type (c) has low strength and stiffness, but high ductility. Ordinary buildings that mainly consist of beams and columns have type (b) and (c) relationships. However, the reactor building of a nuclear power plant, in which seismic walls are the main members that withstand the seismic force, has the type (a) relationship. The deformation of the building is limited so as not to affect the safety of the equipment and facilities installed inside.

Ordinary buildings and structures normally consist of columns, beams, walls, and floor slabs. Such buildings are designed to withstand seismic loads through their bending, shearing, and axial strength. Reactor buildings, however, are designed to withstand seismic loads mainly through their seismic walls. Such buildings withstand seismic loads by the shearing strength of seismic walls. In

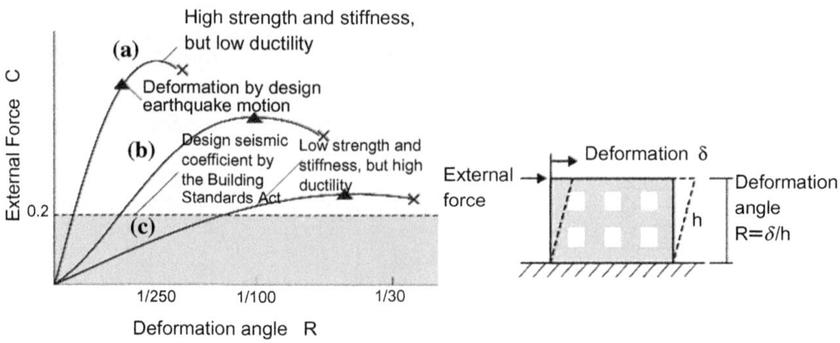

Fig. 4.15 Relationship between an external force and deformation angle of a building [4]

addition to seismic walls, other structural members such as columns and beams are also designed to have relatively high strength, so that the buildings have high strength and stiffness. The following explains the design of the structural members of buildings and structures.

4.5.1 Analysis of Stress Acting on Structural Member

Proper methods should be adopted for the analysis of stress and deformation of structural members depending on the geometry, materials, and other characteristics. The reactor building of an ABWR plant, for example, is designed such that the vertical loads and horizontal seismic loads are borne by seismic walls. The columns and beams at each story level are connected with the seismic walls. The RCCV is connected with the spent fuel pools, adding complexity to the geometry. Frame models are used for the analysis of buildings and structures that mainly consist of beams and columns, while a two- or three-dimensional FEM model is used for massive and continuous components such as the RCCV and foundation slab. Figure 4.16 shows examples of models for the analysis of the stress and deformation.

Based on the results of linear analyses performed using frame models and FEM models, the dimensions of structural members (e.g., walls, columns, beams, roofs, and foundations) are determined to ensure that the stress of the structural members do not exceed allowable stress. Figure 4.17 shows examples of the design of structural components of a reactor building.

Class S buildings and structures are required to have an adequate safety margin for the ultimate strength of their whole structures. The design of the structural members is repeated until this requirement is satisfied.

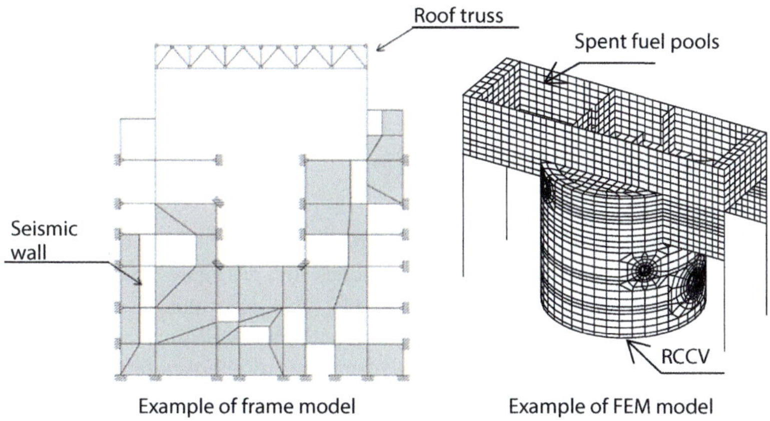

Fig. 4.16 Stress analysis models

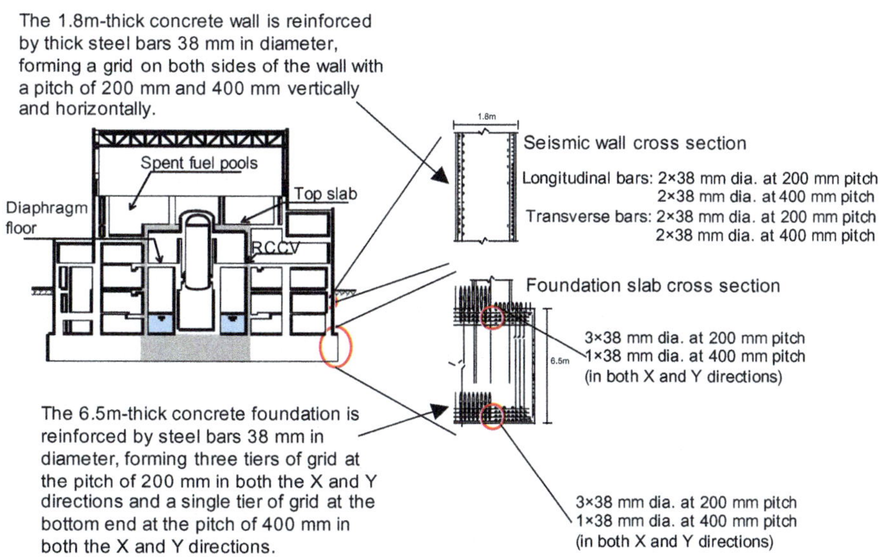

Fig. 4.17 Design of reinforced-concrete components of a reactor building

4.5.2 Design by Allowable Stress

In design employing the allowable stress method, the stress of the structural members generated by the combination of stationary loads, operation loads, and seismic loads. Seismic loads resulting from the dynamic response against the elastic design S_d are calculated under the condition of a linear relationship between stress and strain, and the stress is confirmed not to exceed the allowable stress specified by the design regulatory criteria.

4.5.3 Design by Ultimate Strength

The ultimate strength of a structure is the maximum load that can be withstood by the structure. After the external load reaches the ultimate load which can be withstood by the structure (ultimate strength), the deformation and strain of the structure drastically increase, resulting in the collapse. Buildings and structures that contain class S facilities are required to have a sufficient safety margins to the ultimate load and deformation. In design by ultimate strength, the structural members reach the elastic–plastic range, but the stress and strain of the members do not exceed the ultimate levels, and the members thus maintain their functional requirements. In the ultimate strength design of Class S buildings, the strains acting on structural members are evaluated and the horizontal dynamic bearing capacity of the buildings is assessed.

4.6 Assessment of Functional Requirements

This section outlines the assessment of capabilities of Class S facilities to continue fulfilling their functional requirements.

4.6.1 Assessment of Functionality Against Standard Seismic Motion S_s

The buildings and facilities of S Class are required to have a sufficient deformability as a whole structure and a sufficient safety margin to the ultimate strength against the standard seismic motion S_s.

4.6.1.1 Seismic Wall of Reinforced Concrete

Since the ultimate state of the reactor building is characterized mostly by the shear fracture of reinforced-concrete seismic walls, the safety of the seismic walls is assessed by the magnitude of the shear strain. The maximum allowable shear strain γ_a of reinforced-concrete walls is regulated to be 2.0×10^{-3} by taking into

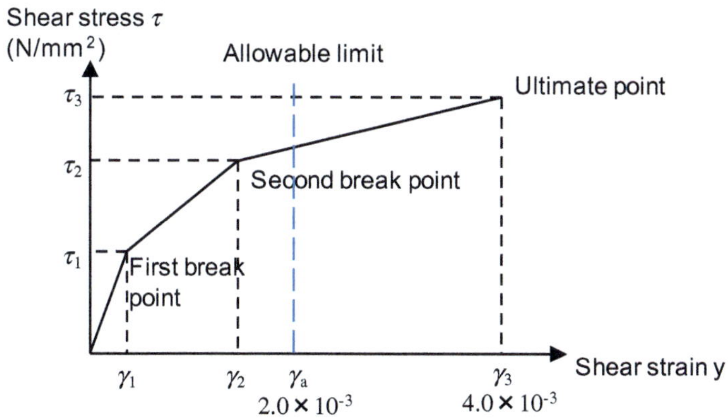

Fig. 4.18 Concept of the maximum allowable shear strain of a reinforced-concrete seismic wall

consideration a safety margin for the ultimate shear strain. Figure 4.18 shows the skeleton curve of the stress–strain relationship of a reinforced-concrete seismic wall and the maximum allowable shear strain.

The concrete reactor containment, such as the RCCV or prestressed-concrete containment vessel, requires an ultimate strength assessment based on the evaluation of the stress arising from the standard seismic motion S_s.

4.6.1.2 Steel Frame Structure

Steel frame structures with braces of Class S buildings and structures require an assessment based on the evaluation of the ductility factor. The ductility factor is a measure of the deformability of buildings and structures. Figure 4.19 shows one

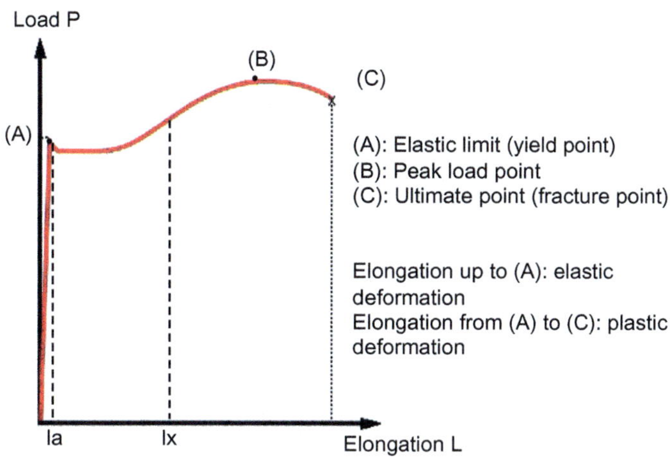

Fig. 4.19 Load–deformation relationship

example of the tensile load and elongation of a steel bar. In the figure, lx is the elongation at a specific load while 1a is the elongation at the elastic limit. The ductility factor is obtained as lx/la. The point B shows the peak load and the point C is the ultimate point (fracture point). The ductility factor of the structural members is shown as lu/la.

4.6.1.3 Load on Foundation Ground

For Class S buildings and structures, it is necessary to perform an assessment to ensure that the stress of ground under the foundation slab induced by the dynamic and static loads in the event of standard seismic motion S_s will not exceed the ultimate vertical bearing capacity of the ground. Given the same weight of the building/structure, a larger foundation slab (i.e., a larger contact surface) will result in a smaller stress of the ground. The stress of the ground at any part under the foundation slab will not exceed the ultimate bearing capacity of the ground to ensure an adequate safety margin for the ultimate bearing capacity.

4.6.2 Assessment of Dynamic Horizontal Bearing Capacity and Examination of Functional Requirement

For the design of Class S buildings and structures, it must be ensured that the dynamic horizontal bearing capacity at each floor exceeds the minimum requirement by a factor of at least 1.5. For Class B and Class C buildings and structures, the dynamic horizontal bearing capacity at each story level exceeds the minimum requirement (Part II, Chap. 15, Sect. 15.4).

In addition to the above, there are structural components that are required to provide particular functions, such as maintaining subatmospheric pressure, preventing leakage, shielding, physically supporting facilities, and preventing consequential impacts. Table 4.2 gives these structural components and their functional requirements.

Table 4.2 Examples of the functional requirements of buildings and structures

Component of building	Functional requirement
Reinforced Concrete containment vessel (including steel liners)	Maintaining subatmospheric pressure and preventing leakage
Some parts of the reactor building (secondary containment facilities, etc.)	Maintaining subatmospheric pressure
Spent fuel pools (including steel liners)	Preventing leakage
Main control room	Shielding

4.7 Verification of Design After Plant Construction

After the construction of reactor buildings, vibration tests and observations of dynamic responses during actual earthquakes will be carried out to verify the integrity of the design and construction.

4.7.1 Shaking Test for Reactor Building

Shaking tests are conducted using vibrators to verify the dynamic characteristics of the constructed reactor building, and the validity of the simulation model for the dynamic response analysis of the buildings and structures is confirmed. Figure 4.20 shows an example of the locations of the vibrators and monitoring sensors.

By the shaking test of the reactor building, the natural period and vibration modes and the damping constant of the building are obtained. The natural period is obtained from the resonance curve. Figure 4.21 shows an example of a resonance curve and natural period (Part II, Chap. 14, Sect. 14.1).

The building's damping constant can be obtained from the amplitude and shape of the resonance curve around the building's natural period.

4.7.2 Earthquake Observation at Nuclear Power Plant

Seismometers are installed in the reactor building and major plant components, foundation ground and bedrock, for the observation of dynamic motions during earthquakes. The validity of the design is reviewed and the integrity of the building and equipment is assessed based on the observed earthquake records.

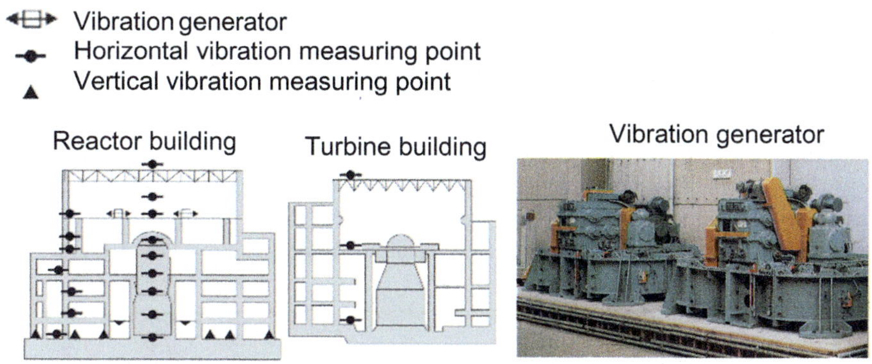

Fig. 4.20 Positions of vibration generator and measuring points in vibration testing

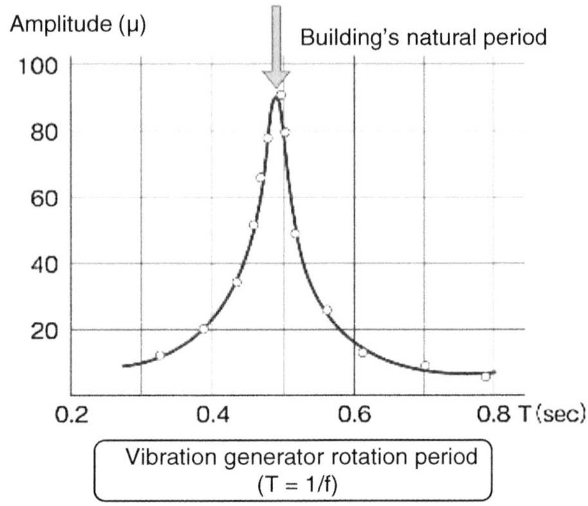

Fig. 4.21 Determination of the natural period from a resonance curve

By analyzing the motions of the building and the foundation ground during earthquakes of small and medium size, it is possible to verify the building's dynamic characteristics and damping properties as well as the characteristics of the building–ground dynamic interactions. When the nuclear power plant experiences large earthquakes, the earthquake observation records are analyzed to find the damage and to confirm the integrity of buildings and facilities. Figure 4.22 shows the flow of the reactor building integrity assessment based on earthquake observation records.

Earthquake motions were observed at the Unit 3 reactor building of the Onagawa Nuclear Power Station during the 2011 Great East Japan Earthquake and the integrity of the plant was verified.

Figure 4.23 shows the seismometer layout on the ground and in the Unit 3 reactor building. Figure 4.24 compares the recorded ground motion with the standard seismic motion S_s that was adopted for the design of the plant. Figure 4.25 shows the model for the dynamic response analysis. The observed records show that the response of the reactor building to the ground motion generated by the earthquake was mostly similar to that simulated for the standard seismic motion S_s.

Figure 4.26 shows the distribution of the maximum dynamic shear strain between two adjacent floors at each level of the reactor building. As shown, the maximum response shearing strain at each story floors obtained by dynamic response analysis based by the observed ground motion is lower than the seismic safety assessment criterion (2.0×10^{-3}).

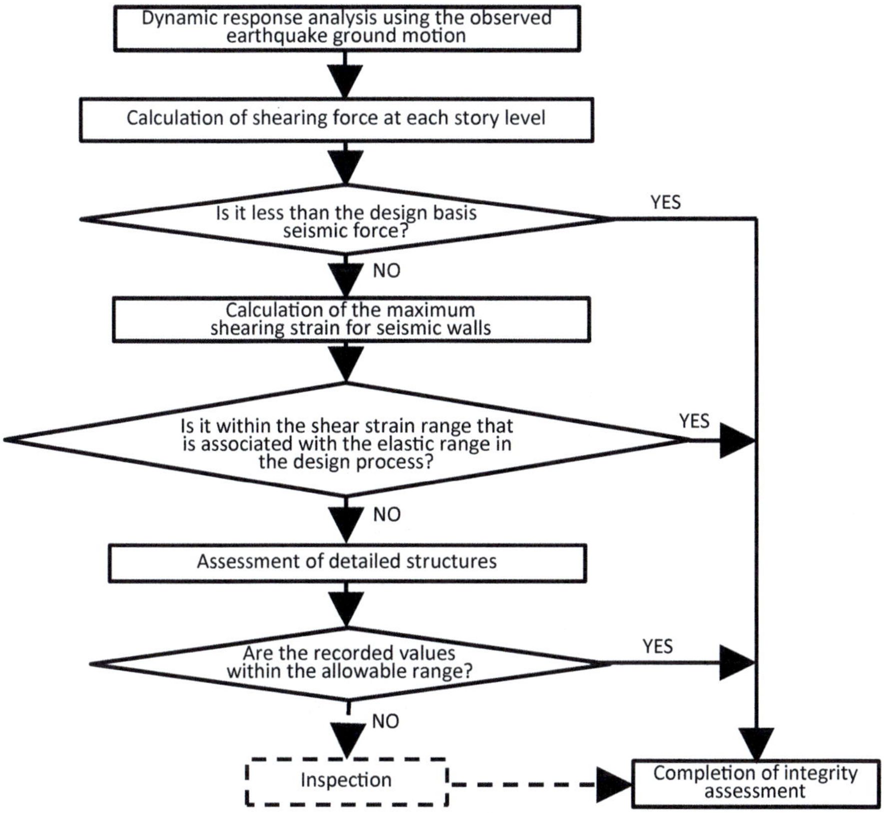

Fig. 4.22 Example of the flow of reactor building integrity assessment based on observed earthquake records

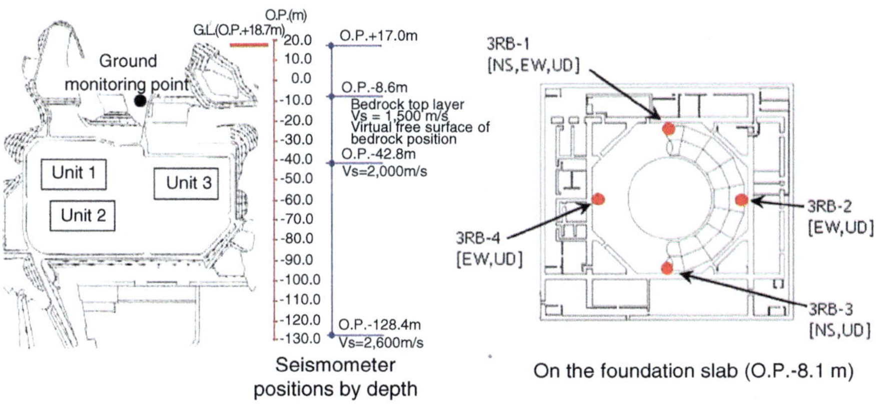

Fig. 4.23 Examples of the seismometer layout on the site ground and in the Unit 3 reactor building of the Onagawa Nuclear Power Station [5, 6]

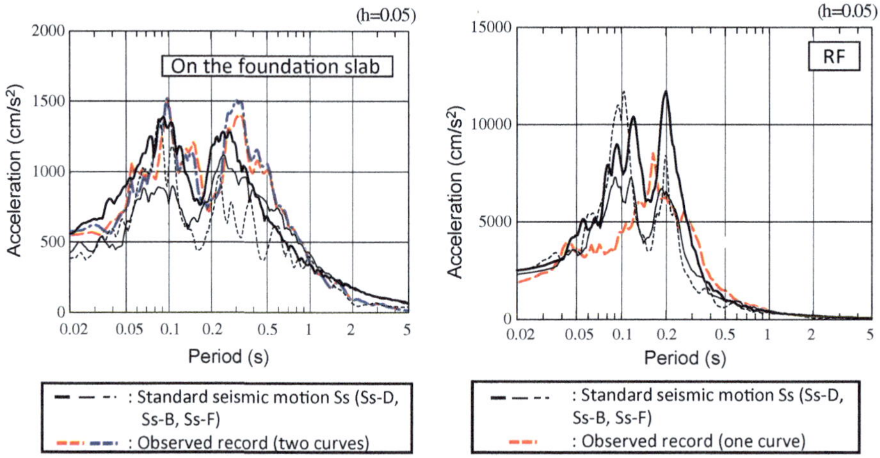

Fig. 4.24 Comparison of the design standard seismic motion S_s with the recorded ground motion (Onagawa Unit 3, north–south direction) [7]

Fig. 4.25 Dynamic response analysis model [7]

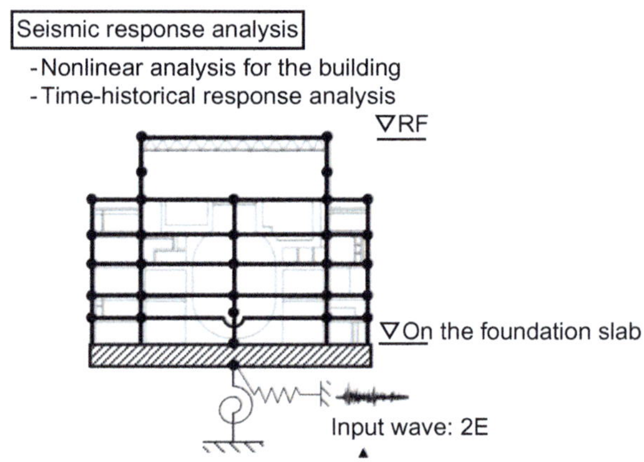

Figure 4.27 shows the shearing force of the seismic walls of the Onagawa Unit 3 reactor building simulated using the observed earthquake ground motion, for comparison with the elastic limit. The maximum shear force acting on the seismic walls was lower than the elastic limit strength, indicating that the seismic walls generally remained within the elastic range.

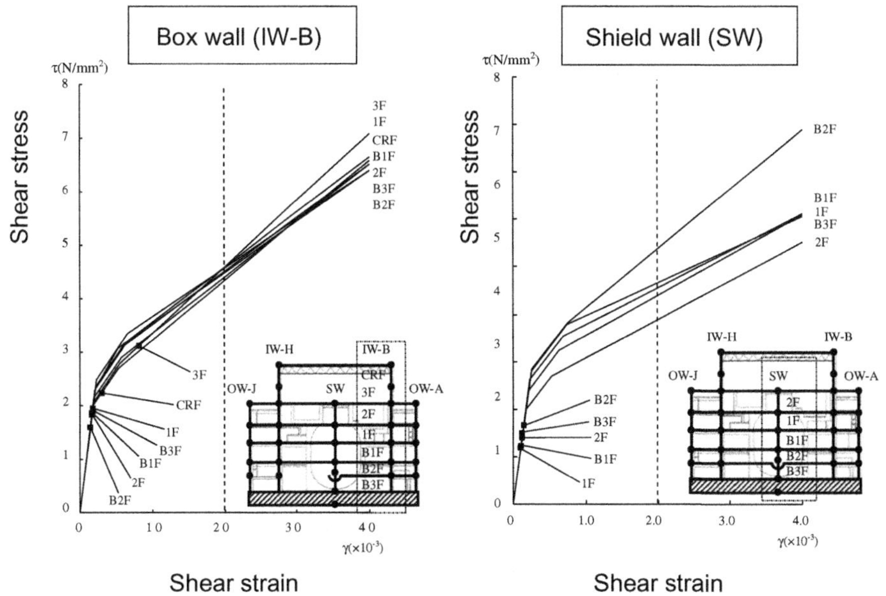

Fig. 4.26 Maximum response shearing strain experienced by Onagawa Unit 3 (north–south direction, where the solid line is the design curve) [7]

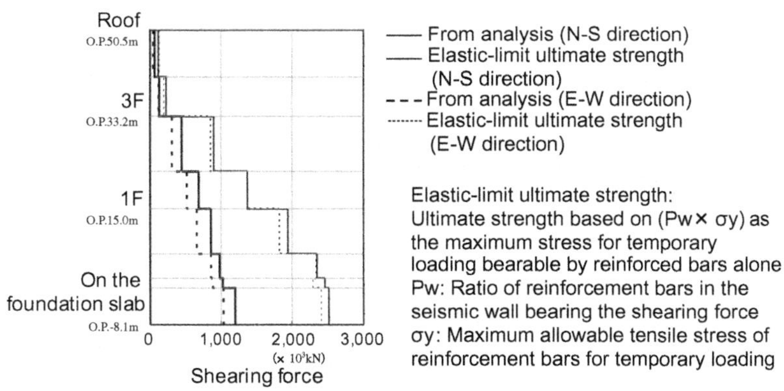

Fig. 4.27 Comparison of the story shearing force experienced by the seismic walls of the Onagawa Unit 3 reactor building with the elastic limit [7]

References

1. "Technical Code for Seismic Design of Nuclear Power Plants (JEAC4601-2008)", The Japan Electric Association, 2008
2. Japanese Architectural Standard Specification "JASS5 N Reinforced Concrete Work at Nuclear Power Plants", and its Explanation, Architectural Institute of Japan, 2013

3. "AIJ Standard for Structural Calculation of Reinforced Concrete Structures for Nuclear Power Facilities, Architectural Institute of Japan, 2013
4. "*Kenchiku no Taishin, Taifu Nyumon* (Introduction of Seismic Resistance and Wind Resistance)", Shokokusha Publishing Co., Ltd., 1995
5. "*Tohoku Chihou Taiheiyou Oki Jishin ni tomonau Onagawa Genshiryoku Hatsudensho ni okeru Jishin Kansoku Kiroku no Bunseki oyobi Tsunami no Chousa Kekka ni tsuite* (Analysis of Earthquake Observation Record and Survey Result of Tsunami caused by the 2011 Tohoku Earthquake at the Onagawa Nuclear Power Plant)", Tohoku Electric Power Co., Inc., April 7, 2011
6. "*Heisei 23 nen Tohoku Chihou Taiheiyou Oki Jishin ni-yoru Onagawa Genshiryoku Hatsudensho oyobi Tokai Diani Hatsudensho no Genshiro Tateya nado eno Eikyou Hyouka ni-tsuite –Chukan Torimatome–* (Interim Report on Impact and Evaluation of Reactor Buildings, etc. at Onagawa Nuclear Power Reactor and Tokai-daini Nuclear Power Reactor by the 2011 Tohoku Earthquake)", Nuclear and Industrial Safety Agency, September 3, 2012
7. "*Genshiryoku Hatsudensho Tateya no Taishin Sekkei –Kako, Genzai, Mirai–* (Seismic Design of Nuclear Power Plant Buildings –Past, Present and Future–)", AIJ Annual Meeting 2012 (Tokai) Structures (Nuclear buildings) Panel Discussion Material, 2012

Chapter 5
Earthquake-Resistant Design of Equipment and Piping

Masanori Hamada and Michiya Kuno

Abstract The equipment and piping systems at nuclear power plants serve not only for generating power but also for sustaining the three important safety-related functions of *shut down*, *cool down*, and *confine*. This chapter provides an overview of earthquake-resistant design of equipment and piping systems, and describes a method for determining seismic loads for use in the design and the design procedure. In addition, an assessment of the functioning of valves and pumps during and after an earthquake and design verification using the shaking table test are described.

Keywords Piping · Equipment · Earthquake-resistant design · Seismic load · Dynamic response analysis · Strength assessment · Allowable stress · Dynamic functionality · Shaking table test

5.1 Equipment and Piping that Require Earthquake-Resistant Design

Nuclear power plants use many equipment and piping systems to enable nuclear power to be generated safely. Figure 5.1 shows an example of such systems for a boiling water reactor (BWR). The plant includes: large components, such as the reactor pressure vessel (RPV) that is loaded with fuel, reactor internals inside the RPV, and a reactor containment vessel that covers the RPV; floor installed components, such as pumps, tanks, heat exchangers, and power distribution panels; and piping systems that interconnect such components and that are used for the circulation and transport of liquid and gas.

M. Hamada (✉)
Waseda University, Shinjuku, Tokyo, Japan
e-mail: hamada@waseda.jp

M. Kuno
Chubu Electric Power Company, Nagoya, Japan

M. Hamada and M. Kuno (eds.), *Earthquake Engineering for Nuclear Facilities*,
DOI 10.1007/978-981-10-2516-7_5

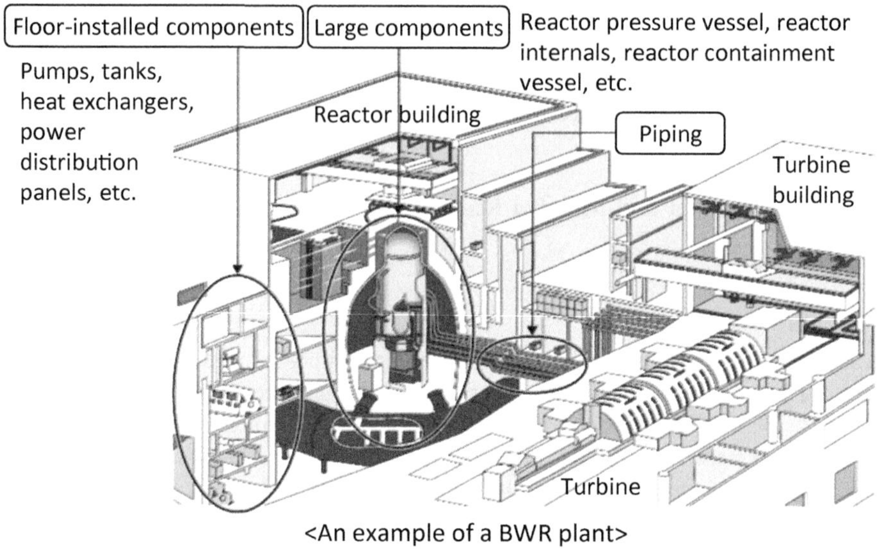

Fig. 5.1 Equipment and piping systems of a boiling water reactor (BWR)

5.2 Earthquake-Resistant Design Methodology

In the earthquake-resistant design of equipment and piping systems, target components are classified by their importance into Classes S, B, and C. Each component is designed to withstand seismic loads of a magnitude that corresponds to its seismic class.

Figure 5.2 shows the process flow for the earthquake-resistant design of equipment and piping systems of a nuclear power plant. This involves designing the basic structures based on their seismic classification and design conditions, and confirming their integrity by analyses. The integrity of dynamic functioning during and after an earthquake is examined, not only by analyses, but also by shaking table tests. As examples, pumps (which are rotating machines) should be designed to remain operational, and valves should maintain their opening and closing capability during and after earthquakes.

(1) Plan for placement and installation of equipment and piping systems

The placement and installation of equipment and piping systems are planned based on seismic classes and design conditions. The design conditions are determined by consideration of their use, the potential risks of hazards, and their specifications, such as capacities.

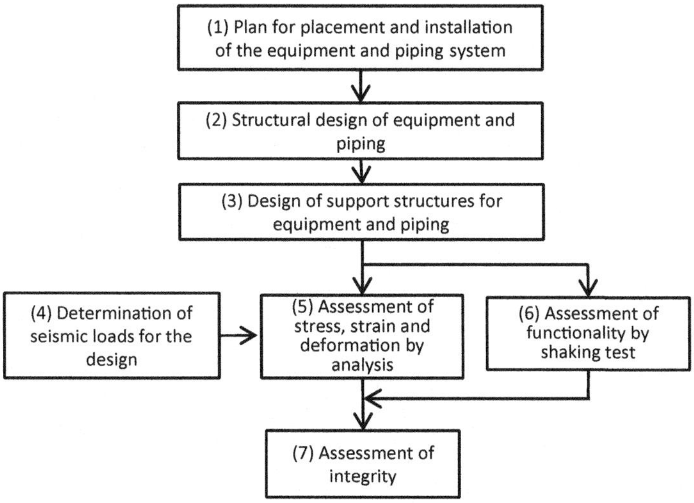

Fig. 5.2 Flow for earthquake-resistant design of equipment and piping systems

(2) Structural design of equipment and piping

Structural features, such as shape, dimensions, and materials, are determined. It is necessary to ensure that the design meets technical standards such as *Codes for Nuclear Power Generation Facilities: Rules on Design and Construction* (JSME SNC1-2010) of the Japan Society of Mechanical Engineers.

(3) Design of support structures for equipment and piping

The support structures are designed to enhance the earthquake-resistance of the equipment and piping systems: earthquake resistance depends largely on the support structures. The positions, directions, and types of supports need to be determined carefully to be able to restrict excessive dynamic responses to earthquake motion while minimizing restrictions of thermal expansion caused by reactor steam heat.

(4) Determination of seismic loads for the design

Dynamic and static seismic loads are defined in accordance with the seismic classification. The dynamic loads are derived based on the dynamic response of the equipment and piping to earthquake motion of each floor on which these are installed. The dynamic response of each floor of the building is obtained from a dynamic response analysis of the building using the standard seismic motion S_S. Static loads are derived from the seismic intensity coefficients in the horizontal and vertical directions, while considering the height of the reactor building, the dynamic characteristics of the foundation ground, and the seismic classification (Sect. 5.3).

(5) Assessment of stress, strain, and deformation by analysis

Stresses, strains, and deformation of the equipment and piping are analyzed by combining the seismic load with the stationary load, the self-weight of the components, and the internal pressure (Sect. 5.4).

(6) Assessment of functionality by shaking test

Acceleration and displacement of the equipment and piping caused by standard seismic motion S_s should not be larger than the values against which their functionalities have been confirmed by vibration tests. In principle, the dynamic responses of the floors of the reactor building on which the equipment and the piping are installed are used as the input motions for the shaking table test. Alternatively, harmonic waves, the acceleration amplitudes of which exceed the maximum dynamic response, can be used. Dynamic functionality assessment using the shaking table test is carried out for Class S components, such as pumps, valves, and electrical devices, which must continue to be functional (capable of rotation, opening, closing, etc.) during and after an earthquake (Sect. 5.5).

(7) Assessment of integrity

Based on the analyses and shaking tests, it is verified that the stress caused by the standard seismic motion S_s does not exceed the allowable limit. If the stress does exceed the limit, then the structural design of the equipment or piping and/or the locations of the seismic supports must be changed. The analyses and shaking tests should then be repeated until it is ensured that the stress will not exceed the allowable limit.

5.3 Dynamic and Static Seismic Loads

This section explains procedures for evaluating dynamic seismic loads based on dynamic response analysis and static loads based on the seismic classification and installation height of the equipment and piping.

5.3.1 Evaluation of Loads

Dynamic loads required for design are evaluated by dynamic response analysis of the reactor buildings, equipment, and piping. Two methods can be used: time–history analysis and modal analysis. In a time–history analysis, the dynamic response of the equipment and the piping against the standard seismic motion S_s is calculated for each time step. In principle, large equipment and piping systems

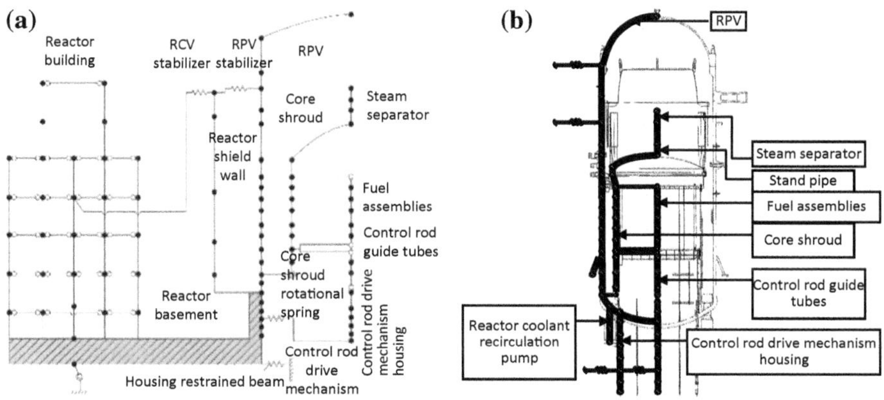

Fig. 5.3 Dynamic analysis model of reactor building and large equipment and piping components, **a** Coupled model for large components, **b** Analytic model for RPV in an ABWR plant

require a time–history response analysis because they are supported by the building at multiple points and the characteristics of those vibrations are complex. Figure 5.3a shows a coupled model of large components with the reactor building; Fig. 5.3b shows an analytical model for an RPV at an advanced boiling water reactor plant.

In a modal analysis, the response of each natural vibration mode is calculated from the response spectra of the standard seismic motion S_s. The dynamic response of the components can be obtained by summing the responses of each vibration mode. Modal analysis is mainly used for the design of floor installed components, such as pumps, containers, and piping (For details of time–history response and modal analyses, Part II, Chap. 14, Sect. 14.2).

Static seismic loads are derived based on the seismic intensity coefficients in the horizontal and vertical directions (Part II, Chap. 15, Sects. 15.1 and 15.2). The seismic intensity coefficient is the ratio of the maximum acceleration in the horizontal or vertical direction to the gravitational acceleration. The static seismic load in the horizontal direction is obtained as the product of the floor shear coefficient Ci, according to the seismic classification (3.6 for Class S, 1.8 for Class B, and 1.2 for Class C). For the design of equipment and piping, the seismic intensity coefficients are 1.2 times the above-mentioned values by taking the resonant vibration of equipment and piping into consideration. The seismic intensity coefficient in the vertical direction is 1.2 times the vertical seismic coefficient Cv.

For Class B facilities, which may experience resonance vibration with earthquake motion, the dynamic response should be analyzed by 1.2 times of the seismic motion for elastic design S_d. The static seismic load is determined based on these analyses.

5.3.2 Damping Constant

The vibrations of buildings, equipment, and piping caused by earthquakes will be damped gradually as the kinetic energy of the vibrating system decreases owing to factors such as frictional heat and air resistance. This damping effect is accounted for in the dynamic response analysis of the equipment and piping.

Table 5.1 lists the damping constants for the design of various components, as determined by shaking tests. The damping constant of piping systems varies between 0.5 and 3% because the damping effect varies with factors such as the number and type of supports, and the presence or absence of insulating materials. Apart from the listed values, specifically validated values based on shaking tests and earthquake observation results are also used.

5.4 Strength Assessment

The stresses of equipment and piping caused by seismic loads and other loads, such as self-weight of the component, should be less than the allowable stresses that are determined according to the seismic classification and materials.

5.4.1 Outline of Strength Assessment

The seismic safety of equipment and piping must be assessed under a combination of seismic loads with loads induced by self-weight, pressure, and thermal expansion of the fluid.

Table 5.1 Values of damping constants of selected components [1]

Component	Design basis damping constant (%)	
	Horizontal	Vertical
Welded structure	1.0	1.0
Bolt and rivet structure	2.0	2.0
Machineries such as pumps and fans	1.0	1.0
Fuel assembly (BWR)	7.0	1.0
Control rod drive mechanism (BWR)	3.5	1.0
Piping	0.5–3.0	0.5–3.0
Cranes (overhead crane and refueling equipment)	2.0	2.0

Technical code for seismic design of nuclear power plant (JEAC 4601-2008), The Japan Electric Association, 2008

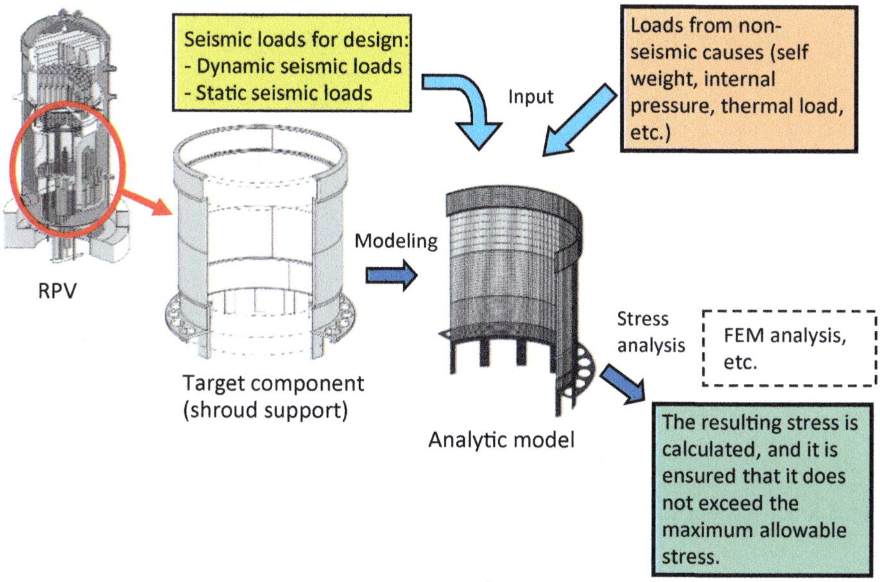

Fig. 5.4 Assessment of strength

Figure 5.4 shows an example of strength assessment of the shroud support inside an RPV. By using an analytical model (usually a finite-element model), the stresses induced at various parts of the shroud are calculated. The stress at each position should not exceed the maximum allowable stress.

5.4.2 *Stress Classification*

The functionality of Class S equipment and piping should be examined by elastoplastic response analysis against the standard seismic motion S_s. Furthermore, the stress caused by the seismic motion for elastic design S_d, should be in the elastic range, i.e., less than the yield stresses.

The stresses of the equipment and piping are classified from the viewpoint of failure modes. Failure modes include ductile failure, which occurs in metals when they are stressed beyond their tensile strength, and fatigue failure, which can occur when a material is stressed repeatedly, even below its tensile strength. The stresses are classified into primary, secondary, and peak stresses. For each classification, it must be ensured that the stress does not exceed the allowable limit. Figure 5.5 shows an example of stresses caused in a slab and beam.

Primary stress is caused by seismic loads, self-weight, and internal pressure. When the primary stresses, which act continuously on the structural members, exceed the yield stress, this may result in ductile failure. Primary stress is further

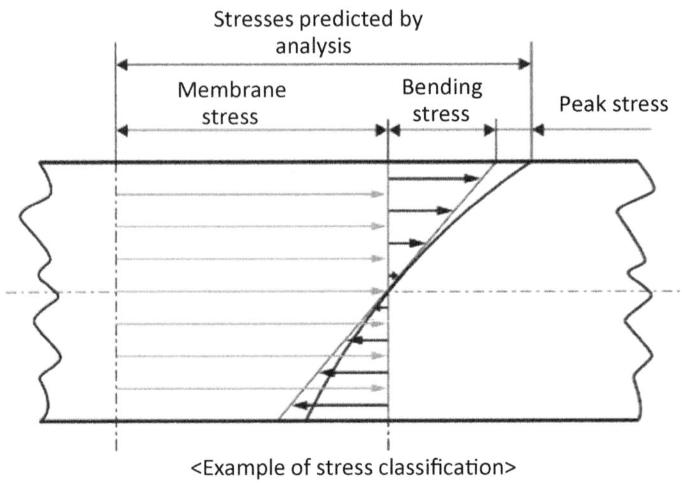

Fig. 5.5 Example of stresses of vessel

classified into membrane stress and bending stress; a maximum allowable limit is defined for each. The membrane stress is uniform within the cross section and the bending stress varies from tensile to compressive within the cross section.

Secondary stress comprises, for example, a stress arising in the structural member from a restraint imposed by the fixture as it expands by heat. If the secondary stress occurs repeatedly—in addition to the primary stress, this may result in fatigue failure, even when the stress is small.

Peak stress is added to the primary or secondary stress as a result of the stress being concentrated at a specific part of the structural member, typically where its cross-sectional shape changes. The peak stress does not cause major failure, but, like secondary stress, its repeated action may lead to fatigue failure. When calculating the peak stress, it is necessary to consider the stress concentration factor.

5.4.3 Allowable Stresses

Allowable stresses are defined according to the seismic classification, type of equipment or piping, and category of stress classification. Table 5.2 shows an example of the allowable stresses for an RPV.

In Table 5.2, *C*s refers to a load condition that requires combining the seismic loads from the seismic motion for elastic design S_d with other loads. *D*s refers to a load condition that requires combining the seismic loads from the standard seismic motion S_s with other loads. The allowable stress is determined for each of the following combinations: the membrane stress only (as part of the primary stress); the primary stress including both membrane stress and bending stress; the primary

Table 5.2 Allowable stresses for a reactor pressure vessel

Service condition	Primary stress (membrane stress only)	Primary stress (membrane stress + bending stress)	Primary stress + secondary stress	Primary stress + secondary stress + peak stress
Cs	min [Sy, (2/3)Su] However, 1.2Sm should be the value for austenite stainless steel and high nickel alloy	1.5 times greater than the value given on the left	3 Sm	Fatigue accumulation factor ≤ 1.0
Ds	(2/3)Su However, min [2.4Sm, (2/3) Su] should be the value for austenite stainless steel and high nickel alloy	1.5 times greater than the value given on the left		

Service condition *Cs*: Assume a combination of loads during normal operation, loads during in-operation transients, and loads during an accident, with seismic loads determined from the seismic motion for elastic design S_d or static seismic forces
Service condition *Ds*: Assume a combination of loads during normal operation, loads during in-operation transients, and loads during an accident, with seismic loads determined from the standard seismic motion S_s
Sy: design yield stress
Su: design tensile strength
Sm: design stress intensity min [(1/3)Su, (2/3)Sy]

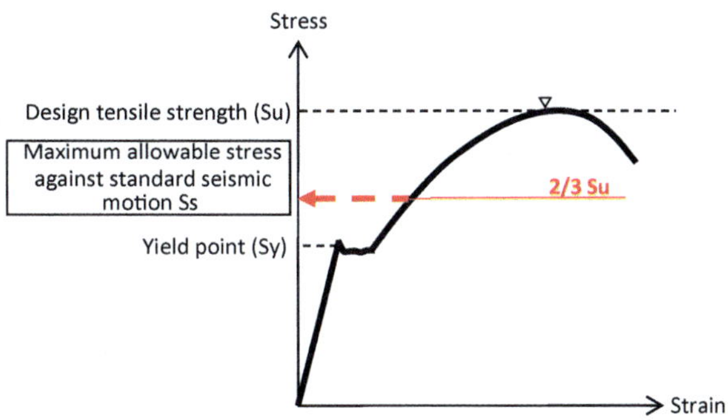

Fig. 5.6 Stress–strain relationship of steel, showing the definition of maximum allowable stress

stress and the secondary stress; and the full combination of the primary, secondary, and peak stresses.

Figure 5.6 shows the concept of allowable stress with reference to the relationship between the membrane stress and the strain occurring in an RPV.

Regarding the stress caused by the standard seismic motion S_s, the allowable stress is defined as 2/3 of the ultimate tensile strength Su. This leaves the safety margin at a full value of the tensile strength Su.

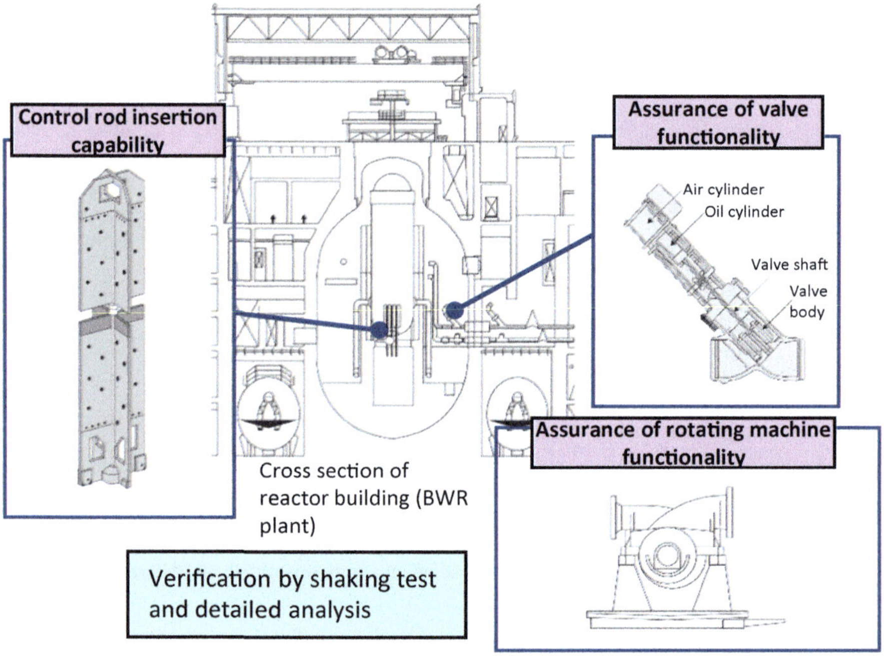

Fig. 5.7 Devices requiring assessment of dynamic functionality [2]

5.5 Assessment of Dynamic Functionality

The dynamic functionality of valves, pumps, and other Class S devices is examined by shaking tests and analyses. These should confirm that the equipment and piping remain functional against acceleration and deformation exceeding the design level. Figure 5.7 shows examples of devices at a BWR plant that require dynamic assessment of dynamic functionality. It is necessary to assess the capability of the control rod drive system to shut down the reactor in an emergency and the capability of the valves to shut off the main steam supply following a piping fracture. The dynamic functionality of the pumps used for cooling the reactor in an emergency must also be assessed.

5.5.1 Control Rod Insertion System

This section explains how the control rod insertion capability at a BWR plant is assessed. The fuel assemblies for BWR plants are approximately 4.5 m long. Bundles of four form a set, and these are placed on fuel supports inside the reactor. Control rods are inserted into the gaps between the fuel assemblies. One control rod

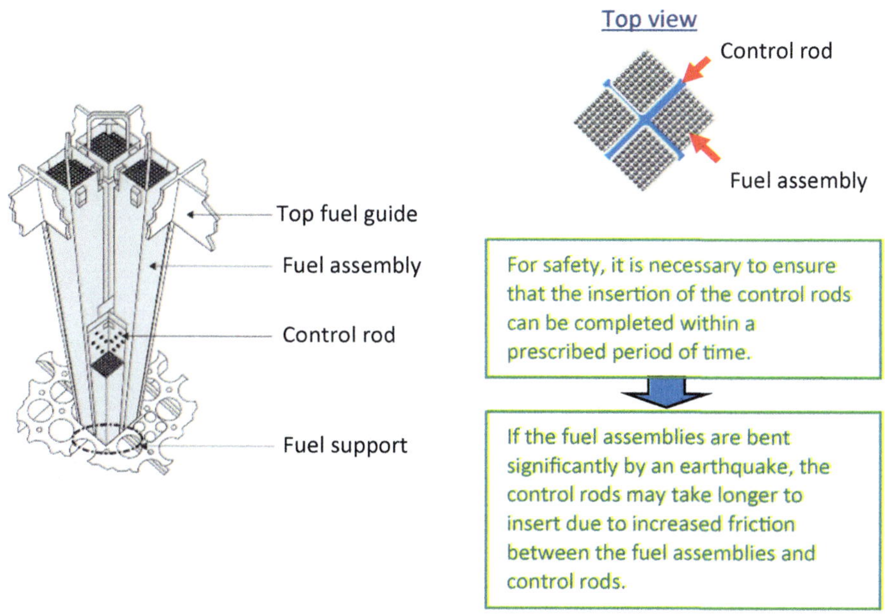

Fig. 5.8 Fuel assemblies

is provided for every four bundles of fuel assemblies. The control rods are withdrawn during reactor operation to permit the nuclear reactions to take place. When the reactor has to be shut down, the rods are inserted to stop the nuclear reactions. In the example of a 1.1 GW BWR plant, there are 764 bundles of fuel assemblies and 185 control rods. Figure 5.8 shows the configuration of the fuel assemblies.

When earthquake ground motion greater than a preset level is detected, the control rods are inserted rapidly into the fuel assemblies from below by the control rod drive mechanism using water pressure. This is termed *scram*. During scram, the insertion of the control rods must be completed within a limited time. If the fuel assemblies are bent significantly during the earthquake, however, the smooth execution of scram may be hindered; it would take longer to insert the control rods because of friction between the fuel assemblies and the control rods. The integrity of the scram function is therefore verified using a shaking test.

Figure 5.9 shows a shaking test used to verify the capability of control rod insertion at a BWR plant. The full-size test set with dummy fuel assemblies is vibrated by the shaking table and the capability for insertion of the control rods is confirmed. The motion of the shaking table is increased stepwise by varying the horizontal displacement to which the fuel assemblies are subjected. The time to complete the insertion of the control rods is measured to determine the allowable horizontal displacement. The capability to insert the control rods during an earthquake is verified by ensuring that the horizontal displacement of fuel assemblies

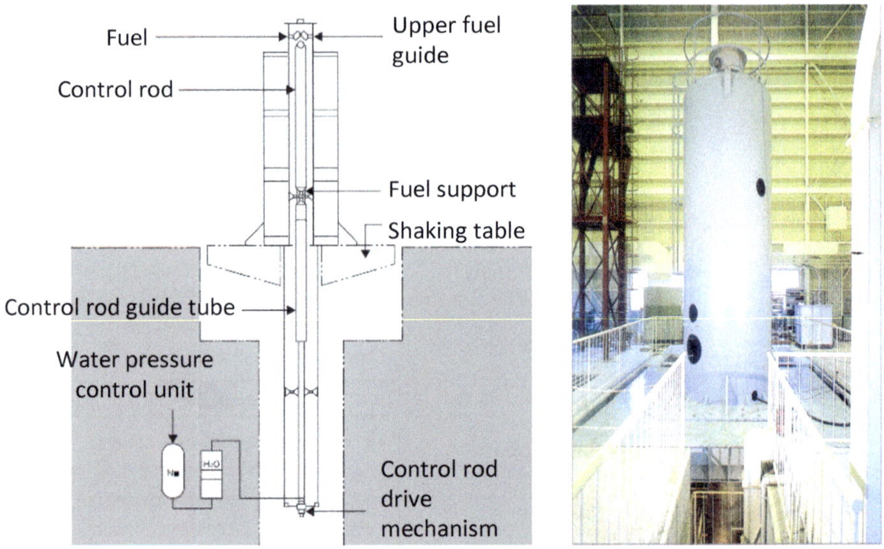

Fig. 5.9 Shaking test to determine the capability to insert control rods during an earthquake

predicted by a dynamic response analysis against the standard seismic motion S_s does not exceed the allowable horizontal displacement determined by the shaking tests.

In the case of scram in a pressurized water reactor, control rods are inserted from above and drop under their own weight. The capability of control rod insertion during an earthquake is similarly tested, and it is ensured that the horizontal displacement that may occur in the fuel assemblies does not exceed the allowable values obtained in the shaking tests. In addition, the capability of control rod insertion is examined to verify that the rods can be inserted completely within the time limit.

5.5.2 *Rotating Machines and Valves*

The functionalities of rotating machines (pumps, fans, etc.) and motor operated valves are also assessed by shaking tests. Figures 5.10 and 5.11 show a horizontal pump and a motor operated globe valve, respectively. Shaking tests are performed with the actual device placed on a shaking table to determine the maximum acceleration up to which the device can remain functional. Functionality is defined as the pump's capability to maintain a specified discharge rate and the valve's capability to open and close as instructed, for example. The maximum acceleration up to which the device has demonstrated its capability to remain functional is referred to as the *maximum functionality-proven acceleration*. In functionality

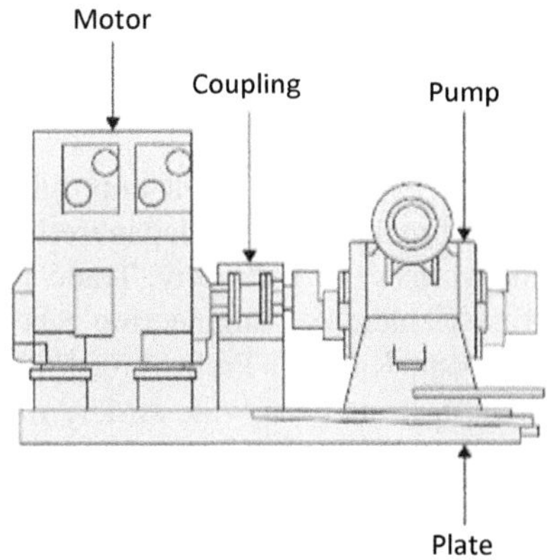

Fig. 5.10 Schematic of a horizontal pump [2]

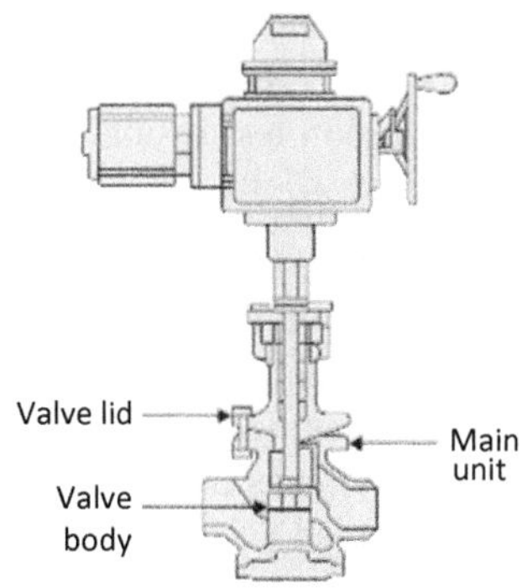

Fig. 5.11 Schematic of a motor operated globe valve [2]

assessments, the acceleration that the device is expected to be subjected, according to the results of dynamic response analysis, should not exceed the maximum functionality-proven acceleration.

5.6 Earthquake-Resistance Demonstration Tests

Earthquake-resistant design specifications for nuclear power plants are stricter than those for normal buildings and structures. Moreover, at nuclear power plants, the responses of equipment, piping, and structures to seismic and non-seismic loads have complex characteristics. Assessment by analysis is therefore followed by verification using a shaking test. Full-size replicas of the plant components important for safety are subjected to vibration tests performed on a large shaking table to determine whether they can maintain their integrity.

5.6.1 Shaking Table Tests at Tadotsu Engineering Laboratory

A joint project aimed at improving the reliability of nuclear power facilities was started in 1975 by the Japanese Ministry of International Trade and Industry (now the Ministry of Economy, Trade and Industry) and the Nuclear Power Engineering Test Center. An engineering laboratory with a large shaking table was built in Tadotsu, Kagawa Prefecture. From 1982, this facility was used for 23 years for shaking tests on a wide variety of large components until its decommissioning in 2005. The purposes of these vibration test programs were as follows:

1. to verify the capability of those components at nuclear power plants that are important for safety to maintain their structural integrity by withstanding earthquakes and to evaluate their safety margins,
2. to verify that those components important for safety could remain functional during and after earthquakes, and
3. to verify the validity of earthquake-resistant design methods by comparing vibration test results with analytical results.

Figure 5.12 shows an overview of the large shaking table, which was 15 m × 15 m, could carry loads of up to 1,000 kN, and could produce accelerations of up to 5 G. The table could produce vibrations simultaneously in the horizontal and vertical directions.

The shaking table was used to perform vibration tests on full-size models and precise replicas of plant components by subjecting them to vibrations stronger than the designed seismic motion. As shown in Fig. 5.13, the earthquake resistance of major nuclear power plant components, including concrete containment vessels, RPV internals, emergency diesel generator systems, and reactor *shut down* cooling systems, were examined. In these tests, vibrations were increased to the highest limit allowed by the shaking table to determine the ultimate strength of the respective components. Table 5.3 summarizes observations from the testing of different components.

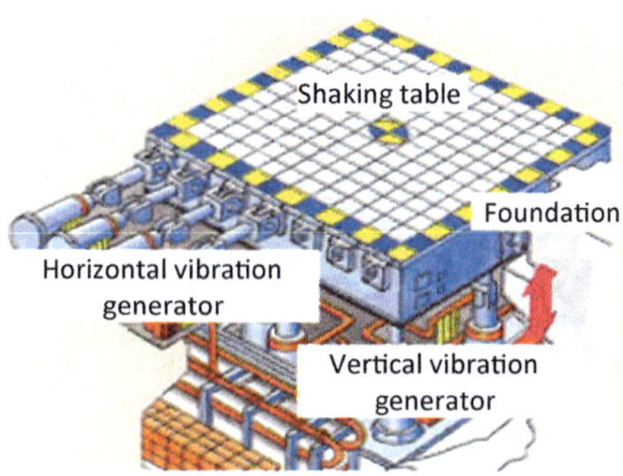

Fig. 5.12 Overview of large shaking table [3]

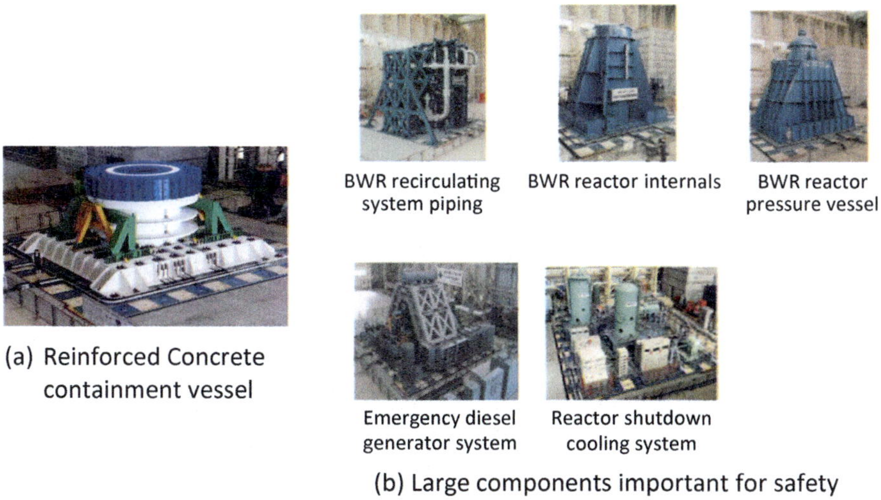

Fig. 5.13 Examples of components tested for their aseismic reliability on the large shaking table [3]

Table 5.3 Observations of shaking table test results [4]

Test name	Summary of observations on test results
Seismic testing of the concrete containment vessel	The leakage prevention capability is maintained until a fracture in the concrete occurs. The fracture level was approx. $6S_2$ in the case of PCCV, and approx. $7S_2$ in the case of RCCV
Ultimate strength testing of piping	The fracture level in terms of the maximum response was at least 8.5 times larger than the assumed maximum allowable level (IV_AS)
Ultimate strength testing of electric distribution panels	The ultimate strength (up to the loss of function) was approx. 5 G or greater (acceleration measured on the panel foundation) for the main control panel, the reactor auxiliary control panel and the logical circuit control panel The ultimate strength was approx. 4 G for protective instrument racks, instrument racks, control center consoles, power center consoles and metal clad switchgears (after redesigning them with additional reinforcement to prevent the malfunctioning of devices)
Ultimate strength testing of horizontal single-stage pumps	The ultimate strength (up to the loss of bearing function) was 8.4 G (acceleration measured on the pump foundation)

(continued)

Table 5.3 (continued)

Test name	Summary of observations on test results
Ultimate strength testing of large vertical pumps	The ultimate strength (up to the yielding of parts/materials) was 12 G in terms of the response acceleration at the motor top and 31 G in terms of the response acceleration at the barrel end
Testing of the ease of control rod insertion	PWR: The capability of smooth insertion was maintained until the response displacement of the fuel assemblies reached just over 40 mm BWR: The capability of smooth insertion was maintained until the response displacement of the fuel assemblies reached just over 80 mm

S_2 refers to the design basis ground motion S_2 (ground motion associated with the design basis for extreme earthquakes) as per the *Earthquake-Resistant Design Technology Guidelines for Nuclear Power Stations* (JEAG4601-1987)
IV_AS refers to the maximum allowable limit assumed in the design based on the ground motion S_2

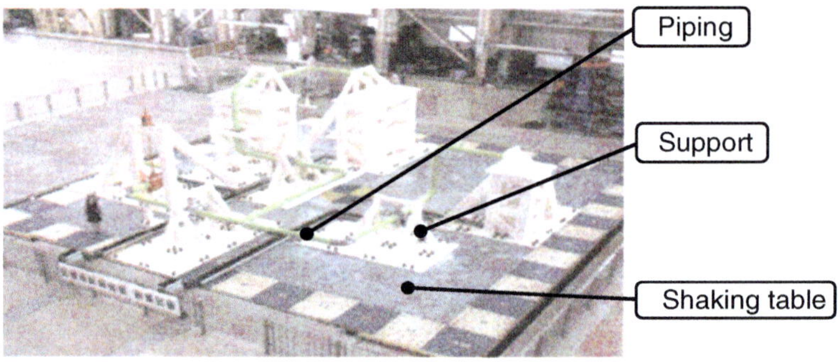

Fig. 5.14 Example of seismic reliability demonstration test (ultimate strength test of piping) [3]

Figure 5.14 shows an example of a vibration test performed on piping. An actual piping system was vibrated by harmonic waves with a magnitude 8.5 times greater than the maximum allowable level assumed by the design until fatigue fracture occurred. This test confirmed that the piping had a sufficient safety margin to the stress level at which the fracture took place.

5.6.2 *Full-Size Three-Dimensional Failure Process Test (E-Defense)*

In 2005, when the shaking table in Tadotsu was decommissioned, the world's largest full-size three-dimensional seismic destructive testing facility (E-Defense) was built in Miki City, Hyogo Prefecture, as one of the initiatives started after the 1995 Kobe earthquake. E-Defense enables the testing of full-size buildings with

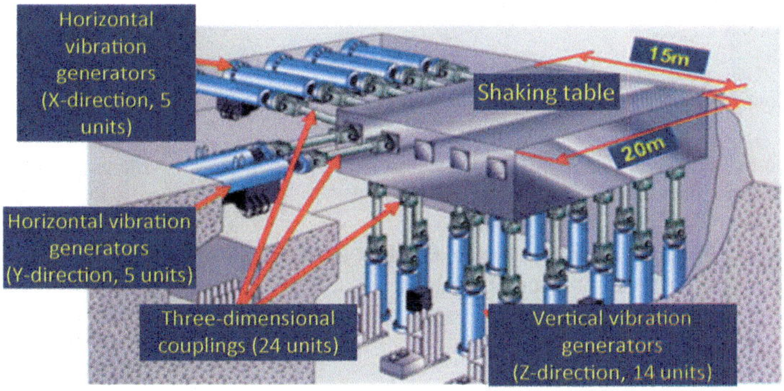

Fig. 5.15 Conceptual overview of the three-dimensional shaking table for testing full-size models (E-defense) [3]

Table 5.4 Specifications of the three-dimensional shaking table for testing full-size models (E-defense)

	E-defense	Tadotsu shaking table
Year of completion	2005	1982
Shaking table dimensions	20 × 15 m	15 × 15 m
Maximum loading capacity (t)	1200	1000
Maximum acceleration (Gal)	900	1900
Maximum velocity (cm/s)	200	75
Maximum displacement (cm)	±100	±20
Direction of vibration	Two horizontal directions + vertical direction	One horizontal direction + vertical direction

three-dimensional vibrations in the horizontal and vertical directions with an intensity comparable with that of the Kobe earthquake. This enables examination of the dynamic characteristics of structures and a study of their failure processes. Figure 5.15 shows an overview of the three-dimensional shaking table for full-size models; its specifications are given in Table 5.4.

The large shaking table at Tadotsu was designed to produce large accelerations with short periods, specifically for testing nuclear power plant facilities. E-Defense was mainly designed for testing buildings and structures with relatively long periods and is capable of producing large displacements. E-Defense has been used to test spent fuel storage casks, the sloshing of contents of large tanks at nuclear power plants, and piping.

References

1. "Technical Code for Seismic Design of Nuclear Power Plants (JEAC4601-2008)", The Japan Electric Association, 2008
2. "Technical Guidelines for Seismic Design of Nuclear Power Plants (JEAG4601-1991, Supplement)", The Japan Electric Association, 1991
3. The Institute of Applied Energy website
4. Japan Nuclear Energy Safety Organization's Material (Material of former Nuclear and Industrial Safety Agency's Council), August 27, 2008

Chapter 6
Earthquake-Resistant Design of Important Civil Engineering Structures

Masanori Hamada and Michiya Kuno

Abstract Nuclear power plants contain many civil engineering structures, such as water-intake and—discharge facilities, tank foundations, and port structures. Those civil engineering structures that support important safety equipment and piping systems and those that assist in cooling water delivery to the reactor and spent fuel cooling facilities in an emergency are referred to as "important civil engineering structures." This chapter describes the earthquake-resistant design of such important civil engineering structures. The earthquake-resistant design methodology, as based on a dynamic response analysis, and a method to model civil engineering structures and nearby ground for dynamic response analysis are explained. Methods for the assessment of seismic safety of civil engineering structures are explained. These assessments are based on an allowable stress and strength of the structural members, and an allowable structure deformation. For an assessment of the allowable structure deformation, an interpretation of the ultimate state of important civil engineering structures in terms of their functional requirements is discussed, and a method to determine the ultimate state that focuses on the failure process of reinforced concrete is described.

Keywords Civil engineering structure · Water intake · Water discharge · Cooling water · Foundation · Support of equipment · Dynamic response analysis · Underground ducts · Ultimate strength

---

M. Hamada (✉)
Waseda University, Shinjuku, Tokyo, Japan
e-mail: hamada@waseda.jp

M. Kuno
Chubu Electric Power Company, Nagoya, Japan

M. Hamada and M. Kuno (eds.), *Earthquake Engineering for Nuclear Facilities*,
DOI 10.1007/978-981-10-2516-7_6

6.1 Civil Engineering Structures in a Nuclear Power Plant

Examples of civil engineering structures at a nuclear power plants include water-intake and—discharge structures; foundations for electrical and mechanical installations, equipment, piping, and tanks; and port structures. Table 6.1 lists civil engineering structures at a nuclear power plant. Figure 6.1 shows a sample layout of these civil engineering structures.

The water-intake and—discharge structures comprise the intake gate, intake channel, intake pit (water cistern), outlet channel, and outlet gate. These are used for sea water-intake and—discharge back to the sea. Sea water is used to cool and condense steam after it has been used to drive turbines, for circulation inside the reactors, and to cool spent fuel. Civil engineering structures also include foundations that support the piping used to transport sea water from the intake gate and to circulate it through cooling systems, foundations for piping that transports fuel oil and service water, and foundations for electric cables and the ducts that house them. Foundations that support tanks that contain fuel for emergency diesel generators and service water for use inside the power plant are also civil engineering structures. Port structures for marine transport such as piers and seawalls, and other infrastructure such as roads, bridges, and drainage channels also exist.

Table 6.1 Civil engineering structures at a nuclear power station

Civil engineering structure	Outline of structures
Water intake and discharge structures	Structures used for intake from the sea and discharge back to the sea of the water used for cooling and condensing the steam after it has been used for driving the turbines Examples: intake gate, intake channel, intake pit, discharge channel and discharge gate
Foundations for electrical and mechanical installations, equipment and piping	Foundations supporting the piping used for circulating inside the power station the sea water from the intake port; the foundations structures supporting the piping used for transporting fuel oil, service water, etc.; and the foundations housing electric and instrumentation cables Examples: sea water pipe ducts and electric cable ducts
Foundations for tanks	Foundations supporting the tanks containing the fuel for emergency diesel generators, the service water for use inside the power station, etc. Example: fuel tank foundations
Port structures	Piers and facilities to load and unload the nuclear fuel, equipment and materials, and the seawalls Examples: seawalls and piers
Other structures	Other facilities required for the management of the power station Examples: roads, bridges and drainage channels

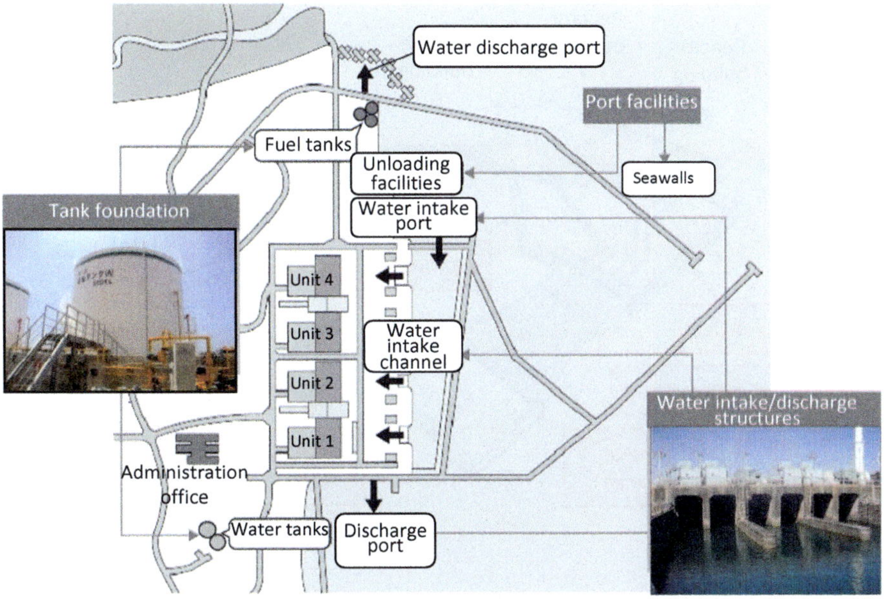

Fig. 6.1 Layout of civil engineering structures at a nuclear power station [1]

6.2 Important Civil Engineering Structures

Important civil engineering structures that have a direct impact on the crucial function of *cool down* include the intake pit and the sea water pipe duct that supports sea water pumps and pipes. These constitute an emergency sea-water-intake system such as an intake gate and channel for the delivery of emergency sea water. Figure 6.2 shows examples of such structures.

The safety of nuclear power plants depends on the integrity of all facilities with main and auxiliary safety functions, including the supporting structures. Important civil engineering structures include the facilities themselves, and those that mediate interactions or interferences between different facilities. The important civil engineering structures should be designed to fulfil these roles.

6.3 Flow of the Earthquake-Resistant Design

Figure 6.3 shows the procedure for designing earthquake-resistance of important civil engineering structures. In principle, an earthquake-resistant design is achieved on the basis of a dynamic response analysis against standard seismic motion S_s as the input motion. The design procedure is described below.

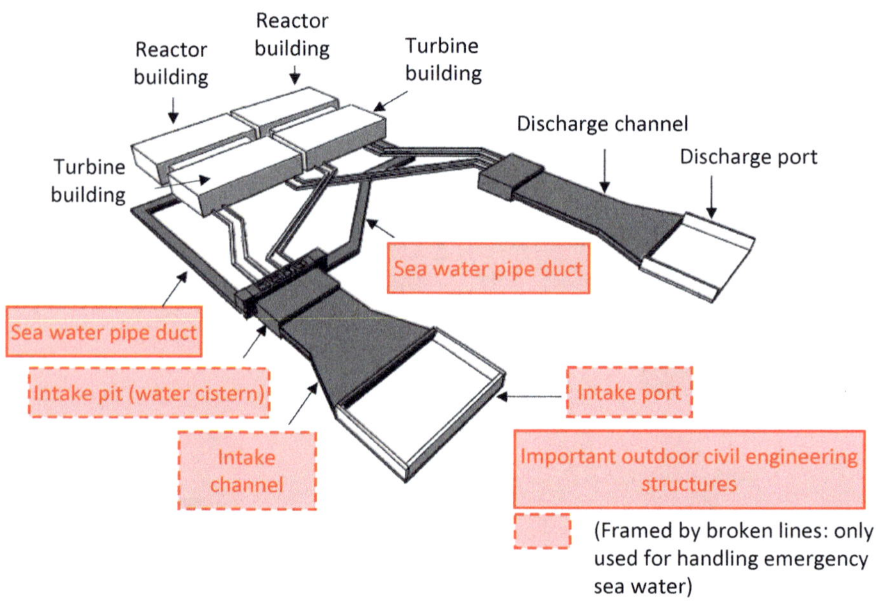

Fig. 6.2 Examples of important civil engineering structures [2]

(1) Structural plan

A structural plan is made in consideration of the functional requirements and seismic classification of the nuclear facilities. Based on the structural plan, basic design conditions are defined, and decisions are made on the physical properties of the materials and the ground.

(2) Structure modeling

The details of structural members, such as the size, materials, required strength, and reinforced bar arrangement are determined based on the structure plan for the design calculations by using static seismic loads.

(3) Load estimation

Stationary loads to be sustained because of the structure's self-weight, the weight of equipment and piping that the civil engineering structures are to support, and soil and water pressure from the surrounding ground are estimated (Sect. 6.4).

(4) Stress and deformation analysis by stationary loads

Deformations of structures, and stresses and strains of structural members induced by stationary loads are calculated using the structure model.

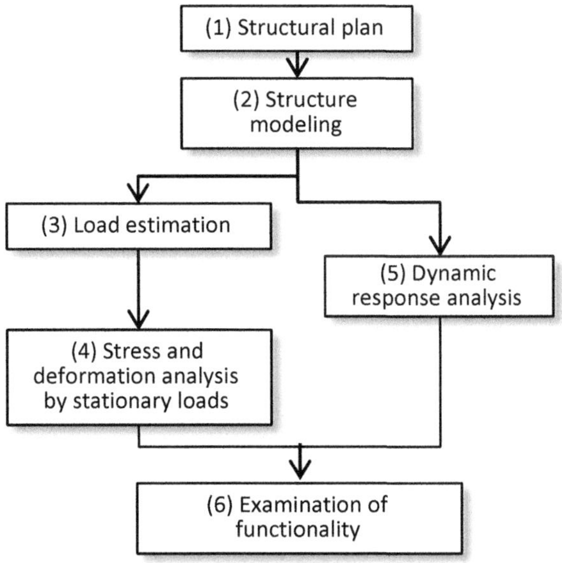

Fig. 6.3 Earthquake-resistant design flow for important outdoor civil engineering structures

(5) Dynamic response analysis

Dynamic response analyses are performed using the standard seismic motion S_s as the input ground motion to determine the seismic loads (e.g., inertial forces, dynamic water pressure, and dynamic soil pressure). The dynamic response (in terms of acceleration and deformation) of the equipment and piping support as calculated by the dynamic response analysis is referred to in the earthquake-resistant design of equipment and piping (Sect. 6.5).

(6) Examination of functionality

By verifying that the deformation, strains, and stresses caused by a combination of stationary and seismic loads calculated by the dynamic response analysis do not exceed the allowable limit, it is ensured that the important civil engineering structures will maintain their required capability, such as providing support and delivering water. If the civil engineering structures do not fulfill the functional requirements, the earthquake-resistant design should be restarted from reexamination of the model such as structural members (Sect. 6.6).

6.4 Loads for the Design

The design of important civil engineering structures involves a consideration of stationary loads, seismic loads, and other loads as follows.

6.4.1 Stationary Loads

As shown in Fig. 6.4a, stationary loads act continuously on the structure under normal circumstances (during operation and plant outage). These loads include the structure's self-weight, the superstructure load, the water and soil pressures, and the reaction forces on the structures from the surrounding ground. The superstructure load refers to loads resulting from the weight of supported objects (equipment, piping, etc.). The soil pressure is the static force from the surrounding ground and the water pressure is the load imposed by the groundwater and sea water around or inside the structures.

(a)

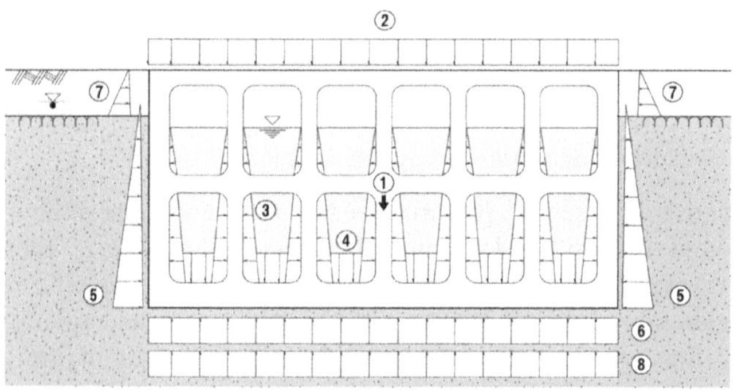

① Structure's self-weight ② Superstructure load ③ Static water pressure from contained water ④ Contained water weight ⑤ Static water pressure from external water ⑥ Uplift pressure ⑦ Soil pressure ⑧ Reaction force from the ground

(b)

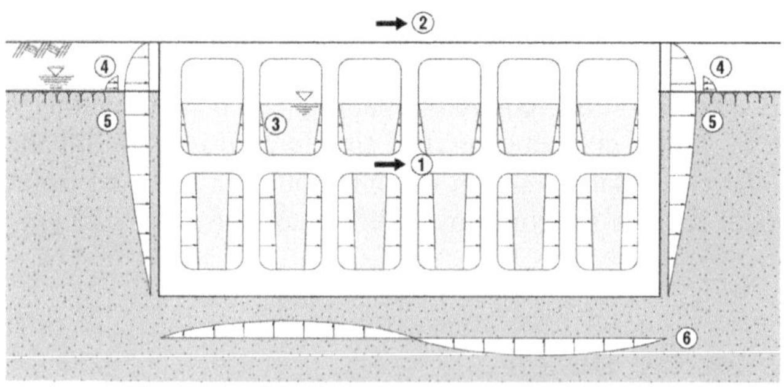

① Inertia force of the structure ② Inertia force of the superstructure ③ Dynamic water pressure from contained water ④ Dynamic water pressure from external water ⑤ Soil pressure during earthquake ⑥ Reaction force from the ground

Fig. 6.4 Load combinations considered in the design, **a** stationary loads, **b** seismic loads

6.4.2 Seismic Loads

As shown in Fig. 6.4b, seismic loads act on structures during earthquakes such as the inertial force of equipment and piping, soil and water pressure, and reaction forces from the ground, which are induced by earthquakes. Although important civil engineering structures are generally considered as Class C facilities themselves, they require Class S level seismic safety because of the importance of the Class S equipment and piping that they support. Therefore, they should be designed to withstand seismic loads that arise from the standard seismic motion S_s.

6.4.3 Other Loads

In addition to the stationary and seismic loads, forces exerted by waves, tsunamis, and wind during typhoons should be taken into consideration. For underground structures, loads from vehicles traveling on the ground surface need to be considered.

6.5 Earthquake-Resistant Design

Earthquake-resistant design of important civil engineering structures is done based on a dynamic response analysis. The response displacement method is also used to design box culverts of underground civil engineering structure, such as when the stress and structure deformation are governed by ground displacement (Part II, Chap. 15, Sects. 15.3, 15.5).

6.5.1 Design Based on Dynamic Response Analysis

Figure 6.5 shows a procedure for the design of important civil engineering structures by dynamic response analysis against the standard seismic motion S_s. Figure 6.6 shows an example of an analytical model by the finite element method (FEM) used in the dynamic response analysis of an underground box culvert.

Deformations, stresses, and strains of underground structures and the ground are analyzed by the FEM using the standard seismic motion S_s as the input motion at the bedrock, which is the bottom of the FEM model. However, the standard seismic motion S_s is usually defined on a virtual free surface of the bedrock. Therefore, input motion at the bottom of the model is redefined from free surface motion by

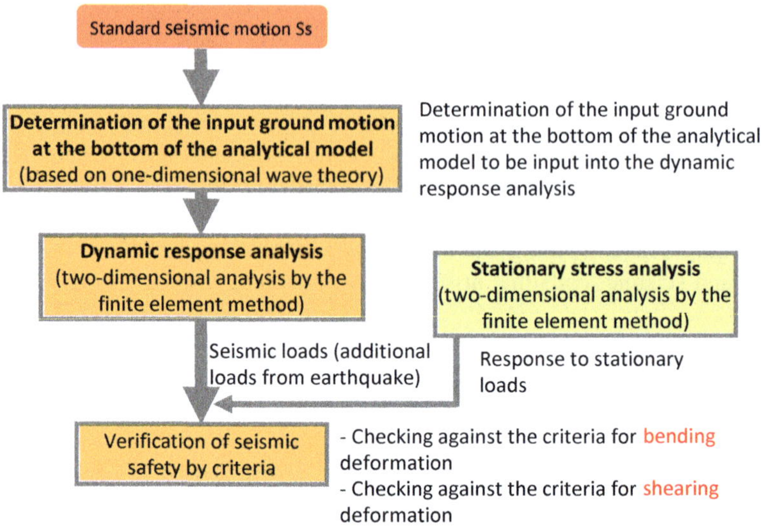

Fig. 6.5 Procedure for earthquake-resistant design by dynamic seismic response analysis

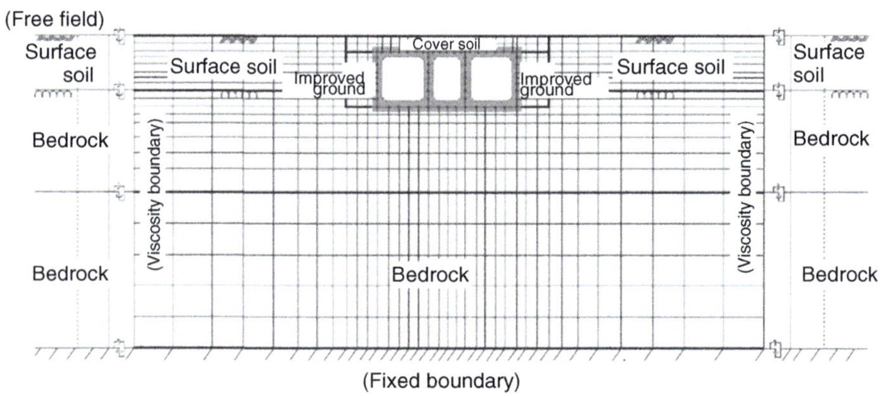

Fig. 6.6 Model for dynamic response analysis of an underground box culvert by the finite element method

one-dimensional wave theory (multi-reflection theory) as explained in Fig. 13.10 (Part II, Chap. 13). The seismic safety of civil engineering structures is examined by deformation, stresses, and strains caused by combining static, dynamic, and other loads.

Four kinds of dynamic response analyses of civil engineering structures exist, depending on a definition of the limit state of the structures for the design. Among them, linear analysis assumes that the stress–strain relationship of the structures and the ground are linear and elastic. This approach is chosen when the structure and ground deformation remain within the linear and elastic range.

In equivalent linear analysis, linear analysis where the stress–strain relationship is assumed to be linear at each calculation step of the dynamic response analysis is repeated by changing soil properties, such as an elastic shear modulus and damping constant depending on the magnitude of the shear strain. This method is often used in the earthquake-resistant design of important civil engineering structures, and enables relatively accurate analysis as long as the response of the structures does not exceed its ultimate state (Fig. 6.7).

Nonlinear dynamic analysis of structures is performed in the time domain (in time history) by using two methods in which the nonlinear characteristics of structural members and structural materials are considered. Nonlinear analysis by structural members is a relatively simple method in which the nonlinearity of the response is accounted for at a structural member level. In nonlinear analysis for materials, the nonlinearity of the stress–strain relationship of the structural members is taken into consideration. In the two kinds of nonlinear analyses of member and material levels, structure deformation beyond the linear and the elastic range can be calculated more precisely and accurately, compared with equivalent linear analysis. However, the computation processes are complex and the analysis is more time-consuming.

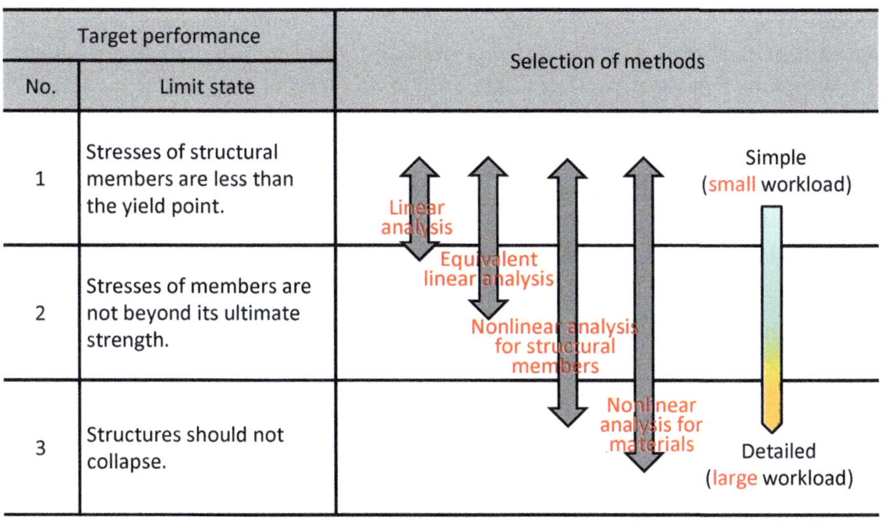

Fig. 6.7 Selection of methods of dynamic response analysis [2]

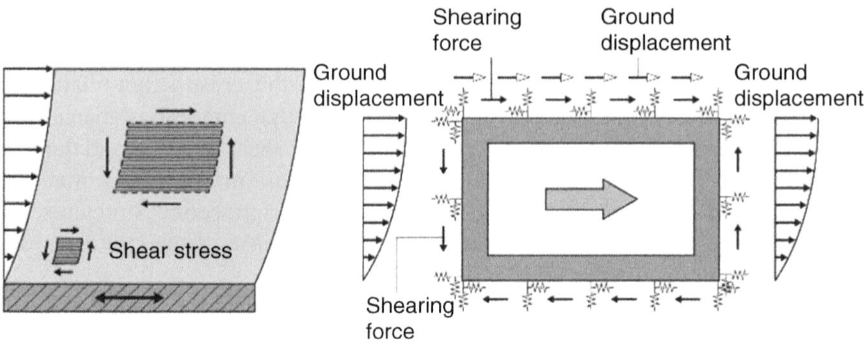

Fig. 6.8 Earthquake-resistant design of a box culvert by the response—displacement method

6.5.2 Earthquake-Resistant Design by Response—Displacement Method

The deformation of underground structures during earthquakes is governed by the displacement of the surrounding ground. Therefore, the response displacement method (Part II, Chap. 15, Sect. 15.5) is used in the earthquake-resistant design of underground civil engineering structures of nuclear power plants. In the response displacement method, the deformation of underground structures is obtained by inputting the displacement of the surrounding ground into the underground structures through ground springs. Figure 6.8 shows an example of where the response displacement method is used to design a box culvert. The maximum ground displacement and shear stresses that are calculated by a dynamic response analysis of a one-dimensional model of the surface ground are input into the culvert. Beside the displacements and shear stresses of the ground, the inertial force of the culvert itself is taken into consideration in the design.

6.6 Examination of Seismic Safety and Functionality of Important Civil Engineering Structure

This section describes methods to examine seismic safety and the functionality of important civil engineering structures.

6.6.1 Methods of Assessment of Seismic Safety and Functionality

Three methods exist to examine seismic safety and functionality of important civil engineering structures. These are the allowable-stress design method, the limit-state design method, and the performance-based design method. The allowable-stress design method, which was used extensively up to the 1980s, was simple because the relationship between the structure deformation and the load is assumed to be linear and elastic. In this design method, the allowable stress of the structural members is specified for the design. Although this method can address the behaviors of structures within a linear range, it cannot quantify the safety margin up to failure of the structures.

The concept of a limit-state design method was introduced in the 1990s and led to the development of techniques to quantify safety margins of structures before they reached a limit state. The limit-state design method involves defining the allowable limit state of the structures, and evaluating the safety margin up to the limit state in terms of multiple safety factors, which are evaluated considering various uncertainties related to external loads and material properties. This method makes it possible to handle the nonlinearity up to the ultimate strength in the design process.

Extensive damage to buildings, bridges, and port harbor structures caused by the 1995 Kobe earthquake led to a proposal for the performance-based design as an approach that can address the deformation of structures up to fracture against very intense ground motions such as those observed in and around Kobe city during this earthquake. This method is used mainly to evaluate the capability of structures to remain functional.

6.6.2 Performance Requirements and Limit State

The performance required by important civil engineering structures against standard seismic motion S_s is to preserve the integrity and ability of equipment and piping to retain emergency cooling water circulation. Therefore, important civil engineering structures should continue serving as an indirect support for equipment and piping, and limit their deformation to enable the normal operation of equipment and piping.

Important civil engineering structures should be examined against standard seismic motion S_s for their own seismic safety from the viewpoint of structural design and for the bearing capacity of the ground foundation. For their own seismic safety, the structure integrity should be examined against criteria that depend on the design methods, which are the allowable limit state and the required performance. When the limit-state design method is applied, the response against the standard seismic motion S_s should not exceed the requirements listed in Table 6.2.

In the allowable-stress design, the stress of structural members against standard seismic motion S_s, should not exceed the allowable stress. In addition, the force acted on the foundation ground from the buildings and structures should not exceed the ultimate bearing capacity of the ground.

Table 6.2 Performance requirements and limits for important civil engineering structures

Performance requirement	Limit state of the structure (specifications)
Sustaining the functionality of equipment/piping	Sustaining the functionality of equipment – Keeping the piping intake – Keeping the internal space intact
Sustaining the delivery of water through the water-intake channel	– Keeping the internal space intact (enabling the intake of the required quantity of sea water)

6.6.3 Limit State of Reinforced Concrete Structures

A reinforced concrete structural member can be subjected to two major types of failure: bending failure and shearing failure shown in Fig. 6.9. Figure 6.10 shows two examples of failure from a loading test. Figure 6.9a gives an example of bending failure: a compression failure has occurred on the interior side of the wall. Figure 6.9b provides an example of shearing failure: the wall has been sheared and a crack runs diagonally from left to right.

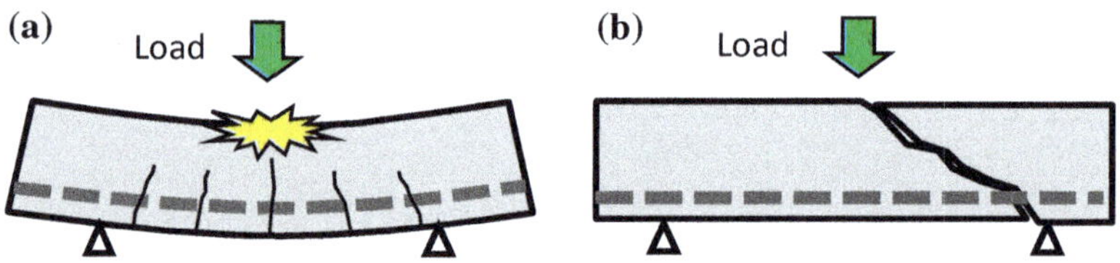

Fig. 6.9 Failure modes of reinforced concrete structures, **a** bending failure, **b** shearing failure

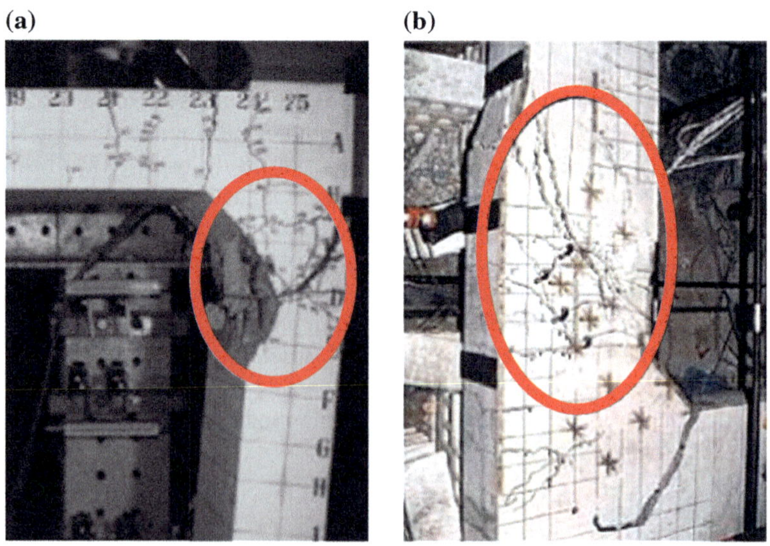

Fig. 6.10 Reinforced concrete failure examples from a loading test, **a** bending failure, **b** shearing failure [2]

6.6.3.1 Examination of Bending Failure

Figure 6.11 shows a typical relationship between the bending moment and deformation of a reinforced concrete member. As the bending moment increases, a crack develops in the concrete. As the bending moment increases further, the reinforcement bars yield. When the bending moment exceeds the ultimate moment, the concrete begins to soften and the cover concrete comes off. This is followed by buckling of the reinforcement bars and finally by a collapse of the structure.

The ultimate bending strength is determined from the cross-sections of the members, and the volume and arrangement of reinforcement bars. The capability to retain functionality is judged according to the following four criteria, which are focused on the deformation of structural components:

(1) The bending strain of concrete at the compression end should be 1 % or less.
(2) The curvature does not exceed the level that is associated with condition 1.
(3) The story deformation angle does not exceed the level that is associated with condition 1 (i.e., the maximum allowable story deformation angle).
(4) The story deformation angle should be 1/100 or less.

The condition for cover concrete exfoliation is the *nearly total absence of stress in the concrete*. Experimental results have shown that concrete retains some residual compressive strength as long as its strain at the compression end remains at ~1 % or less. As the strain at the compression end increases beyond 1 %, the residual compressive strength decreases gradually, which makes the cover concrete more prone to exfoliation. Therefore, the maximum allowable strain at the compression end is found to be 1 %.

As shown in Fig. 6.12, the story deformation angle is the ratio of the deformation of the structure in the horizontal direction to its height H, and is a measure of the deformation of the structure. Experimental results have shown that a story

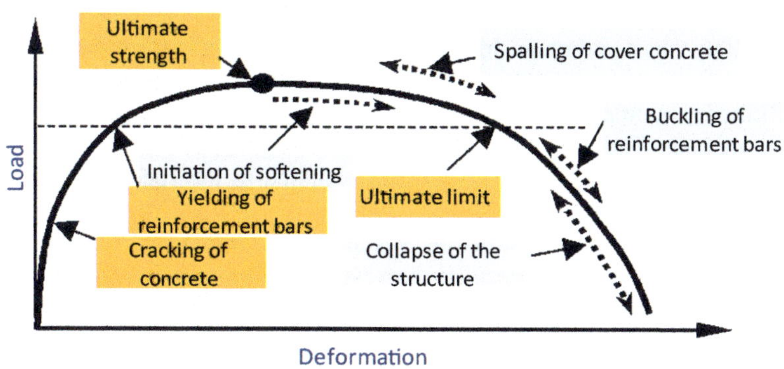

Fig. 6.11 Relationship between the load and deformation of bending members of reinforced concrete structures [2]

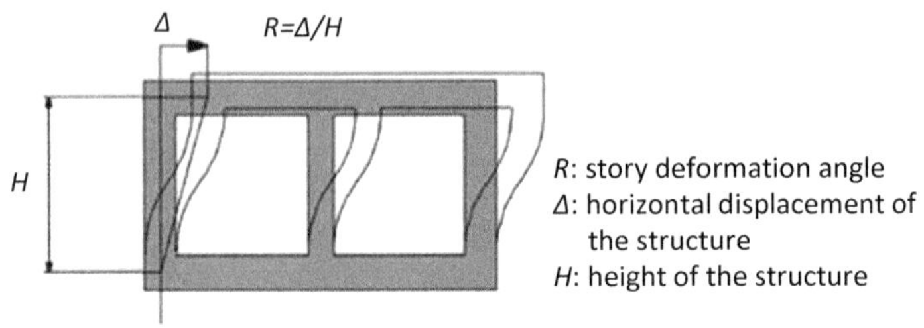

Fig. 6.12 Story deformation angle [2]

deformation angle of less than 1/100 provides a sufficient safety margin for cover concrete exfoliation.

6.6.3.2 Examination of Shear Failure

Two methods exist for determining the limit values used in an examination of shear failure:

(1) One method refers to the design shear strength.
(2) The other method involves performing a nonlinear analysis for materials (assessment on material failure).

Method (1) allows for a simple examination because it is based on the use of an evaluation formula derived from the results of loading tests. Method (1) is usually used in structural design to ensure that structural members do not reach their shear failure.

However, when a structure has complex geometry and a high degree of statical indeterminacy, a failure of some structural members reach the limit state does not lead to an immediate loss in stability of the entire structure. In such cases, an examination may be conducted using Method (2), which is more accurate and takes the failure process into account. Figure 6.13 provides an example of evaluating the seismic safety of a structure by nonlinear analysis performed on a structural component. By using an analytical model of the structural component, an analysis is performed assuming earth pressure and other loads derived from dynamic response analysis. The limit point at which the structural component shear fails can be determined by testing, in which the load applied to the specimen is increased gradually.

When the capability of a structure to remain functional is evaluated, Method (2) provides more accurate results; however, the approach is more complex. Therefore, a common approach to evaluate all structural components is by using Method (1) and then applying Method (2) to some components that require a more detailed evaluation.

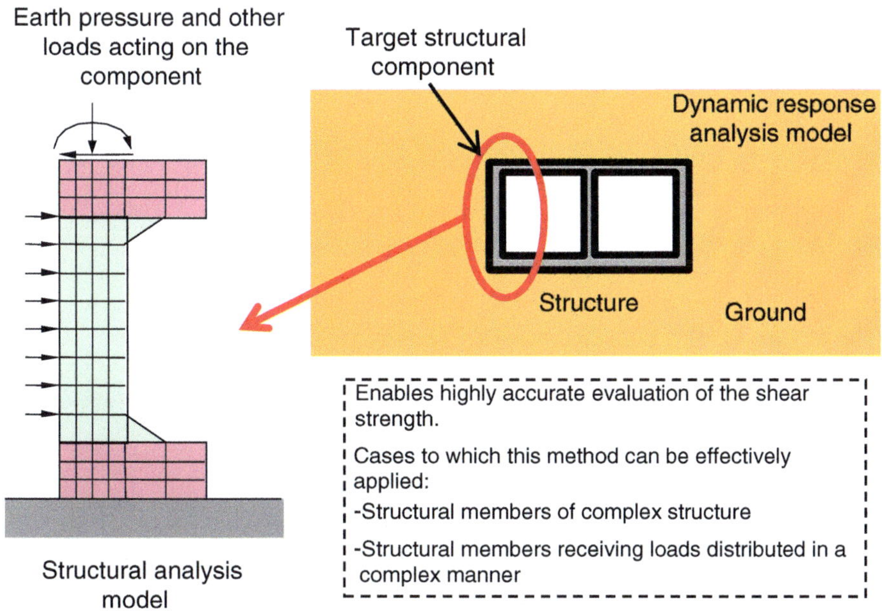

Fig. 6.13 Evaluation by analysis that considers the nonlinearity of structural materials

References

1. Japan Nuclear Energy Safety Organization website
2. "*Genshiryoku Hatsudensho Okugai Juuyou Doboku Kouzoubutsu no Taishin Sekkei Shousa Shishin* (Verification Guidelines for Seismic Design of Outside Important Civil Facilities in Nuclear Power Plants)", Japan Society of Civil Engineers

Chapter 7
Tsunami-Resistant Design

Masanori Hamada and Michiya Kuno

Abstract Nuclear power plants need large quantities of sea water to cool steam that has been used to drive turbines. All nuclear power plants in Japan are thus located by the sea. The design of nuclear power plants ensures that important safety functions of the nuclear power plant will not be lost in the event of supposed tsunamis that may occur in the future. The first step in developing a tsunami-resistant design is to identify tsunamis that may greatly affect the nuclear power plant and their potential sources according to literature reviews and geological surveys. Next, a numerical simulation is performed to investigate how a tsunami arising at the source will propagate over the sea, reach the coast near the plant, and run up the shore, to predict the water level, flow velocity, and flooded zone. The tsunami inundation height, wave force, and other values derived from this analysis are used in the tsunami-resistant design of structures and facilities, and the functionality of structures and facilities is assessed. This chapter explains the tsunami assessment method and the tsunami-resistant design of structures and facilities.

Keywords Tsunami · Tsunami deposit · Tsunami source · Tsunami-resistant design · Tsunami propagation · Tsunami protection wall · Tsunami forces · Tsunami warning

7.1 Flow of Tsunami-Resistant Design

Figure 7.1 shows the flow of tsunami-resistant design.

(1) Surveys for tsunami assessment

M. Hamada (✉)
Waseda University, Shinjuku, Tokyo, Japan
e-mail: hamada@waseda.jp

M. Kuno
Chubu Electric Power Company, Nagoya, Japan

M. Hamada and M. Kuno (eds.), *Earthquake Engineering for Nuclear Facilities*,
DOI 10.1007/978-981-10-2516-7_7

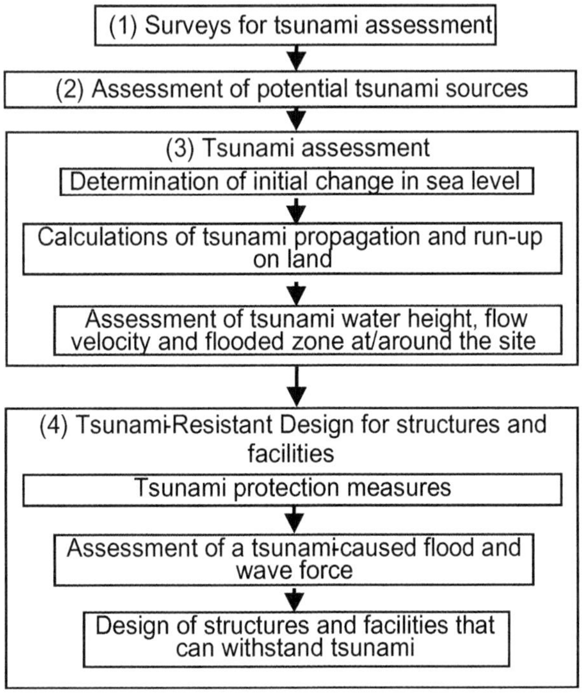

Fig. 7.1 Flow of the design of tsunami resistance

Various surveys are performed on the causes and source areas of tsunamis that may occur in the future and greatly affect the plant (Sect. 7.2).

(2) Assessment of potential tsunami sources
Potential tsunami sources are identified according to the survey results (Sect. 7.3).

(3) Tsunami assessment
The initial change in sea level is determined using a tsunami source model. Numerical analyses are then performed on the propagation of the tsunamis on the sea surface and run-up on land. These analyses are followed by an assessment of the tsunami height expected at the plant (Sect. 7.4).

(4) Tsunami-resistant design for structures and facilities
According to the tsunami height, flood areas, and wave force predicted in the numerical analysis, tsunami protection measures are developed and the tsunami-resistant design of structures and facilities is conducted (Sect. 7.5).

7.2 Survey for Tsunami Assessment

Tsunamis are caused by not only earthquakes but also other events such as volcanic eruptions and seabed landslides. However, earthquake-caused tsunamis account for 90 % of all past tsunamis. Various surveys are performed to identify the relevant causes and sources of past tsunamis. To determine the potential causes of tsunamis (e.g., faults, landslides, and massif failures), surveys on tsunami deposits are performed in addition to the surveys described in Sect. 7.2.2.

7.2.1 Survey on Traces of Past Tsunamis

7.2.1.1 Literature Survey

By reviewing the existing literatures, including historical records such as ancient manuscripts, tsunami monitoring records, and archeological survey documents, information is collected on the year of occurrence, return period, scale, and cause of past tsunamis that may have struck areas around the target site. Figure 7.2 shows an example description of a tsunami in *Nihon-shoki,* the oldest chronology for Japan.

7.2.1.2 Survey of Tsunami Deposits

To clarify the tsunamis that have struck areas around a plant in the past, the deposits left by tsunamis in the geological stratum are surveyed. The 2011 Great East Japan Earthquake showed the importance of such surveys on tsunami deposits as sources of information about past tsunamis. However, even in areas that have been struck by a tsunami in the past, tsunami deposits are preserved only under topographically favorable conditions. Moreover, tsunami deposits can be lost by subsequent erosion

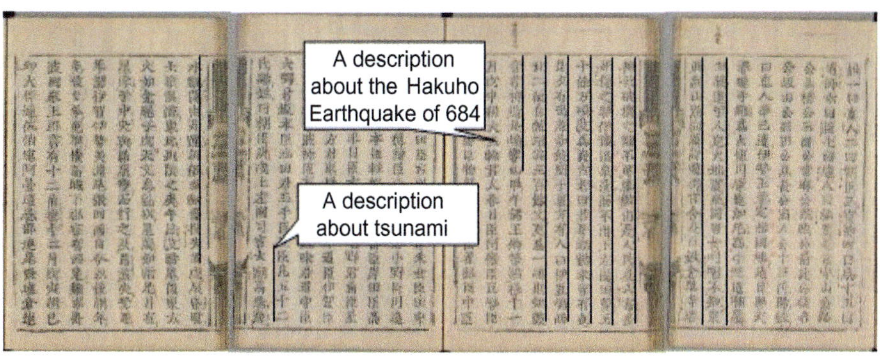

Fig. 7.2 A description of a tsunami in *Nihon-shoki* [1]

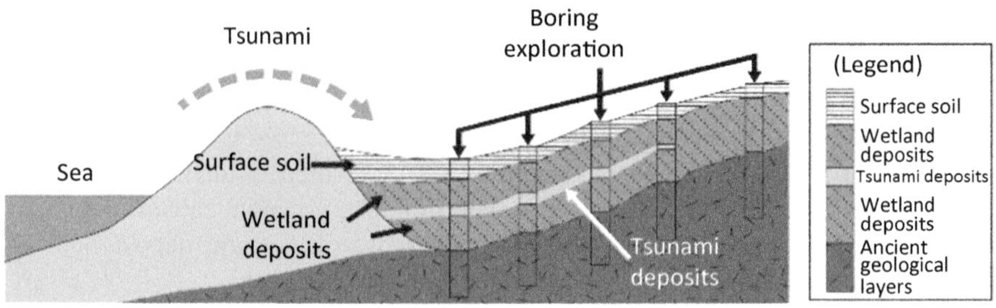

Fig. 7.3 Survey on tsunami deposits [2]

of the ground and artificial changes to the land, which limit the application of surveys on tsunami deposits.

Specifically, explorations are made employing borings and excavated trenches to find geological strata that were formed by past tsunamis and to understand their distribution. These surveys normally address the geological strata formed in or after the Jomon Transgression period (i.e., the Holocene glacial retreat period), during which sea levels rose. As to whether any geological stratum was formed by a tsunami, a judgment should be made after comprehensively evaluating survey results and test results of specimens, as well as the geological and geomorphic features of the given area. Figure 7.3 shows an example of a tsunami deposit survey conducted through boring exploration.

7.2.2 Survey of Causes of Tsunamis

Most tsunamis are caused by earthquakes, but there are other causes as mentioned earlier.

The causes of tsunamis including earthquakes are categorized as

- interplate earthquakes,
- oceanic intra-oceanic plate earthquakes,
- crustal earthquakes caused by an offshore active fault,
- landslide or slope failures on land or under the sea, and
- volcanic eruptions.

Considering that tsunamis affect larger areas than earthquakes, surveys on the causes of tsunamis must cover a wider area than surveys on earthquake ground motion. Various survey techniques are combined in an appropriate manner considering the geomorphic and geological conditions of the given area.

Surveys conducted on earthquakes as a cause of tsunamis include not only interplate earthquakes but also a type of oceanic intra-oceanic plate earthquake known as the subduction oceanic plate earthquake, which has its epicenter on or close to a sea trench axis. A typical example of a subduction oceanic plate earthquake is the 1933 Sanriku earthquake in Japan. This earthquake caused significant tsunami damage even though the earthquake was of low magnitude.

Past tsunamis caused by onshore and offshore slope failures and landslides should be studied by conducting an examination of geomorphic and geological features over a wide area around the plant. An example of a landslide-caused tsunami is the 1958 Lituya Bay mega-tsunami (Alaska, USA).

As for volcanic activity (e.g., eruptions) as a cause of tsunamis, surveys should be conducted on volcanic activity around the site, to evaluate the risks from landslides and massif failures due to a volcanic eruption or volcanic earthquake.

Surveys on tsunami propagation routes involve factors that affect the propagation of tsunamis, such as the onshore and offshore geomorphic features of the areas along the route from the tsunami source to the target site. In addition, surveys should be conducted on the sand drift that may be caused by a tsunami around the water intake gate to judge its potential effect on water intake facilities.

7.3 Assessment of Potential Tsunami Sources

According to the results of tsunami surveys, the causes of tsunamis that may happen in the future and greatly affect the site are identified, and the tsunami source is determined for each such potential case of tsunami. Figure 7.4 shows a tsunami source model prepared by the Cabinet Office of the Japanese Government based on the 1677 Boso Coast earthquake (an interplate earthquake) that struck during the Eiho Period.

7.4 Tsunami Assessment

A tsunami assessment based on numerical simulation analysis is performed for identified causes of tsunamis using a tsunami source model. The numerical simulation of a tsunami involves determining the initial change in sea level and calculating the tsunami propagation over the sea and run-up on land. The inundation height, flow velocity of the run-up tsunami, flooded zone, and depth at/around buildings, structures, and facilities are then assessed.

7.4.1 Determination of Initial Change in Sea Level

7.4.1.1 Tsunamis Caused by Earthquakes

In the case of a tsunami caused by the seabed rising or dropping owing to fault movement (i.e., an earthquake-caused tsunami), upward/downward displacements

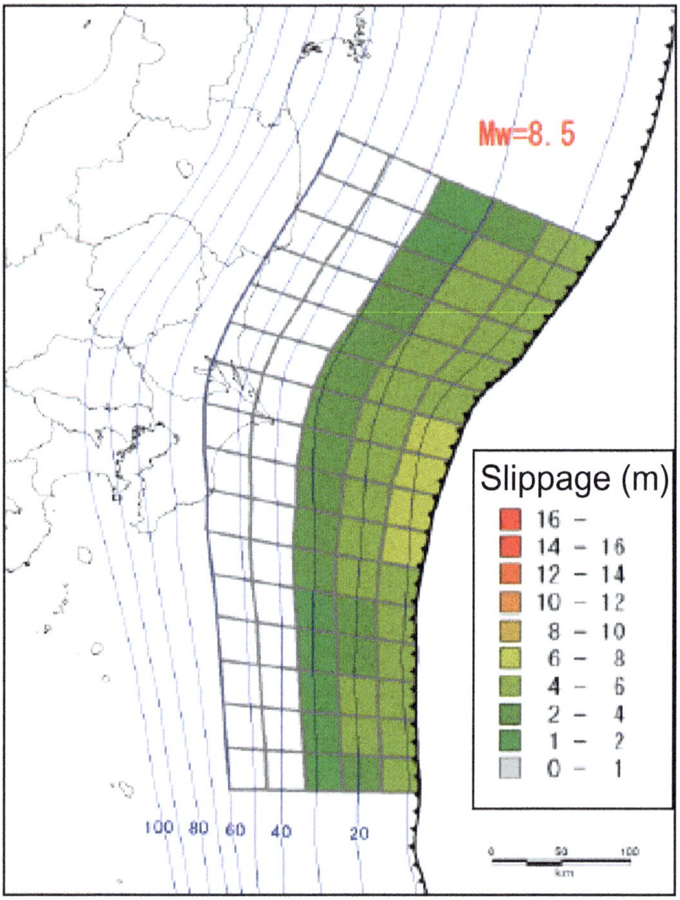

Fig. 7.4 Example of a tsunami source model (interplate earthquake) [3]

of the seabed are determined. The initial change in sea level is assumed to be same as the seabed displacement. This is based on the assumption that the sea level will follow the seabed-level displacement within a short time. Figure 7.5 shows the fault parameters defining a fault movement and presents an overview of seabed deformation analysis. The seabed deformation analysis is performed using semi-infinite elastic theory proposed by Manshinha and Smylie [6].

7.4.1.2 Tsunamis by Other Causes

As mentioned in Sect. 7.2, tsunamis can be produced by causes other than earthquakes, such as onshore/offshore landslides and slope failures and volcanic activity. In each case of such tsunamis, the initial change in sea level and other parameters are determined using an appropriate cause-specific analysis method. When

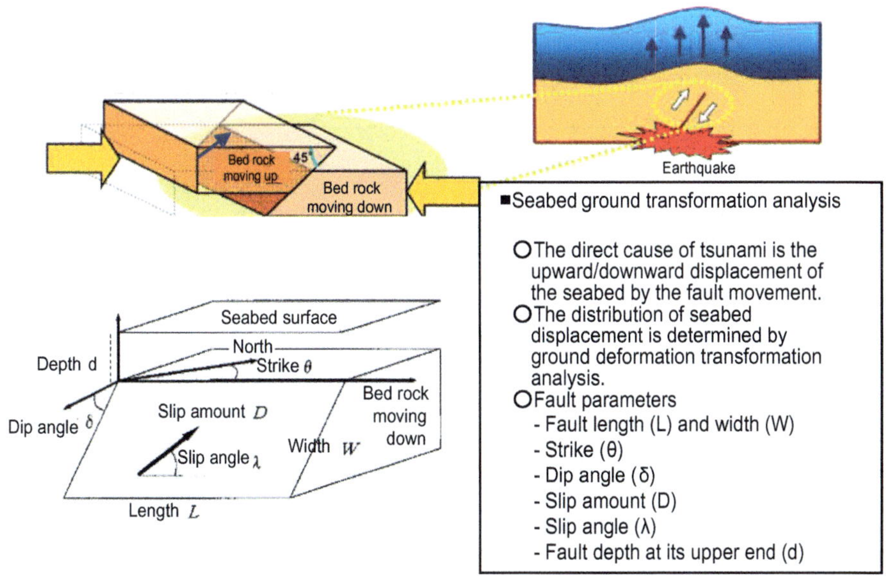

Fig. 7.5 Overview of ground deformation analysis [4, 5]

considering landslides and slope failures, an analysis can be conducted using a two-layer flow model of soil and water.

7.4.2 Calculations of Tsunami Propagation and Run-up on Land

The wavelength of a tsunami ranges from several tens of kilometers to several hundred kilometers, which is a length much greater than the sea depth of several kilometers. Therefore, long-wave theory is used for tsunami propagation analysis. Long-wave theory makes it possible to consider the leaning forward of a waveform, as shown in Fig. 7.6, which happens during the propagation of tsunamis owing to the effect of the submarine topography.

Calculations of tsunami propagation and run-up on land are made using a two-dimensional model in a horizontal plane. According to the initial change in sea level, the propagation of the tsunami over the sea is analyzed using the differential method. The analysis accounts for the changes in the sea depth due to crustal deformation and the change in the sea surface due to the tide. The maximum water level is evaluated at the time of high tide, and the minimum water level is evaluated at low tide. Figure 7.7 shows an example of the numerical simulation analysis of a tsunami.

With the evolution of computers, it has become possible to directly conduct three-dimensional analysis when the three-dimensional seabed topography affects

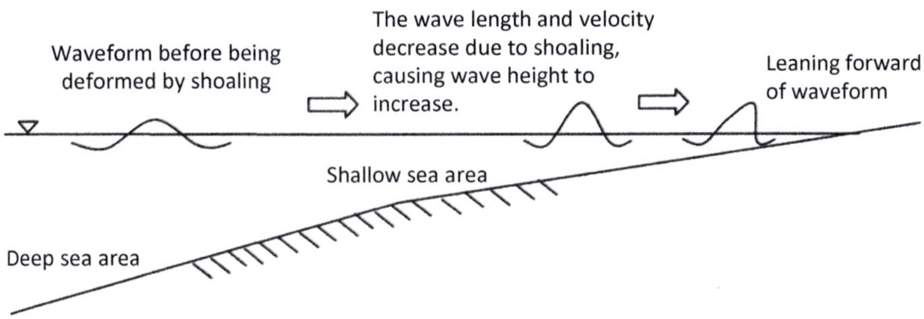

Fig. 7.6 Tsunami propagation over the sea and the leaning forward of a wave form [7]

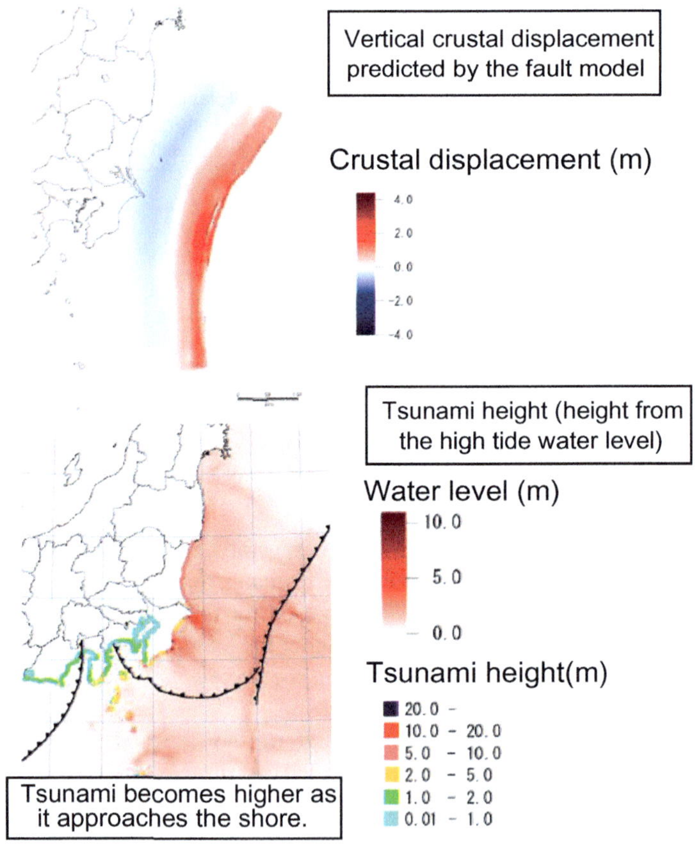

Fig. 7.7 Numerical simulation analysis of tsunami propagation [3]

the propagation of a tsunami. In addition, it is useful to perform experiments on tsunami propagation using water tanks to verify the validity of the numerical simulation.

Furthermore, it is important to compare the water levels and flooded zones of past tsunamis with the results of the numerical simulation.

7.5 Tsunami-Resistant Design for Structure and Facility

According to the assessment of tsunami effects on the site, protective measures are developed and the tsunami-resistant design of facilities and structures is conducted for the wave force and height predicted by the tsunami simulation. Furthermore, the effect of the change in the seawater level at the water intake is studied.

7.5.1 Tsunami Protection Measures

The functionality of the facilities for cooling the reactors and spent fuel (water injection, heat removal, and power supply components) should be ensured by taking tsunami protection measures. Specifically, a tsunami protection wall may need to be built to prevent the flooding of the site, watertight doors may need to be installed in buildings housing components important for safety, and severe accident management facilities in the event of a loss of cooling capability due to a tsunami may need to be located at an elevated site. Figure 7.8 shows an example of tsunami protection measures including a tsunami protection wall, watertight doors of buildings, and the preparation of fire engines and pipes at an elevated location.

7.5.2 Assessment of a Tsunami-Caused Flood and Wave Force

Tsunami protection measures should be designed and constructed to ensure the safety functions of the nuclear power plant against inundation, wave forces, and other effects of tsunamis. It is therefore necessary to determine the inundation height, wave force, and other consequences of a tsunami. For the inundation height, flow velocity, and flooded zone, the results obtained from the tsunami assessment described in Sect. 7.4 are used. For the wave force that acts upon structures when a tsunami runs up the shore, the magnitude and distribution of the wave force can be estimated employing the following methods.

Evaluation formulas derived from hydraulic model experiments are proposed to determine the magnitude and distribution of wave forces acting on structures. When

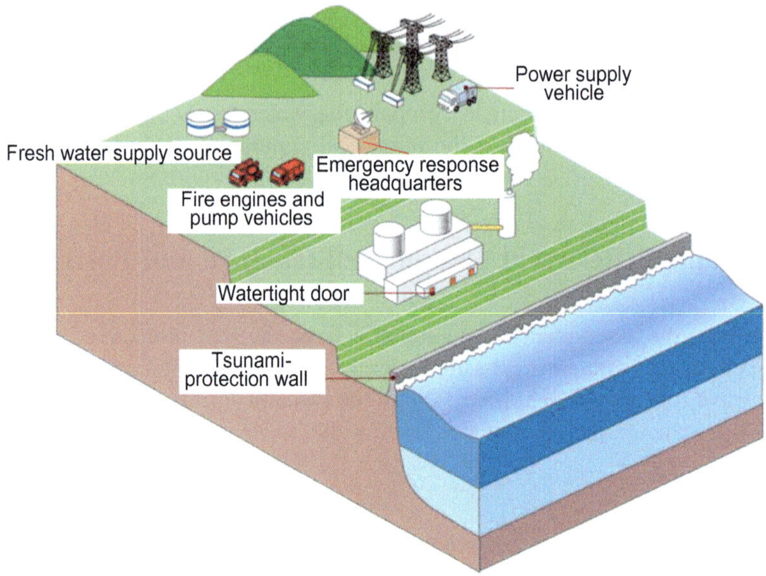

Fig. 7.8 Tsunami protection measures [8]

assessing the tsunami wave force, the evaluation formula to be used must be chosen in consideration of the locations of the target structure and other structures around it, and the local geomorphic features and the characteristics of the tsunami.

A tsunami may behave in a complex manner owing to interactions with local geomorphological features and nearby buildings and structures. In that case, it is desirable to perform the assessment by considering also the results of model experiments and numerical analyses. Examples of an evaluation formula proposed by Asakura et al. [9] and an evaluation formula taken from guidelines published by the Ministry of Land, Infrastructure, Transport and Tourism (MLIT 2011) are presented below. Asakura et al. proposed Eq. (7.1) as an evaluation formula that can be used to determine the maximum wave pressure acting on an upright wall in the event of a tsunami according to the results of two-dimensional model experiments. Figure 7.9 illustrates the concept of the evaluation of tsunami wave pressure:

$$p(z) = (3h - z)pg \tag{7.1}$$

The MLIT published *Provisional Guidelines for the Structural Requirements of Tsunami Evacuation Buildings*, which prescribes a tsunami load calculation method

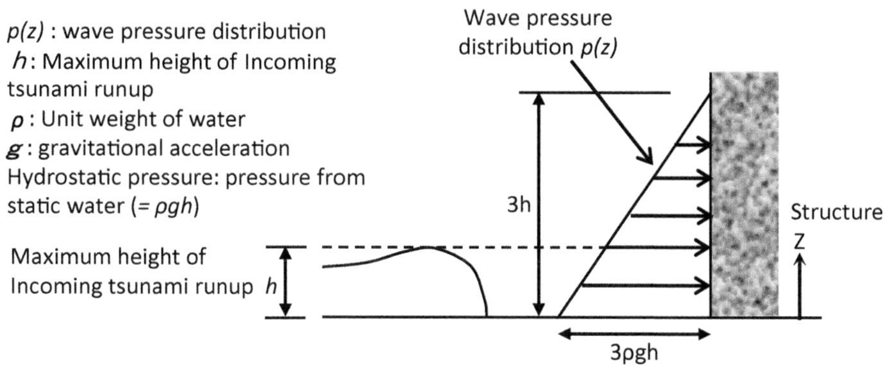

Fig. 7.9 Concept of tsunami wave pressure [9]

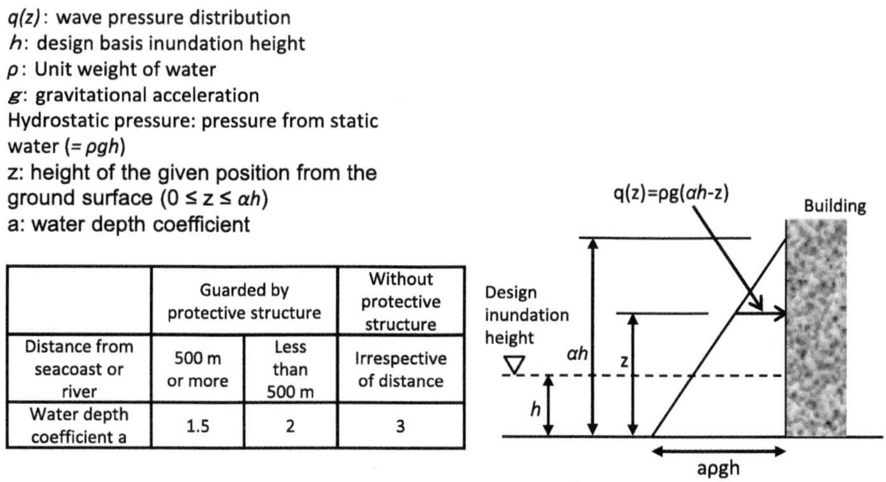

	Guarded by protective structure		Without protective structure
Distance from seacoast or river	500 m or more	Less than 500 m	Irrespective of distance
Water depth coefficient a	1.5	2	3

Fig. 7.10 Tsunami wave force evaluation according to the guidelines of the MLIT (2011) [10]

based on the surveys of tsunami damage to buildings as a consequence of the 2011 Great East Japan Earthquake. The guidelines recommend the use of the tsunami pressure (shown in Fig. 7.10) expressed as

$$q(z) = (ah - z)pg \tag{7.2}$$

7.5.3 *Design of Structure and Facility that Can Withstand Tsunamis*

7.5.3.1 Prevention of Flooding by Tsunami Protection Wall

A tsunami protection wall prevents nuclear power plants from flooding. Such a wall must maintain its integrity by resisting both the tsunami wave force and earthquake ground motions. Some tsunami protection walls are designed to withstand the tsunami wave force and seismic load by having their foundations deeply embedded into the ground. Important facilities can be sited at an elevated location to ensure that they are not flooded. Figure 7.11 shows an example of the prevention of flooding employing tsunami protection walls and embankments.

The tsunami protection wall and other structures that protect plants from flooding should be strictly designed to withstand the wave force of a tsunami, ensuring that they maintain their integrity even after being struck by a tsunami that passes above them.

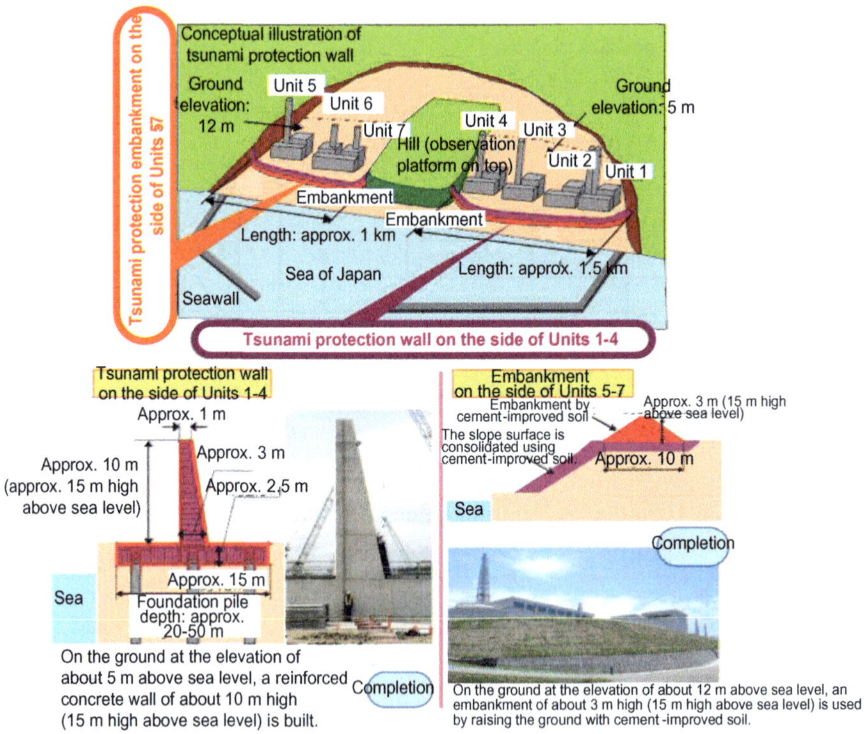

Fig. 7.11 Prevention of the flooding of a plant by wall and embankment [11]

7.5.3.2 Prevention of Overflow from Water Intake and Discharge Facility

The water intake and discharge facilities of a nuclear power plant are connected with the sea. If there are openings in the intake and discharge facilities, the sea water may overflow into the plant. It is therefore important to take necessary measures to prevent such overflow. As measures, walls higher than the rise in water level may be built around openings, or openings may be made capable of closing. Figure 7.12 shows an example of walls that prevent the sea water overflowing from the water intake pit into the plant. Figure 7.13 shows a closing plate installed at the opening of water discharge facilities.

7.5.3.3 Prevention of Water Penetration into Building

To protect important facilities that have a safety function for the nuclear power plant in the event of a tsunami, the doors of buildings should be watertight and penetration spaces of piping through walls should be perfectly sealed. Moreover, considering that water may still enter buildings, measures should be taken to prevent the penetration of water into rooms where important facilities having a safety function are installed. Figure 7.14 shows an example of watertight doors attached to external building walls. These doors may be exposed to the wave force of a tsunami. Accordingly, a double-door configuration (combining a wave force-withstanding door and a watertight door) should be employed to ensure protection against the wave force of a tsunami.

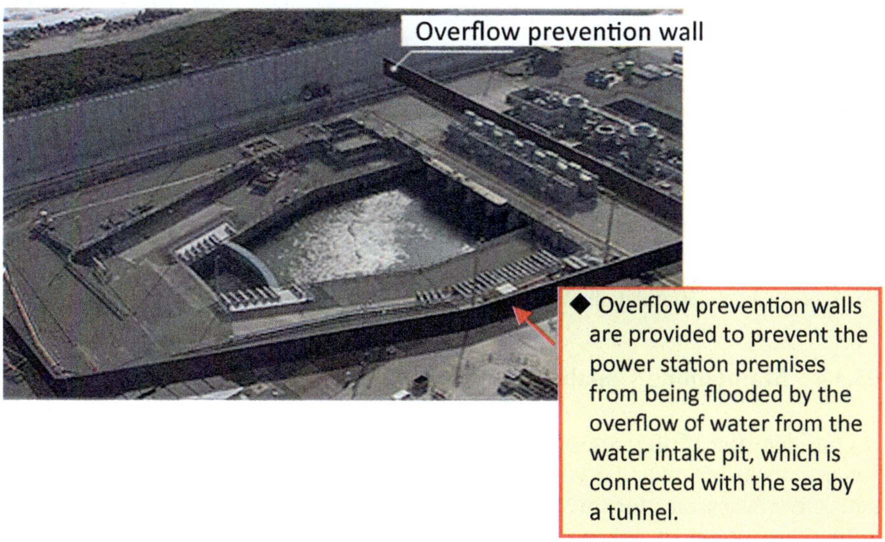

Fig. 7.12 Overflow prevention walls

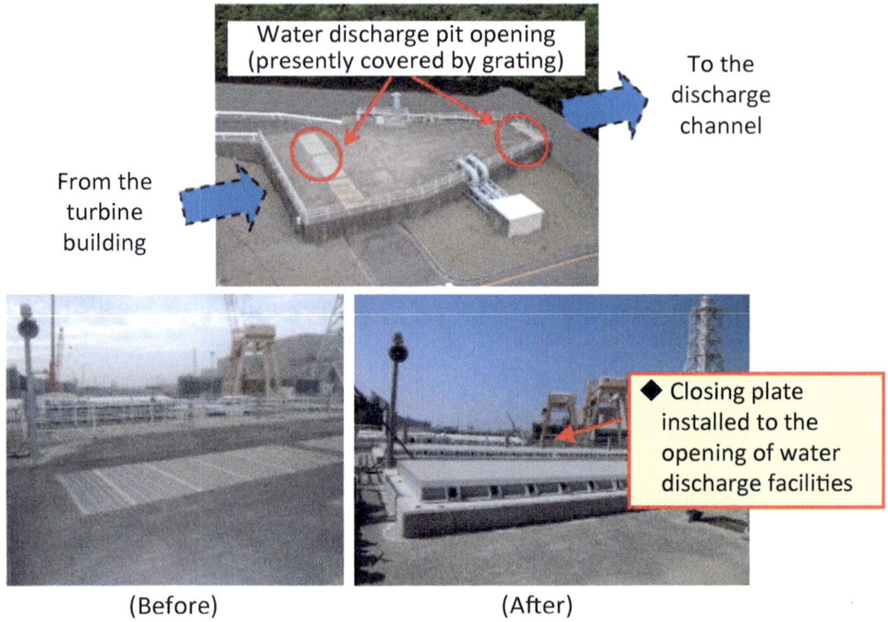

Fig. 7.13 Example of a closing plate installed at the opening of water discharge facilities [2]

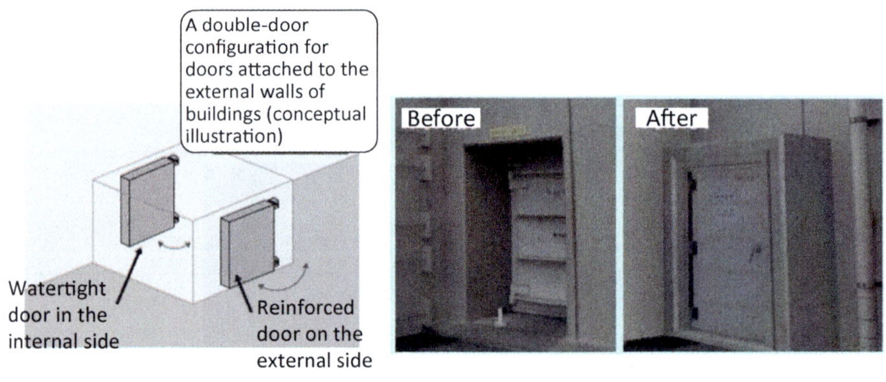

Fig. 7.14 Example of watertight doors attached to the external walls of buildings [2]

7.5.3.4 Maintaining Capability of Water Pumps

In consideration of the consequence of a drop in the sea level in the event of a tsunami, measures must be taken to preserve the water intake capability of pumps that deliver sea water for the emergency cooling of reactor cores and spent fuel.

Measures should be taken to ensure the sea water intake capability by employing a structural design that allows the intake of cooling water (sea water) at a certain

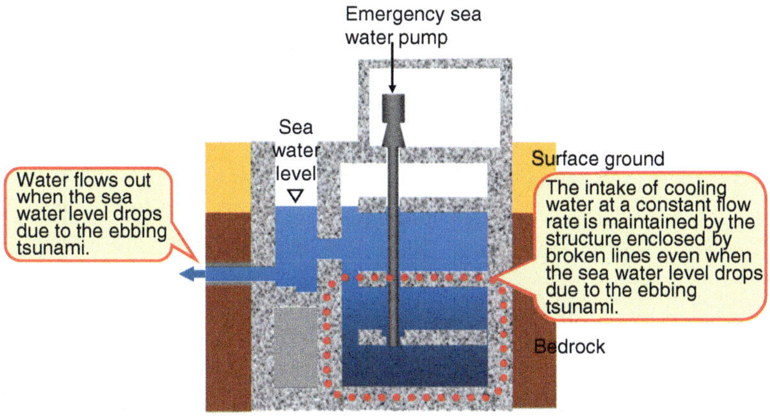

Fig. 7.15 Maintaining the water intake capability of sea water pumps [2]

flow rate even if the sea level drops. The water intake gate and water channel should continue to allow the passage of water by resisting a blockage due to tsunami-induced sand drift, sedimentation, and floating debris. The seawater pumps should be able to continue to take in water without being affected by the mixing of sand with water. Figure 7.15 shows a configuration designed to maintain the intake of cooling water by the structure framed with broken lines even when the sea level drops owing to an ebbing tsunami.

7.5.3.5 Other Tsunami Protection Measures

An early warning of a tsunami attack allows actions to be taken in preparation against a tsunami at a nuclear power plant. Measures that can be taken for early tsunami detection include the use of a wave gage that employs the Global Positioning System (GPS). Employing this method, a tsunami is detected by observing the vertical movement of an offshore buoy with GPS satellites. Figure 7.16 shows tsunami detection employing a GPS device.

Examples of tsunami detection measures that can be employed at nuclear power plants include the installation of tide gages and the monitoring of tsunamis with outdoor surveillance cameras. Outdoor surveillance cameras can be used to warn of a tsunami attack by directly observing the abovementioned GPS buoy.

It is important that tsunami monitoring devices such as outdoor surveillance cameras are installed at locations that are unlikely to be damaged by the tsunami wave force and floating debris in order for the devices to effectively serve the purpose of tsunami monitoring. Figure 7.17 shows a high-sensitivity camera as an example of a tsunami monitoring device.

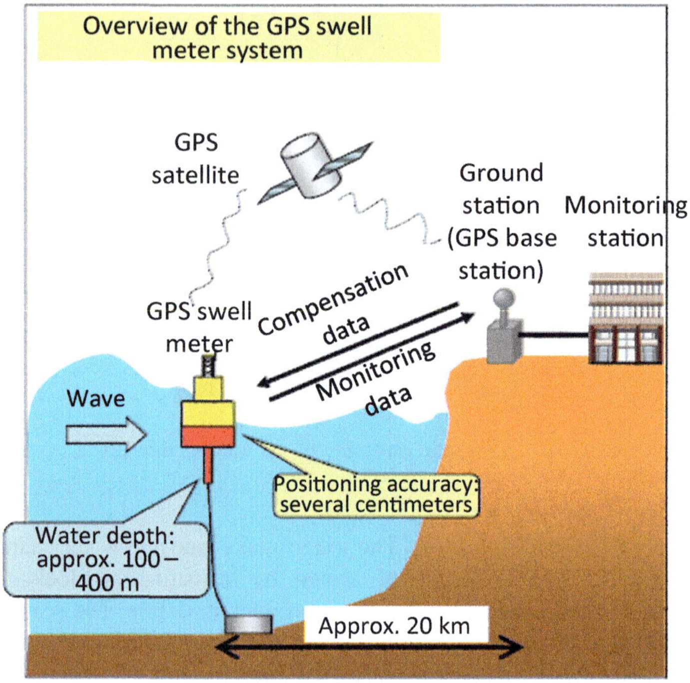

Fig. 7.16 Overview of tsunami detection using a GPS swell meter [12]

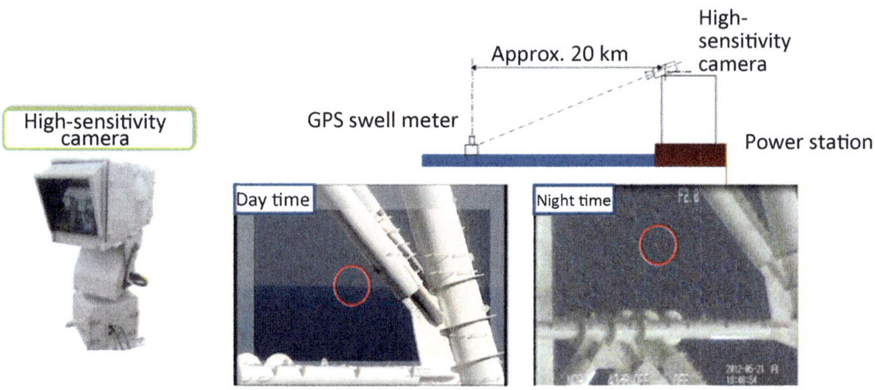

Fig. 7.17 Example of a tsunami monitoring device (high-sensitivity camera) [13]

References

1. "Nihon-shoki Vol. 29", National Diet Library Digital Collections
2. Chubu Electric Power website
3. "*Shuto Chokka no M7 Kurasu no Jishin oyobi Sagami Torafu zoi no M8 Kurasu no Jishin nado no Shingen Dansou Moderu to Shindo Bunpu, Tsunami-daka nado ni Kansuru Houkokusho* (Report about Hypocenter Fault Model, Distribution of Seismic Intensity and Tsunami Height of M7 Class Earthquake directly below Tokyo and M8 Class Earthquake beside Sagami Trough, etc.)", Cabinet Office, December, 2013
4. Japan Meteorological Agency website
5. Electrical Review, June, 2005
6. "The displacement fields of inclined fault." Manshinha, L., and Smylie, D.E. Bull. Seism. Soc. Amer., 61, 1971
7. "Manual for Analyzing Tsunami Run-up in Rivers (draft)", Japan Institute of Country-ology and Engineering, May, 2007
8. FEPC's Web Magazine "Enelog" Vol. 3, June, 2013
9. "An Experimental Study on Wave Force Acting on On-Shore Structures due to Overflowing Tsunamis", Ryosuke Asakura, Koji Iwase, Takeshi Ikeya, Makoto Takao, Toshimichi Kaneto, Naoki Fujii, Masanori Omori, Proceedings of Coastal Engineering, Vol. 47, Japan Society of Civil Engineers, 2000
10. "*Tsunami Hinan Biru Nado no Kouzou-jou no Youken no Kaisetsu* (Explanation of Structural Requirements for Tsunami Evacuation Buildings, etc.)", National Institute for Land and Infrastructure Management, Building Performance Standardization Association, Assisted by Building Research Institute, February 2012
11. Tokyo Electric Power website
12. Ministry of Land, Infrastructure, Transport and Tourism website
13. "*Hamaoka Genshiryoku Hatsudensho no Tsunami Kanshi ni tsuite* (Tsunami Monitoring in the Hamaoka Nuclear Power Plant)", The Disaster Preparedness and Nuclear Council in Shizuoka Prefecture, Material 2-1 for the first meeting of Tsunami Protection Subcommittee in FY2012, August 9, 2012

Further Readings

14. "Tsunami Assessment Method for Nuclear Power Plants in Japan", Japan Society of Civil Engineers, The Tsunami Evaluation Subcommittee, The Nuclear Civil Engineering Committee, February, 2002
15. "*Aichi-ken Tokai Jishin, Tonankai Jishin nado Higai Yosoku Chousa Houkokusho* (Investigation Report of Damage Estimation caused by Tokai Earthquake and Tonankai Earthquake)", Earthquake Department, Disaster Prevention Council of Aichi Prefecture, March 2013
16. "Interpretation of Ordinance on the Standards for Locations, Structures, and Installations of Commercial Power Reactors and Auxiliary Facilities for These Reactors", Decided on June 19, 2013, Nuclear/Regulation/Technology No. 1306193, NRA's decision
17. "Review Guide for Geological and Geological Structure Surveys within and around the Site", Decided on June 19, 2013, Nuclear/Control/Earthquake No. 1306191, NRA's decision
18. "Review Guide for Design Basis Tsunami and Anti-Tsunami Design Policy", Decided on June 19, 2013, Nuclear/Control/Earthquake No. 1306193, NRA's decision
19. "Regulatory Guide for Reviewing Tsunami Resistant Design on Construction Plan", Decided on June 19, 2013, Nuclear/Control/Earthquake No. 1306196, NRA's decision
20. "Ordinance on the Standards for Locations, Structures, and Installations of Commercial Power Reactors and Auxiliary Facilities for These Reactors", Ordinance of the Nuclear Regulation Authority No. 5, June 28, 2013

Chapter 8
Seismic Probabilistic Risk Assessment

Tatsuya Itoi, Michiya Kuno and Masanori Hamada

Abstract This chapter presents an overview of the seismic probabilistic risk assessment (PRA) for nuclear power plants. First, the basic concept of seismic PRA and its application to continuous improvement of seismic safety of nuclear power plants are explained. Then, the methodology of seismic PRA for nuclear power plants is explained. Seismic PRA process can be divided into probabilistic seismic hazard analysis, fragility analysis of structures/components, and accident sequence analysis. As for the probabilistic seismic hazard analysis, modeling of earthquake occurrence, a method for earthquake ground motion prediction, and logic tree model are introduced. Then, the fragility analysis of structures/components is described. This includes an explanation of the selection of failure modes, and the realistic strength and realistic response of buildings/components.

Keywords Probabilistic risk assessment · Accident scenario · Fragility analysis · Seismic hazard analysis · Fragility curve · Event tree · Fault tree

8.1 Probabilistic Risk Assessment (PRA)

The accident that took place at the Fukushima Daiichi Nuclear Power Plant as a result of the 2011 Great East Japan earthquake and consequent tsunami raised awareness of the risks associated with nuclear facilities. To achieve high level of safety for nuclear facilities, probabilistic risk assessment is considered useful because it provides insight into the weakness by identifying as many possible accident scenarios as possible.

T. Itoi (✉)
The University of Tokyo, Tokyo, Japan
e-mail: itoi@n.t.u-tokyo.ac.jp

M. Kuno
Chubu Electric Power Company, Nagoya, Japan

M. Hamada
Waseda University, Shinjuku, Tokyo, Japan

M. Hamada and M. Kuno (eds.), *Earthquake Engineering for Nuclear Facilities*,
DOI 10.1007/978-981-10-2516-7_8

Risk is defined as combination of consequence of an accident and its associated likelihood, i.e., occurrence probability.

When a severe accident happens at a nuclear power plant, leading to damage to the reactor core and the reactor containment vessel, large amount of radioactive material are released into the environment, exposing the public in the vicinity of the plant with respect to radiation exposure as well as environmental contamination.

As a result of probabilistic risk assessment, contributors to accident risk can be quantitatively assessed, which are potential causes of an accident. Moreover, PRA can estimate quantitatively the effectiveness of countermeasures to be taken to reduce risks.

Quantitative safety goal and subsidiary objectives, so-called performance goals, are referred to as criteria that represent the societal acceptability of the risks (Appendix 8.1).

Safety goals usually concern the radiation risk of the public as well as environmental contamination, while performance goals concern, for example, equivalent plant condition such as core damage frequency as well as large early release frequency.

8.2 Categorization of Initiating Events

Events that initiate accidents are classified into (a) random failure of components, (b) internal hazards such as internal flooding and an internal fire, and (c) external hazards. External hazards are classified into natural hazards such as earthquakes and tsunamis, and human-induced hazards such as an airplane attack and an intentional act.

8.3 Overview of Seismic PRA

In seismic PRA, attention is focused on earthquakes that may happen in the future and affect the safety of nuclear power plant. The failure probabilities of structures (e.g., buildings) and components are assessed. Uncertainty associated with the earthquake ground motion, the responses of buildings and components, and strength of structures and components are considered. Then, the occurrence probabilities (or frequencies) of various accident sequences and the magnitude of their consequences are estimated.

Furthermore, in seismic PRA, the core damage frequency is estimated on the basis of seismic hazard analysis, structures/components fragility analysis, and accident sequence analysis. The seismic hazard analysis estimates the probability of occurrence, within a certain period of time (e.g., 50 years), of an earthquake for which the ground motion, e.g., at the hypothetical free surface of bedrock, exceeds a certain level. The fragility analysis gives the failure probability of structures/components (for example, buildings, structures, and component piping systems) as a function of the seismic ground motion level. Figure 8.1 illustrates the overview of seismic PRA.

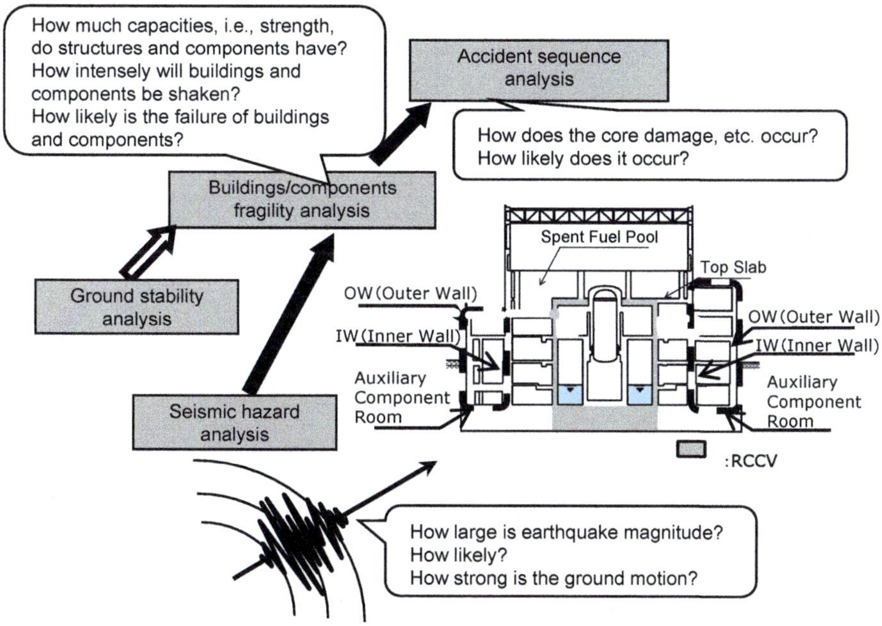

Fig. 8.1 Overview of seismic PRA

Figure 8.2 shows the seismic PRA procedure.

(1) Collection and analysis of site/plant information, and field surveys
To start with, the information required for seismic PRA is collected and compiled. Field surveys (site/plant walkdown surveys) are performed as required. Through field surveys, the type of information is obtained which cannot be acquired sufficiently from documents alone. They may include the information whether components and piping systems have been installed exactly as designed, the presence of any facility that may affect another facility, and factors that may complicate post-earthquake recovery actions.

(2) Identification of accident scenarios
According to plant information and field survey results obtained in Step (1), possible scenarios of an accident, for example, leading to core damage, are identified.

(3) Seismic hazard analysis
Based on information on the seismic source locations and magnitudes, occurrence frequencies of earthquakes that may affect the nuclear power plant, the probability of ground motion, e.g., at virtually free surface of bedrock, exceeding a certain level are estimated.

(4) Ground stability analysis
The potential effects of hazards such as sliding of foundation ground under the buildings/structures and slope failure are studied. Their potential impacts on structures and components require attention.

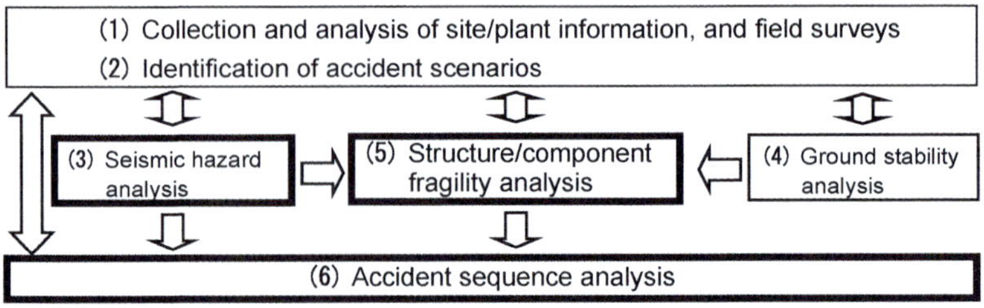

Fig. 8.2 Seismic PRA procedure

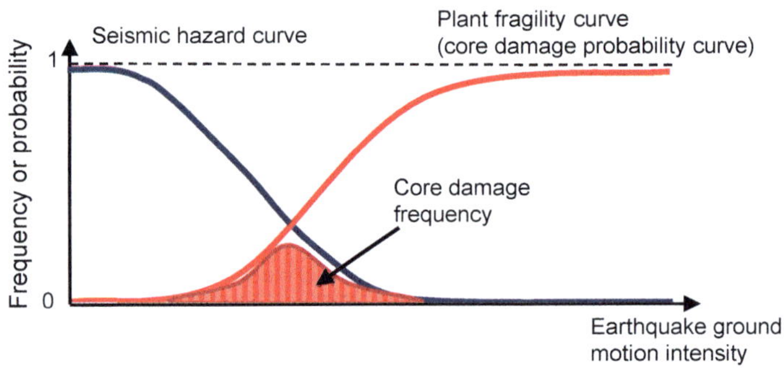

Fig. 8.3 The core damage frequency obtained from the seismic hazard curve and the plant fragility curve

(5) Structure/component fragility analysis
On the basis of information on the dynamic response analysis and strength of structures/components, failure probabilities are estimated as a function of the earthquake ground motion intensity. The failure should not be assessed based on conservative design assumptions, but based on realistic response as well as realistic strength of structures/components.

(6) Accident sequence analysis
On the basis of the seismic hazard analysis, the fragility analysis and information about random failures of safety functions and the scenarios of accidents leading to core damage are analyzed (in the form of event trees and fault trees). The conditional probability and occurrence frequency of core damage are calculated. This leads to the identification of components and facilities that are likely to serve as a cause of core damage.

Figure 8.3 shows how the core damage frequency is calculated from the seismic hazard curve and the plant fragility curve (representing conditional probabilities of core damage as a function of earthquake ground motion intensity).

8.4 Probabilistic Seismic Hazard Analysis

A relationship between the earthquake ground motion intensity and its likelihood of occurrence is obtained by the probabilistic seismic hazard analysis (PSHA). This section describes the conventional method of PSHA for a nuclear facility.

8.4.1 Probabilistic Seismic Hazard Analysis Flow

As shown in Fig. 8.4, a seismic hazard curve, a relationship between the earthquake ground motion intensity and its annual exceedance frequency, is obtained by taking account of various uncertainties. The uncertainties are due to the location and characteristics of seismic sources, and the characteristics of the seismic wave propagation and ground motion amplification. Typically, amplification up to reference bedrock surface (virtually free surface of bedrock) is considered in seismic hazard analysis. Amplification from reference rock to surface ground is considered in fragility analysis.

Figure 8.5 shows the seismic hazard analysis flow.

(1) Identification of the seismic source model
In seismic hazard analysis, a probabilistic model is employed to model the possibilities of earthquake occurrence that may affect the target site. The magnitudes and occurrence of such earthquakes are considered probabilistically and statistically.

(2) Ground motion prediction model
A ground motion prediction model, i.e., attenuation formula, is used to estimate the probability distribution of earthquake ground motion at the target site for each seismic source identified in Step (1).

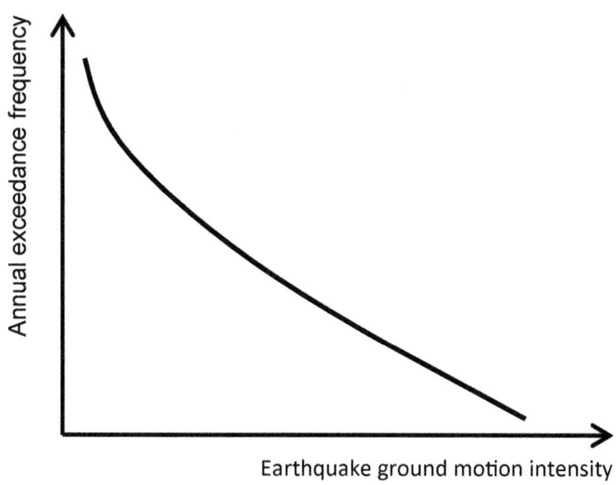

Fig. 8.4 Probabilistic seismic hazard curve

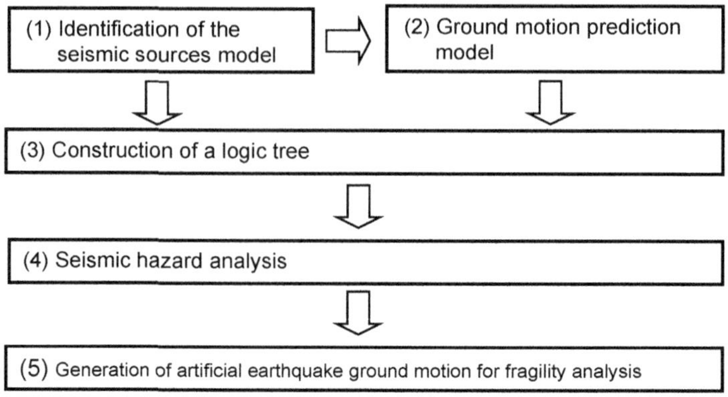

Fig. 8.5 Seismic hazard analysis flow

(3) Construction of a logic tree
For the source model and ground motion prediction model, a logic tree is employed to account for epistemic uncertainty, i.e., knowledge-oriented uncertainty, in parameters that differ in assumptions concerning, for example, the fault length and the earthquake magnitude.
(4) Seismic hazard analysis
As a result of analysis, the seismic hazard curve, i.e., the relationship between the earthquake ground motion intensity and the annual exceedance frequency, is obtained. Uncertainties in the estimated hazard curve are analyzed using the logic tree.
(5) Generation of artificial earthquake ground motion for fragility analysis
A seismic hazard curve is calculated for response spectra of the ground motion for different vibration periods. A value representing a specified level of exceedance probability is selected at each vibration period for the response spectrum. Then, a uniform hazard spectrum is obtained by connecting these points. From the uniform hazard spectrum, simulated ground motion for use in dynamic response analysis is generated.

8.4.2 Probabilistic Seismic Hazard Analysis Method

8.4.2.1 Seismic Source Model

Considering all earthquakes that may affect the target site, seismic sources are modeled and the occurrence probabilities of such earthquakes are estimated. Usually, the potential seismic sources within a distance of approximately 100–150 km from the target site are considered for seismic hazard analysis.

Earthquakes of all seismogenic mechanisms, including inland crustal earthquakes, interplate earthquakes, and intra-oceanic plate earthquakes, should be addressed.

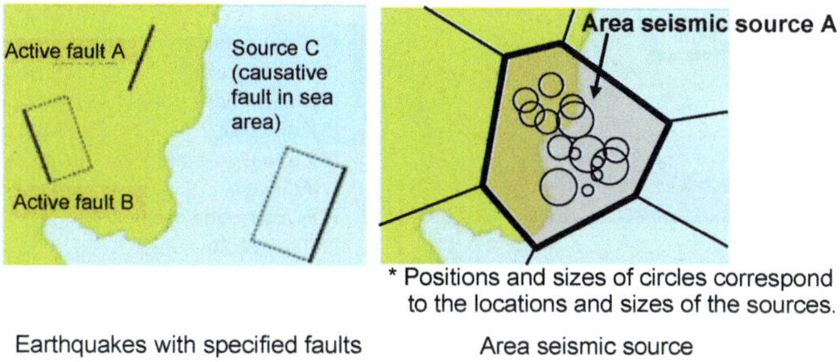

Fig. 8.6 Earthquakes with specified faults and area seismic sources

As shown in Fig. 8.6, the seismic sources are classified into two categories: earthquakes with specified faults, for which the magnitude and location can be specified in advance, and area seismic sources (i.e., earthquakes without specified faults or diffuse seismicity), where seismic activities of unspecifiable magnitudes at unspecifiable locations within a certain zone are modeled statistically.

An example of interplate earthquakes is earthquakes that are located off the Pacific coast of Tokai/Tonankai/Nankai area in Japan. The Japanese Government's Headquarters of Earthquake Research Promotion has released estimation about the probabilities of their occurrence in the regions as shown in Fig. 8.7.

The probability of occurrence of a particular type of earthquake is modeled using a time-dependent probability model to address the periodicity of earthquake occurrence, if the time and the frequency of past occurrences are known. In the case that the time of past occurrences is unknown, a homogeneous probability model such as a Poisson model is used alternatively. A homogeneous probability model assumes an unchanging probability of occurrence in time. Figure 8.8 shows three options (a–c) to estimate annual occurrence frequency or probability of earthquake.

As for the types of earthquakes that are difficult to be specified in advance in terms of magnitude, location, and their occurrence, the occurrence frequencies are modeled using the Gutenberg–Richter law as shown in Fig. 8.9. The spatial distribution of earthquakes is assumed to be uniform within the same seismotectonic zone. A seismotectonic zonation proposed for Japan is shown in Fig. 8.10.

8.4.2.2 Ground Motion Prediction Model

Earthquake ground motion is estimated using a ground motion prediction equation, i.e., attenuation formula. Attenuation formula is usually a function of the magnitudes and fault-to-site distance and site amplification characteristics. The predicted earthquake ground motion is an average value. Attenuation formula is derived by means of regression analysis from ground motion records at many observation stations, and usually exhibits large variability. Referring both to the average value and the variability, the probability of ground motion exceeding a certain level is

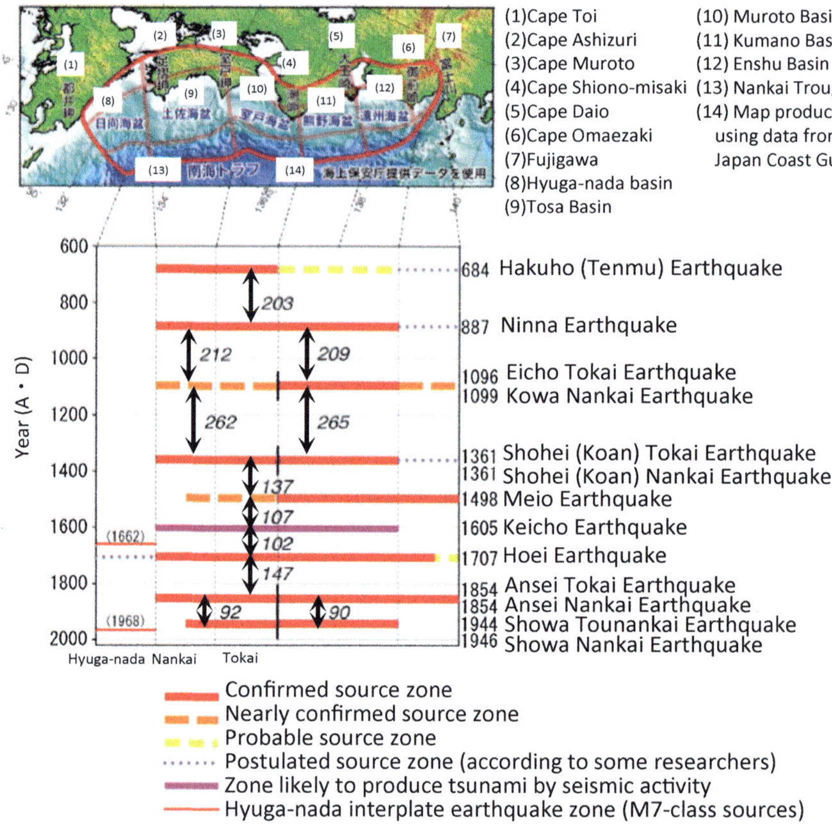

Fig. 8.7 Temporal–spatial distributions of earthquakes occurrence around the Nankai Trough [1]

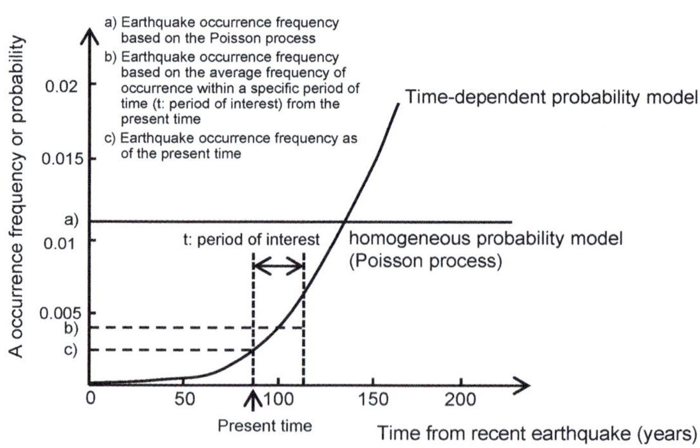

Fig. 8.8 Three options to obtain earthquake occurrence frequency

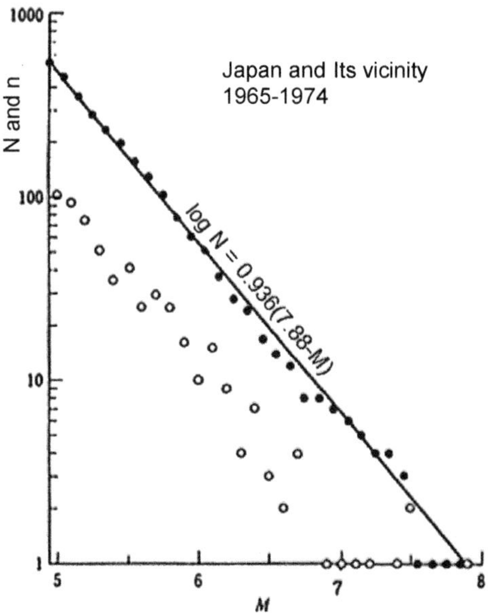

Fig. 8.9 Gutenberg–Richter law [2]. *White dots* correspond to $n(M)$ which is the number of earthquakes in which the magnitude is M. *Black dots* represent $N(M)$ which is the total number of earthquakes in which the magnitude is M or above

estimated. As shown in Fig. 8.11, the variability of attenuation formula, i.e., the variability of logarithm of the ground motion, is modeled as the normal distribution.

8.4.2.3 Logic Tree

Uncertainty in assumptions concerning the parameters is called epistemic uncertainty. In seismic hazard analysis, the epistemic uncertainty in parameters, e.g., the fault length, earthquake magnitudes, and earthquake occurrence frequency, is modeled as possible different opinions among experts. Therefore, a logic tree, like the one shown in Fig. 8.12, is constructed to account for all possible opinions. The logic tree is a tree diagram showing differences in expert opinions. With the weighing by confidence in different opinions, it contributes to the estimation of uncertainty in seismic hazard analysis.

8.4.2.4 Seismic Hazard Curve

As shown in Fig. 8.13, the seismic hazard curves for individual sources are unified to produce a seismic hazard curve. The result of seismic hazard analysis may change in the future as more seismological knowledge are accumulated. It is advisable to examine the validity of seismic hazard analysis from time to time for continuous updating.

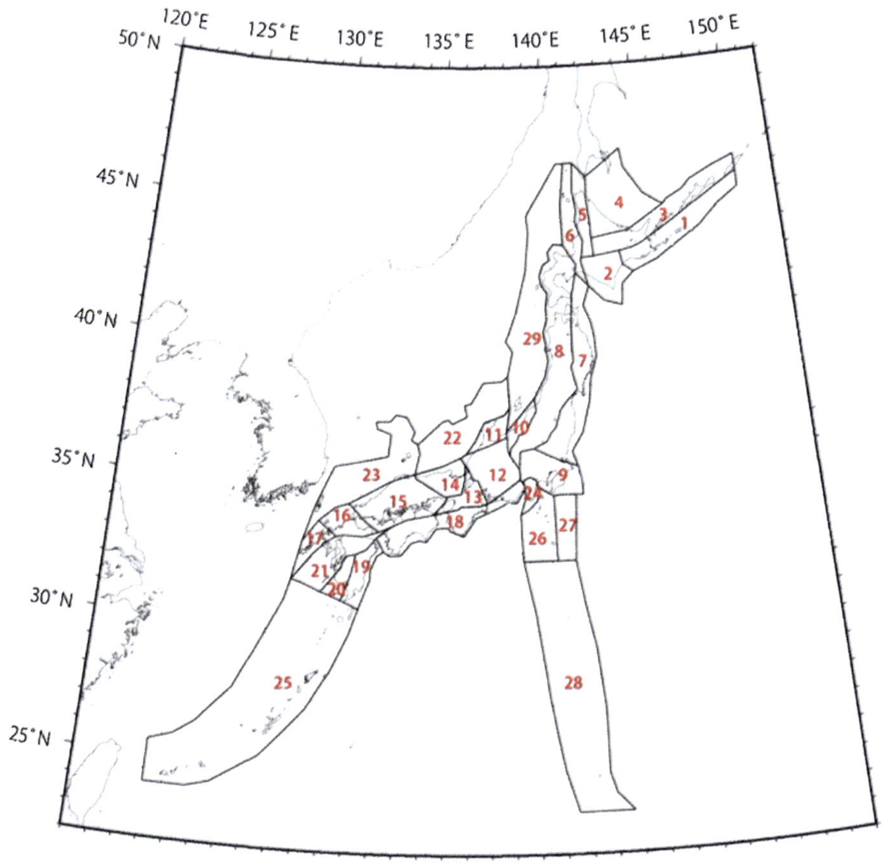

Fig. 8.10 Seismotectonic zonation for earthquake without specified faults [2]

8.4.2.5 Generation of Artificial Earthquake Ground Motion for Fragility Analysis

A seismic hazard curve is produced for acceleration response spectrum, i.e., the maximum responses to the ground motion of one-mass-spring-damper system (Part II, Chap. 14, Sect. 14.1) with different natural vibration periods. For the same exceedance frequency, the response values are plotted at each vibration period to produce a uniform hazard spectrum as shown in Fig. 8.14. The fragility analysis is performed using the artificial time history earthquake ground motion generated so that its response spectrum fits the uniform hazard spectrum.

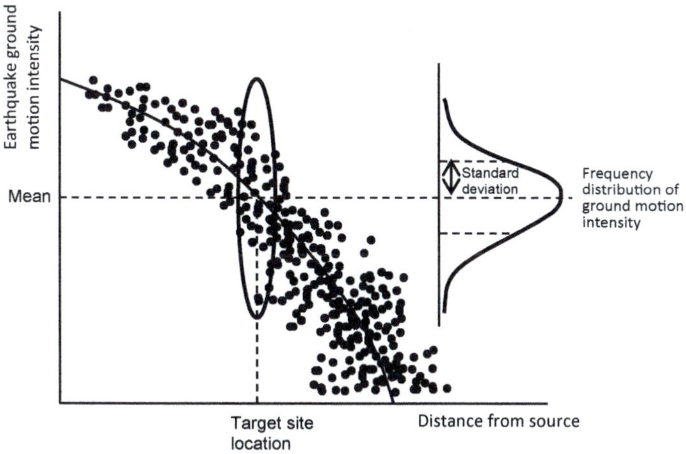

Fig. 8.11 Mean and variability of ground motion estimated using the attenuation formula

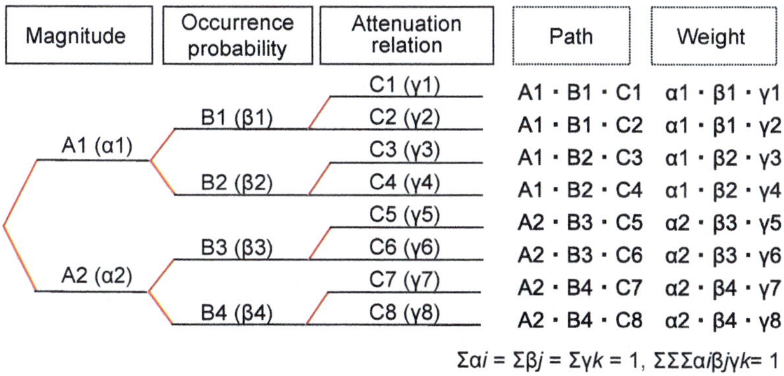

Fig. 8.12 An example of a logic tree

8.5 Structure/Component Fragility Analysis

Relationships between the earthquake ground motion intensity and probability of failure of structure/component are obtained by the seismic fragility analysis. This section describes the conventional method of the seismic fragility analysis.

8.5.1 Fragility Analysis Flow

A fragility curve shows conditional failure probabilities for different ground motion intensity levels. On the basis of information concerning the dynamic response and strength of structures and components, a fragility curve is developed. A fragility

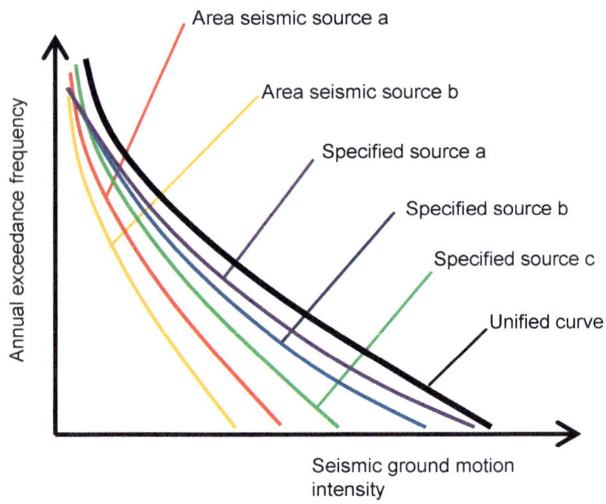

Fig. 8.13 Seismic hazard curve

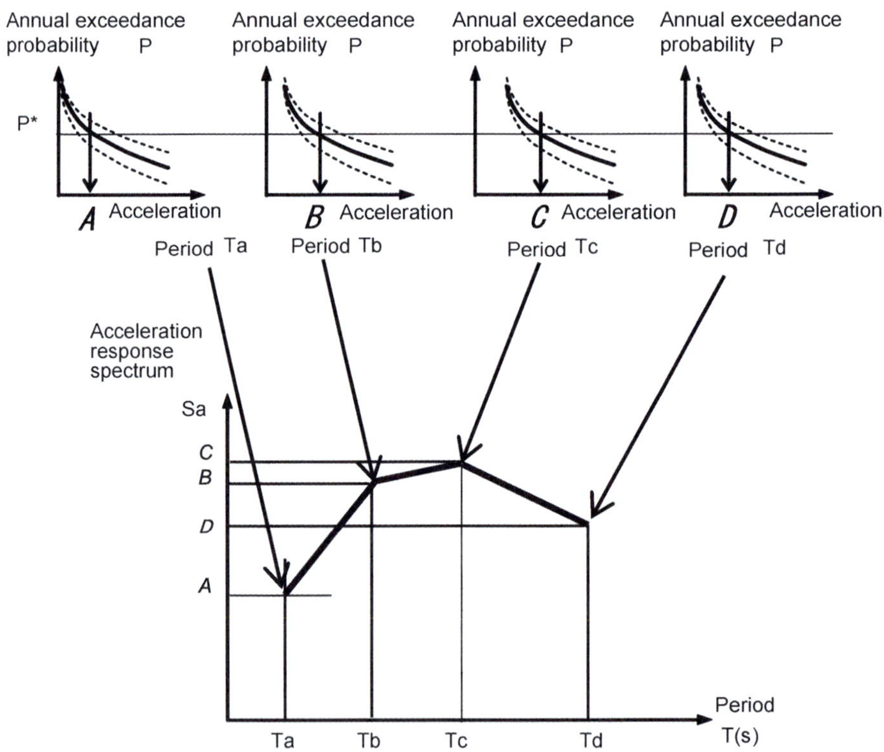

Fig. 8.14 Uniform hazard spectrum

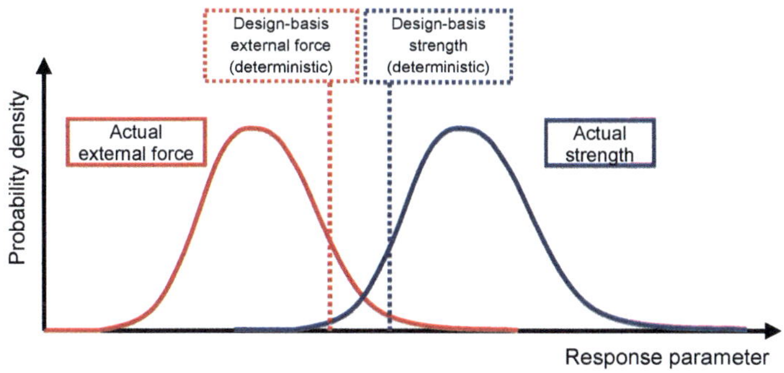

Fig. 8.15 External force and strength relationship

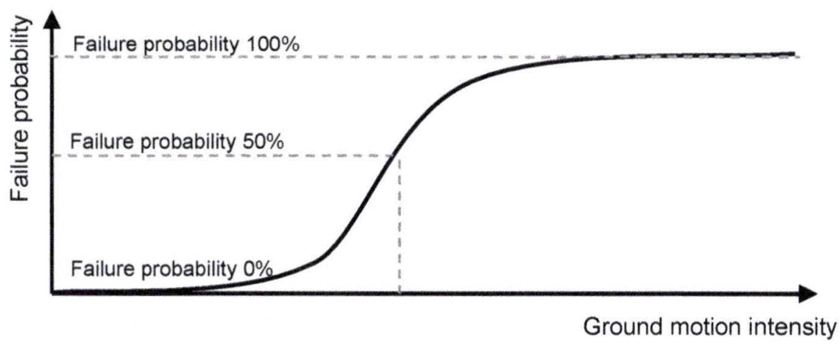

Fig. 8.16 Fragility curve

curve is estimated for structures and components that are significant in assessing the nuclear power plant risks.

For earthquake-resistant design, a deterministic approach is employed and structural components are designed to have an enough safety margin for the deformation and stress produced by the design basis external forces. On the other hand, the actual external forces from an earthquake and the actual material strength inherently exhibit scatter. In other words, the external forces and the strength of structures and components are random variables as shown in Fig. 8.15, and the failure probability of structures and components is represented by a fragility curve like the one shown in Fig. 8.16.

Figure 8.17 shows the structures/components fragility analysis flow.

(1) Selection of target structures and components, and the determination of their failure modes
After selecting target structures and components, their failure and associated response parameters to be used for determining their failure are defined.
(2) Selection of an analysis method
The analysis method is chosen in view of required accuracy and the objectives of the analysis.

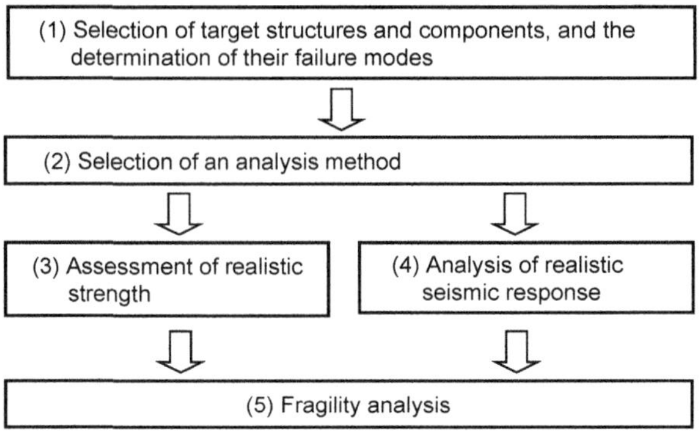

Fig. 8.17 Structure/component fragility analysis flow

(3) Assessment of realistic strength
For each of the target structures and components, the limit beyond which it will fail (i.e., the strength of the structures or components) is assessed probabilistically. The assessment method that may be employed is either an empirical method using experiments, a numerical method based on analysis, or a method that makes use of engineering judgments integrating both empirical and numerical methods.

(4) Analysis of realistic seismic response
For each of the target structures and components, the dynamic response is estimated stochastically, i.e., as a random variable. Conventionally, the input ground motion for this analysis is that derived from the uniform hazard spectrum (Sect. 8.4.2).

(5) Fragility analysis
On the basis of the result of the assessment of the probabilistically estimated strength and response, the fragility curves for the structures and components are obtained.

8.5.2 Building Fragility Analysis

8.5.2.1 Selection of the Failure Mode

First, the building's failure mode and the elements of the building prone to failure are determined. This is followed by the selection of response parameters for assessment of failure and the ground motion intensity.

Failure modes of structure that may lead to the damage of buildings may include sliding, overturning, story collapse, failure of local elements, and the failure of nonstructural members like partitions, ceilings, and doors. The dominant failure mode is selected from among them, and the safety critical elements of the building are identified. Conventionally, a failure mode like story collapse may be assumed to lead directly to core damage.

8.5.2.2 Selection of the Response Parameter

A parameter of the seismic response (e.g., stress, acceleration, strain, or deformation) is chosen for assessment of failure. Note that a failure is assessed probabilistically for the chosen response parameter.

When assessing the probability of shear failure of seismic walls leading to story collapse, the shearing force or shearing strain is chosen as the response parameter.

8.5.2.3 Assessment of Realistic Strength

A typical process of earthquake-resistant design includes the calculation of the dynamic response of building to design input earthquake ground motion. This process is to verify that the dynamic response does not exceed the allowable limit (design strength) depending on the material property. A fragility curve is derived considering the probability distribution of the realistic dynamic response and the probability distribution of realistic strength.

The probability distribution of strength is obtained by statistically analyzed experimental data. The dominant failure mode of buildings considered for a reactor building is the shear failure of a seismic wall made of reinforced concrete.

Table 8.1 summarizes failure limit values of shearing stress and shearing strain of reinforced concrete seismic wall. Figure 8.18 shows the relationship between the shear stress and the shear strain of seismic wall. The failure limit point for seismic walls is probabilistically determined on the basis of high confidence value for experimental results. However, such values may change in the future as more data are collected in experiments. It is advisable to examine the validity of experimental data from time to time and to update the failure limit point values.

8.5.2.4 Analysis of a Realistic Seismic Response

The dynamic response analysis is conducted probabilistically in consideration of uncertainty in the input earthquake ground motion and material property values.

In the analysis of the realistic seismic response, the factors that may significantly affect the response must be accounted for by considering the causes of the

Table 8.1 Shearing stress and strain settings [3]

Failure limit point (assessment index)		Mean	Coefficient of variation (including both aleatory and epistemic uncertainties)
Shearing stress τ_u		Strength equation	0.15
Shearing strain γ_u	Box wall	5.36×10^{-3}	0.24
	Ring wall	9.77×10^{-3}	0.33

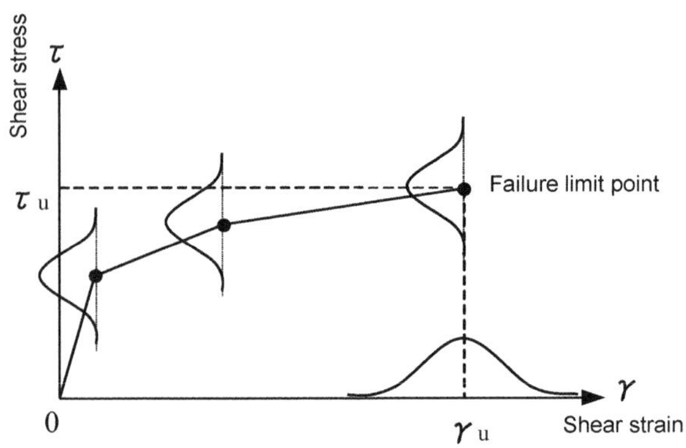

Fig. 8.18 Shearing stress and shearing strain relationship of seismic wall (conceptual)

Table 8.2 Characteristics of different sampling methods [3]

Method	Strength	Weakness
First-order second-moment method	– Simple and highly applicable – Sensitivity of input parameter can be estimated	– Limited applicability – Estimation at the tail of probability distribution is relatively inaccurate because the method focuses on values close to the mean value
Multiple point estimate (two-point estimate, etc.)	– Simple and highly applicable	– Requires assumption about probability distribution – The number of trial increases with the number of input parameters – Includes certain errors
Monte Carlo method	– Very highly applicable – The detail including probability distribution can be estimated	– Large number of trial is required in order to achieve accuracy of a satisfactory level
Experimental design method (Latin hypercube method, orthogonal array method, etc.)	– Highly applicable – Sensitivity to input parameter can be estimated – Possible to decrease the number of trial for the Monte Carlo method, two-point estimate, etc.	– Significant errors may arise when there exists a strong nonlinear relationship between input and output

uncertainties. The probability distribution of the dynamic response can be obtained efficiently by means of sampling performed using, for example, the Monte Carlo simulation or the two-point estimate. Table 8.2 summarizes the characteristics of different sampling methods.

In addition, it should be noted that there are two methods for the evaluation of realistic seismic response: a method based on response analysis and a method based on the response factors (i.e., an assessment of the probabilistic dynamic response using the response factors).

8.5.2.5 Determination of the Fragility Curve

Figure 8.19 shows how the failure probability $F(\alpha)$ for ground motion level α is calculated. As shown in the equation below, $F(\alpha)$ is a conditional probability that $f_R(\alpha, x)$, the probability density function of response to a ground motion intensity level α, exceeds $f_S(x)$, the probability density function of realistic strength. $F(\alpha)$ is calculated as follows:

$$F(\alpha) = \int_0^\infty f_S(x_R)\left(\int_{x_R}^\infty f_R(\alpha, x)\mathrm{d}x\right)\mathrm{d}x_R = \int_0^\infty f_R(\alpha, x_R)\left(\int_0^{x_R} f_s(x)\mathrm{d}x\right)\mathrm{d}x_R \quad (8.1)$$

The large computation time is required when calculation of Eq. (8.1) is done for large number of ground motion level α. Therefore, as shown in Fig. 8.20, discretization is made appropriately and the fragility curve is obtained by interpolation. Typically, the lognormal distribution is assumed for the fragility curve.

8.5.3 Components Fragility Analysis

The fragility of plant components is assessed in consideration of their required functions, paying attention to both structural failures and functional failures.

When assessing the fragility of passive, i.e., static, components such as a tank or heat exchange, the probability of losing required functions due to structural failure in the form of, for example, ductile fracture or brittle fracture is examined. As for active components such as an electrical board, pump, or valve, attention is given not

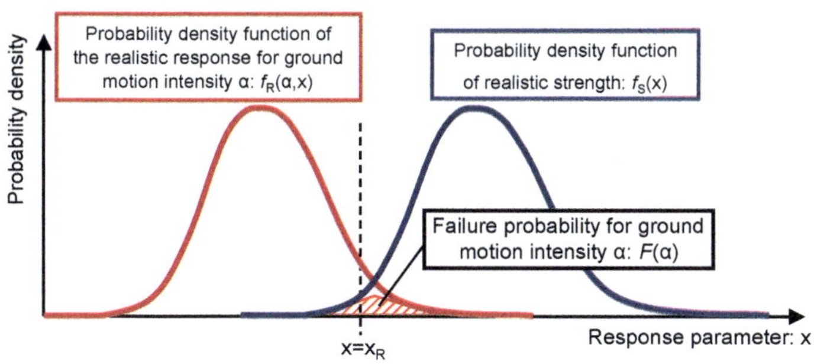

Fig. 8.19 Calculation of failure probability for ground motion intensity α

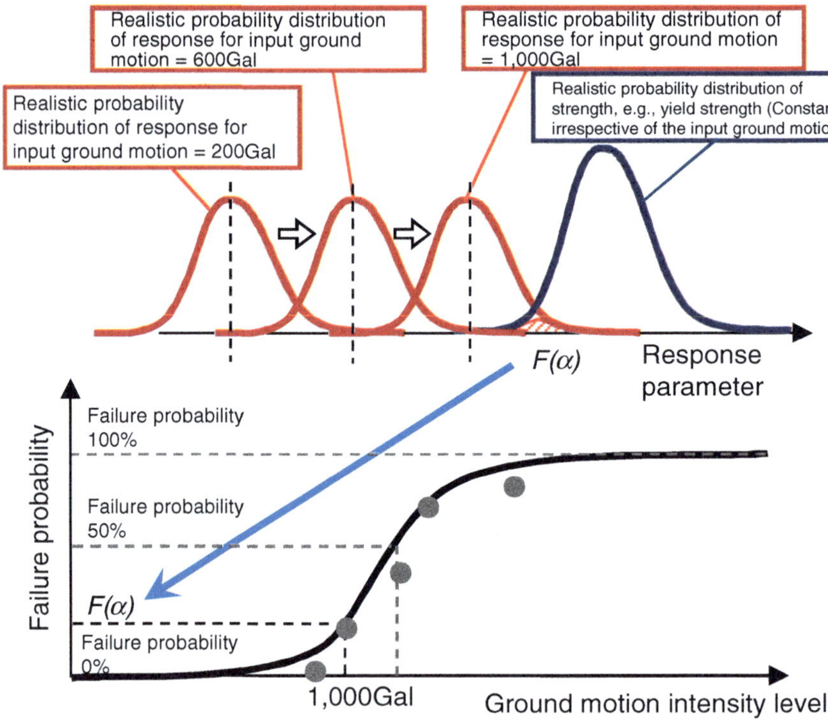

Fig. 8.20 Interpolation of $F(\alpha)$ to obtain fragility curve

only to the possibility of their structural failure, but also to the possibility of their limit for active or electrical functioning being lost.

To identify the functioning limit for active components, vibration testing is performed using a shaking table as described in Chap. 5, Sect. 5.6.

Table 8.3 summarizes component strength assessment methods. Figure 8.21 shows photos from the vibrating testing of active components.

A component fragility analysis is conventionally performed using the safety factor method. With the safety factor method, fragility is determined by referring to the result of response analysis along with the strength factors and response factors.

The fragility in relation to structural failure is determined in reference to material strength. The fragility in relation to functional failure is determined in reference to vibration test results. The median and uncertainty, i.e., logarithmic standard deviation, are estimated to obtain fragility curve.

The value of the response factor is determined from response analysis, accounting for building-related and component-related uncertainties.

Major assumptions (regarding, for example, uncertainty and the response factor) made in fragility analysis are as follows.

Table 8.3 Component strength assessment [4]

Strength assessment	Target components	Assessment procedure and its characteristics
Strength in relation to structural failure mode	Passive (Static) components: – Tanks – Piping, etc.	– The component strength is assessed on the basis of outputs from the seismic design process – The strength normally varies from plant to plant due to dependency on component geometry (support positions, routing, etc.)
Strength in relation to functional failure mode	Active components: – Pumps – Electrical boards, etc.	– The capability to withstand earthquake and remain functional is assessed by the vibration testing of the given component on a shaking table – The capability of a dynamic component to withstand earthquake and remain functional is normally component-specific

The fragility of a component is measured according to the maximum ground acceleration that the component may tolerate without failing. This maximum tolerable ground acceleration (referred to as the fragility acceleration) is given as a random variable as the following equation

$$A = A_{\mathrm{m}} \cdot \varepsilon_R \cdot \varepsilon_U. \tag{8.2}$$

A_m Median of A, which is the ground acceleration beyond which the component will fail (i.e., median fragility acceleration).

ε_R Random variable, i.e., variable with probability distribution, representing scatter caused by aleatory uncertainty inherent to physical randomness; the mean value is 1.0 while the logarithmic standard deviation is denoted as β_R.

ε_U Random variable representing scatter caused produced by epistemic uncertainty; the mean value is 1.0 while the logarithmic standard deviation is denoted as β_U.

A fragility curve, used for the representing component fragility, is obtained as a cumulative distribution function of the fragility acceleration A.

Figure 8.22 shows examples of three fragility curves with 5, 50, and 95 % confidence level. In the figure, β_R is represented as the gradient of the fragility curve, while β_U contributes to the width between fragility curves of different confidence levels. The fragility curve considering both aleatory and epistemic uncertainties, so-called composite fragility curve, is also shown in Fig. 8.22.

Typically, the component strength can be represented according to the high confidence and low probability of failure (HCLPF) value derived from the fragility curve and then compared with the design basis ground motion. The HCLPF value is the strength at which failure probability of 5 % can be assured with 95 % confidence level.

Fig. 8.21 Photos from the vibration testing of active components [4]

Transverse-type pump

Electrical board

8.6 Accident Sequence Analysis

8.6.1 Accident Sequence Analysis Flow

On the basis of the seismic hazard analysis and buildings/components fragility analysis, the probability of failure is determined for different facilities to enable the modeling of accident sequences leading to core damage. The accident sequence

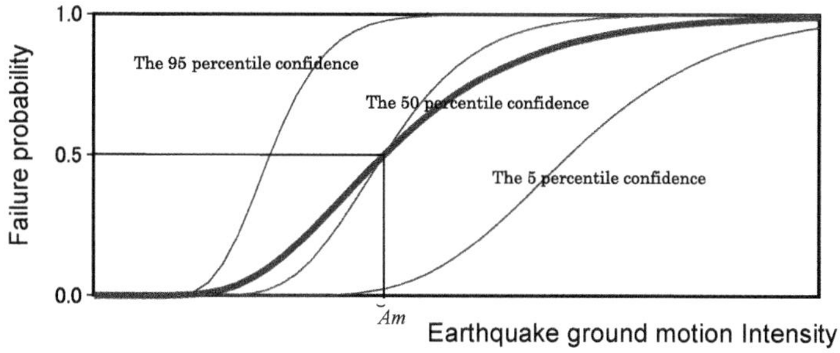

Fig. 8.22 Fragility curve

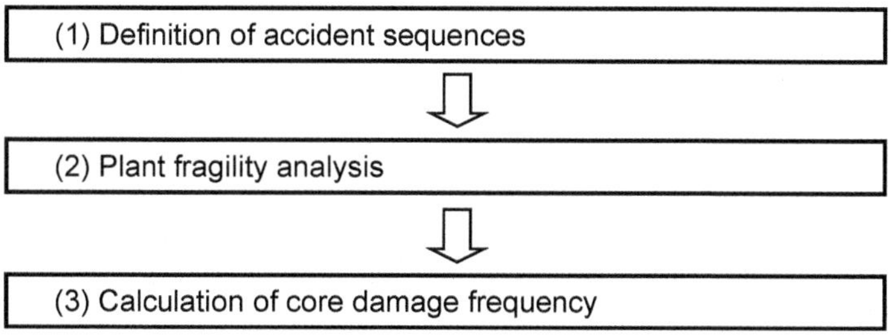

Fig. 8.23 Accident sequence analysis flow

analysis starts with the analysis of accident scenarios and to calculate core damage frequency. Figure 8.23 shows the accident sequence analysis flow.

(1) Definition of accident sequences
Accident sequences are defined using event trees and fault trees as a systematic representation of scenarios leading to core damage. The event tree shows how progressive failures of safety functions may lead to an accident including core damage, while the fault tree shows how the combinations of component failures may be a cause of a safety function failure.

(2) Plant fragility analysis
On the basis of the structure/component fragility analysis and the accident sequence analysis, the fragility of the entire plant is obtained.

(3) Calculation of core damage frequency
The core damage frequency is calculated according to the seismic hazard analysis and the plant fragility analysis. This is followed by the identification of the dominant accident sequences leading to core damage, and the components and facilities for which failure would contribute significantly to core damage are also identified.

8.6.2 Accident Sequence Analysis Method

Figure 8.24 shows the technical components of accident sequence analysis. Examples of the initiating events for an accident are the loss of off-site power, damage to buildings and structures (including the fracture of the reactor pressure vessel and/or the containment vessel), and piping fracture. After defining how the accident sequences progress depending on the success and failure of maintaining safety functions after the occurrence of the initiating event, the total core damage frequency is calculated as the sum of the probabilities of different accident sequences that lead to core damage.

Following the calculation of core damage frequency, a cut-set analysis is performed to identify the dominant sequences leading to core damage and to identify the components and facilities for which failure contributes significantly to core damage. On the basis of the results of such analyses, introduction of measures for safety improvement (including management-oriented measures) can be decided for the reduction of the core damage frequency.

Seismic Level-2 and Level-3 PRA (Appendices 8.2 and 8.3) are considered to be a future challenge, and researches and development are being conducted toward their realization. Other important challenges include the assessment of risks due to earthquakes combined with other hazards (e.g., tsunamis, floods, and fires) and earthquake risks at multiple-unit plants and multiple sites.

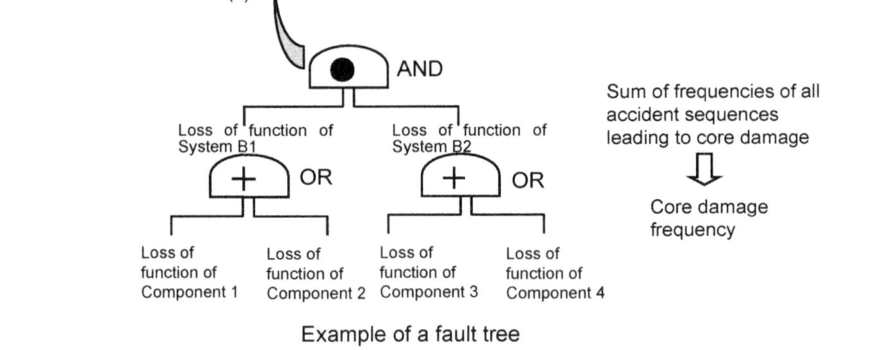

Fig. 8.24 Technical components of accident sequence analysis

Appendix 8.1: Safety Goals of Countries/Agencies Around the World

See Table 8.4.

Appendix 8.2: Levels of Probabilistic Risk Assessment (PRA)

PRA is a method for analyzing and assessing events (i.e., accidents and failures) that can happen for components and systems at a plant in a comprehensive and systematic manner, and it enables the quantitative assessment of the probability of occurrence of each event as well as the magnitude of consequence for each event.

PRA estimates three levels of risk, depending on different stages in accident progression.

Level-1 PRA develops events up to the occurrence of core damage and the core damage frequency is estimated.

Level-2 PRA further develops events leading to a release of radioactive material, and the frequency of occurrence of such release is obtained. PRA up to the assessment of the risk of containment failure is referred to as Level 1.5 PRA.

Level-3 PRA consists of assessing the societal risk including the health risks to the public in the nearby area including its occurrence frequency based on the results of Level-2 PRA.

Figures 8.25 and 8.26 show overview of Level-1, Level-2, and Level-3 PRA.

A methodology for seismic Levels-2 and Level-3 PRA are considered to be a future challenge to be developed.

Appendix 8.3: Overview of Level-2 PRA for Earthquake-Initiated Events

Seismic Level-2 PRA requires the quantification of accident leading to release of radioactive materials including occurrence frequencies and the determination of the source term. But it is basically considered that seismic Level-2 PRA can be implemented by applying Level-2 PRA procedures that address internal events.

Seismic Level-2 PRA identifies and estimates accident scenarios leading to release of radioactive materials. Attention is given not only to the possibility of damage of containment vessel by the earthquake, but also to possible damage of SSCs which support the capacity of the containment vessel. The accident scenarios leading to core damage are classified into several groups according to their types. When classifying, attention is given to the particularities of and similarities among different scenarios. Then, for each group of accident sequences, representative plant

Table 8.4 Safety goals of countries/agencies around the world

Country/agency	Safety goal	Performance goals
IAEA INSAG12 [5]	*General nuclear safety objective* • "To protect individuals, society and the environment by establishing and maintaining in nuclear power plants an effective defense against radiological hazard" *Radiation protection objective* • "To ensure in normal operation that radiation exposure within the plant and due to any release of radioactive material from the plant is as low as reasonably achievable, economic and social factors being taken into account, and below prescribed limits" • "To ensure mitigation of the extent of radiation exposure due to accidents" *Technical safety objective* • "To prevent with high confidence accidents in nuclear plants" • "To ensure that, for all accidents taken into account in the design of the plant, even those of very low probability, radiological consequences, if any, would be minor" • "To ensure that the likelihood of severe accidents with serious radiological consequences is extremely small"	Existing plants core damage frequency $<10^{-4}$/plant operating year Future plants core damage frequency $<10^{-5}$/plant operating year Practical elimination of large early releases Severe accident management and mitigation measures could reduce by a factor of at least ten the probability of large off-site releases requiring short term off-site response
United States NRC [6, 7]	*Qualitative safety goals* • Individual members of the public should be provided a level of protection from the consequences of nuclear power plant operation such that individuals bear no significant additional risk to life and health • Societal risks to life and health from nuclear power plant operation should be comparable to or less than the risks of generating electricity by viable competing technologies and should not be a significant addition to other societal risks *Quantitative safety goal*	Core damage frequency (existing plant, New plant) $<10^{-4}$/reactor year Large early release frequency (Existing plant) $<10^{-5}$/reactor year Large release frequency (new plant) $<10^{-6}$/reactor year

(continued)

Table 8.4 (continued)

Country/agency	Safety goal	Performance goals
	• The risk to an average individual in the vicinity of a nuclear power plant of prompt fatalities that might result from reactor accidents should not exceed 0.1 % of the sum of prompt fatalities risks resulting from other accidents to which members of the US population are exposed • The risk to the population in the area near a nuclear power plant of cancer fatalities that might result from nuclear power plant operation should not exceed 0.1 % of the sum of cancer fatality risks resulting from all other causes	

INSAG The International Nuclear Safety Advisory Group
NRC Nuclear Regulatory Commission

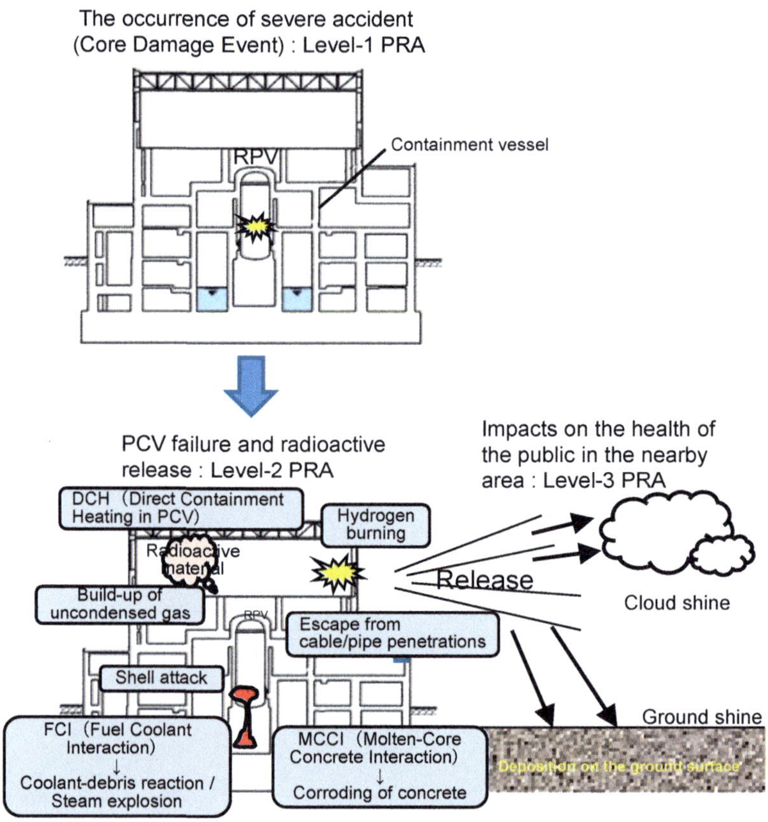

Fig. 8.25 Illustration of PRA

damage states are assigned. When classifying accident sequences into the plant damage states, accident sequences after core damage are carefully studied in terms of how accident progresses and effects on the characteristics of source term. Due considering is given to the fact that the characteristics of accidents due to earthquake are different from those due to random failure of components. This process enables the appropriate characterization of accident sequences. To appropriately estimate the source term in seismic Level-2 PRA, it is necessary to review the condition that is simplified in conventional seismic Level-1 PRA. One of examples is an assumption that failure of reactor building directly results in core damage. An event tree is constructed for the modeling of accident sequences showing how events may develop and lead to release of radioactive materials. Accident sequences in the event tree start from the initial states (the classified plant damage states). Then, this is followed by the determination of the source term for each containment failure mode.

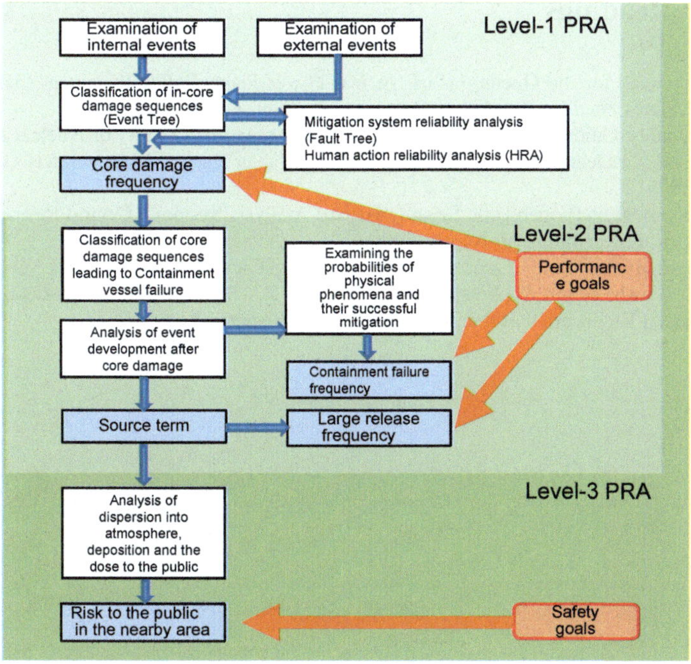

Fig. 8.26 Overview of Level-1, Level-2, and Level-3 PRA

Seismic level-3 PRA, which is to follow the determination of source term, can be implemented basically by applying Level-3 PRA procedures that address internal events.

References

1. "Long-term Evaluation of Nankai Trough Earthquakes (Revised version)", The Headquarters for Earthquake Research Promotion, May 2013
2. "National Seismic Hazard Map for Japan (2014) –Overview of nationwide seismic motion hazard –", The Headquarters for Earthquake Research Promotion, December 2014
3. "AESJ Standard: Seismic PRA Implementation Standards: 2015", Atomic Energy Society of Japan, September 2015
4. "Tadotsu Kogaku Shikenjo no Rekishi to Yakuwari (History and Role of Tadotsu Engineering Laboratory)", The Institute of Applied Energy website
5. "Basic Safety Principles for Nuclear Power Plants 75-INSAG-3 Rev.1 INSAG-12", International Atomic Energy Agency, October 1999
6. "Use and Development of Probabilistic Safety Assessment CSNI (NEA/CSNI/R(2007)12)", Organization for Economic Co-operation and Development Nuclear Energy Agency, November 2007
7. "Probabilistic Risk Criteria and Safety Goals (NEA/CSNI/R(2009)16)", Organization for Economic Co-operation and Development Nuclear Energy Agency, December 2009

Further Readings

8. "Safety Goals for the Operations of Nuclear Power Plants;Policy Statement (51FR28044)", United States Nuclear Regulatory Commission, August 1986
9. "The Quality Guidelines for Probabilistic Safety Assessment (PSA) of Nuclear Installations (trial use)", Nuclear and Industrial Safety Agency, Japan Nuclear Energy Safety Organization, April 2006
10. "Development of Probabilistic Seismic Hazard Analysis System", Central Research Institute of Electric Power Industry, August 2006
11. "Standard for Level 1 / Large Early Release Frequency Probabilistic Risk Assessment for Nuclear Power Plant Applications (ASME/ANS RA-S-2008)", The American Society of Mechanical Engineers and the American Nuclear Society, 2008

Chapter 9
International Standards and National Regulation on Seismic Safety Assessment

Tatsuya Itoi, Michiya Kuno and Masanori Hamada

Abstract Many nuclear power plants have been or will be constructed around the world, including countries and regions with middle to high seismicity. Safety standards published by the International Atomic Energy Agency (IAEA) reflect an international consensus. Each nation, e.g., the United State or Japan, develops its own codes and standards on its own responsibility. IAEA safety standards are bases for codes and standards for each nation. This chapter outlines the standardized practice with respect to seismic safety including earthquake-resistant design. IAEA safety standards and US nuclear safety regulations are introduced mainly focusing on the structures and hierarchy, as well as methods and techniques related to seismic safety assessment.

Keywords Nuclear regulation · Seismic category · Safety requirement · Nuclear safety guide · Seismic design criteria · Seismic design ground motion

9.1 International Atomic Energy Agency (IAEA) Standards

This section describes the hierarchy of the IAEA safety standards, followed by the introduction of some of the requirements and recommendations in the standards related to seismic safety of nuclear facilities.

9.1.1 Structures and Hierarchy of IAEA Standards

The IAEA Safety Standards Series are composed of the following three levels of documents:

T. Itoi (✉)
The University of Tokyo, Tokyo, Japan
e-mail: itoi@n.t.u-tokyo.ac.jp

M. Kuno
Chubu Electric Power Company, Nagoya, Japan

M. Hamada
Waseda University, Shinjuku, Tokyo, Japan

M. Hamada and M. Kuno (eds.), *Earthquake Engineering for Nuclear Facilities*,
DOI 10.1007/978-981-10-2516-7_9

- Safety Fundamentals,
- Safety Requirements, and
- Safety Guides.

Figure 9.1 shows the hierarchy of the IAEA Safety Standards. "Safety Fundamentals [1]" is the policy document of the IAEA Safety Standards series. They state the basic safety objectives, safety principles, and concepts. What is stated at the beginning is the fundamental safety objective. It is "to protect people and the environment from harmful effects of ionizing radiation." Then, ten safety principles have been formulated to achieve the fundamental safety objective. The ten principles are as follows [1]:

- Responsibility for safety (including the prime responsibility of the licensee),
- Role of government,
- Leadership and management for safety,
- Justification of facilities and activities,
- Optimization of protection,
- Limitation of risks to individuals,
- Protection of present and future generations,
- Prevention of accidents,

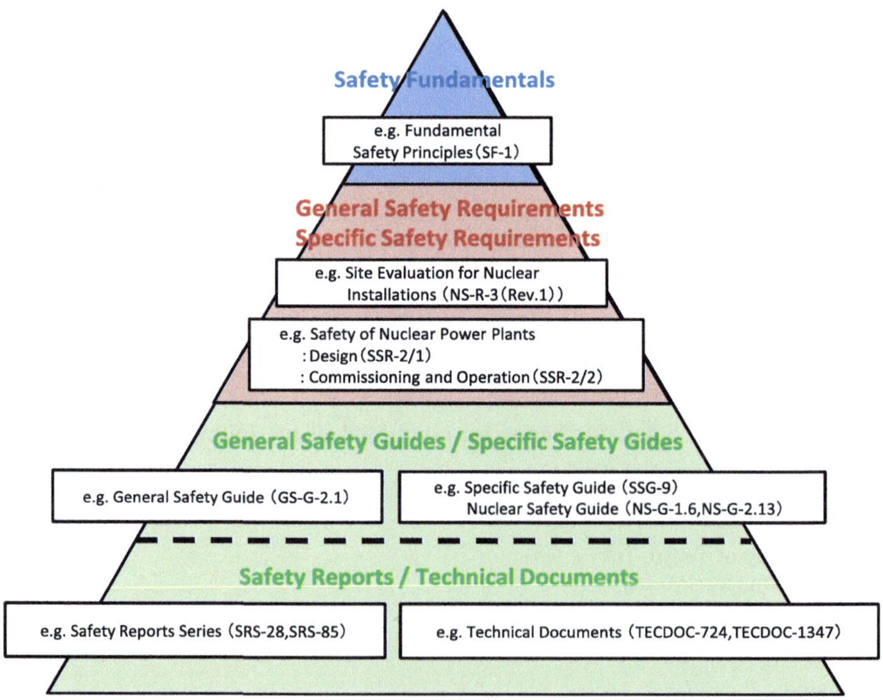

Fig. 9.1 Hierarchy of IAEA safety standards [1–12]

- Emergency preparedness and response, and
- Protective actions to reduce existing or unregulated radiation risks.

The concept of "defense in depth" is also introduced as the primary means to prevent and mitigate the consequences of accidents.

"Safety Requirements" and "Safety Guides" provide technical details on the basis of these principles. "Safety Requirements" come below "Safety Fundamentals," which provides requirements ("shall" statements). "Safety Requirements" are grouped into two: "General Safety Requirements" and "Specific Safety Requirements." The latter includes requirements for specific facilities and activities. "Safety Guides" provide recommendations ("should" statements) regarding measures that ensure that the "Safety Requirements" are met.

IAEA Safety Standards have been developed on the basis of an international consensus. Recently, these documents were reviewed and some of them have been revised based on the action plan in response to the Fukushima Daiichi accident. These documents are considered to be important also from the viewpoint of international cooperation with respect to nuclear safety, because the risks of nuclear power plant are transboundary.

9.1.2 Specific Safety Requirements No. SSR-2/1 (Rev.1) [2]

IAEA Specific Safety Requirements "*Safety of Nuclear Power Plants*: *Design* [SSR-2/1 (Rev.1)]" provide 82 requirements regarding design. Requirements include the application of defense in depth as well as requirements for comprehensive deterministic and probabilistic assessment of safety. It should be noted that the requirement for the design extension condition is prescribed in Requirement 20. Design extension condition (DEC) is a postulated plant state due to multiple failures of safety systems. The concept of DEC was not considered in the conventional design. It is, however, an important concept to enhance the capabilities to withstand severe accidents. External hazards like earthquake are not included in the current definition of DEC. Similar requirements for external hazards, however, have been introduced in some nations after the Fukushima Daiichi accident. The discussion on the issues related to design considerations for beyond design external hazards is initiated by, for example, Draft Technical Document entitled "*Considerations on the Application of the IAEA Safety Requirements for design of Nuclear Power Plants.*"

9.1.3 General Safety Requirements No. NS-R-3 (Rev.1) [3]

In IAEA Safety Guide "*Site Evaluation for Nuclear Installations* [NS-R-3 (Rev.1)]," the general safety criteria for site evaluation of external hazards to the nuclear installation are provided.

Specific requirements for evaluation of earthquakes and surface faulting are also stated. Requirements for earthquakes prescribed in IAEA NS-R-3 (Rev.1) are as follows:

- Evaluation of seismological and geological conditions in the region,
- Evaluation of geological and geotechnical aspects of the site,
- Collection and documentation of information on prehistoric, historical, and instrumentally recorded earthquakes,
- Determination of earthquake hazards by means of seismotectonic evaluation and use of the information collected,
- Assessment of ground motion hazards considering seismotectonic characteristics and site conditions, and
- Thorough uncertainty analysis as part of the evaluation of seismic hazards.

Requirements for surface faulting prescribed in IAEA NS-R-3 (Rev.1) are as follows:

- Assessment of the potential for surface faulting for the site,
- Assessment of capability of a fault on the basis of geological, geophysical, geodetic, or seismological data, and
- Consideration of alternative site in case that reliable evidence shows the existence of a capable fault.

9.1.4 Specific Safety Guide No. NS-G-1.6 [4]

In IAEA Safety Guide "*Seismic Design and Qualification for Nuclear Power Plants* (NS-G-1.6)," a generally accepted technical detail for seismic design of a nuclear power plant is provided. It includes the general safety concept, the design process, the design principles, the concept of periodic safety review, seismic qualification, as well as seismic instrumentation and monitoring. Among them, descriptions on seismic categorization and design basis earthquake are summarized below for the purposes of comparison with previous chapters.

9.1.4.1 Design Basis Earthquake

Two levels of ground motion (Seismic Level (SL)-1 and Seismic Level (SL)-2) should be evaluated for each plant as design basis earthquakes. Two levels are evaluated according to the target probability level. SL-1 is often termed the operating base earthquake. SL-2 is often denoted as the safe shutdown earthquake. The procedures to define these earthquakes are outlined in IAEA SSG-9.

9.1.4.2 Seismic Categorization

Seismic categorization is a categorization of items important to safety during and after an earthquake. Safety classification is also considered for safety design in IAEA SSR-2/1 (Rev.1). Seismic categorization is not necessary related to safety classification. Four levels of seismic categories are introduced in IAEA NS-G-1.6 as shown in Table 9.1. Seismic Category 1 is the highest category, and items classified in Category 1 should be designed to maintain their function in the event of earthquake corresponding to SL-2.

9.1.5 Specific Safety Guide No. 9 (SSG-9) [5]

IAEA published "*Seismic Hazards in Site Evaluation for Nuclear Installations* (SSG-9)" in 2010. As an IAEA safety standard, SSG-9 supersedes "*Evaluation of Seismic Hazards for Nuclear Power Plants* (NS-G-3.3)" and provides general recommendations, guidance, and procedures for conducting a deterministic/probabilistic seismic hazard assessment for a nuclear installation site. The concepts of SL-2 and SL-1 stated in IAEA NS-G-1.6 are also adopted in IAEA SSG-9. "SL-2 is associated with the most stringent safety requirements," while "SL-1 corresponds to a less severe, more probable earthquake level which normally has different implications for safety."

It is preferred that the ground motion hazard is assessed using both probabilistic and deterministic methods. Table 9.2 summarizes the complementary role of deterministic and probabilistic methods for earthquake ground motion hazard assessment. "*Consideration of external events in the design of nuclear facilities other than nuclear power plants, with emphasis on earthquakes* (TECDOC-1347)"

Table 9.1 Seismic categorization [4]

Category	Descriptions
1	• Items whose failure could directly or indirectly cause accident conditions as a consequence of an earthquake • Items required for shutting down, maintaining the reactor in a shutdown condition, removing residual heat, and monitoring parameters essential to these functions • Items required to prevent or mitigate radioactive releases for any postulated initiating events considered in the design
2	• Items which may have interactions with items in categories 1 and 3 • Items needed to prevent and mitigate the accident originated by events other than earthquake • Items which support getting into the site, and items necessary for the evacuation plan in case of emergency
3	• All items that are not installed to reactors, but could cause radiological hazard
4	• All items which are not in categories 1–3

Table 9.2 Earthquake ground motion hazard assessment in IAEA SSG-9 [5]

	Deterministic method	Probabilistic method
Role	Check against probabilistic assessments	Evaluation of deterministic values Input to seismic probabilistic safety/risk assessment
Uncertainty treatment	By conservative process at each step	Realistic assessment and explicit incorporation of all uncertainties
Procedure	1. Evaluate the seismotectonic model 2. Evaluate the maximum magnitude 3. Select the attenuation relationships 4. Perform the hazard calculation 5. Take account of the site response 6. Consideration of conservativeness by taking account of all uncertainties	1. Evaluate the seismotectonic model for the site region including uncertainty 2. Assess the maximum magnitude, occurrence frequency rate with uncertainty of each seismic source model 3. Select the attenuation relationships and assess uncertainty 4. Perform hazard calculation 5. Take account of the site response including uncertainty
Results	Deterministic response spectrum, etc.	• Mean or median seismic hazard curve for horizontal and vertical ground motion • Fractile hazard curves (e.g., mean, 16th, 50th and 84th percentile hazard curves) • Uniform hazard spectra

can be referred to for a method of defining design basis ground motions for nuclear facilities other than nuclear power plants.

9.1.6 Safety Report Series No. 28 (SRS-28) [6]

IAEA published "*Seismic Evaluation of Existing Nuclear Power Plants* (SRS-28)" in 2003. Evaluation of seismic safety of an existing facility is needed because of evidence of greater seismic hazard than expected before, regulatory requirements such as periodic safety review, or new technical findings from, for example, earthquake experiences. This report provides guidance for conducting seismic safety evaluation programs. The program contains three topics:

- Seismic hazard assessment,
- Safety analysis of a nuclear power plant using a seismic margin analysis and/or a site-specific seismic probabilistic safety/risk assessment, and
- Evaluation of the plant-specific seismic capacity for making a decision on upgrading.

9.1.7 Safety Report Series No. 85 (SRS-85) [7]

IAEA published "*Ground Motion Simulation Based on Fault Rupture Modelling for Seismic Hazard Assessment in Site Evaluation for Nuclear Installations* (SRS-85)" in 2015. This report explains the principles that underlie strong ground motion simulation, describes various methods for simulating strong ground motions, and gives examples of strong ground motion simulations using fault rupture modeling.

9.1.8 Technical Document No. 724 (TECDOC-724) [8]

IAEA published "*Probabilistic Safety Assessment for Seismic Events* (TECDOC-724)" in 1993. This report describes a general method for seismic probabilistic safety/risk assessment of nuclear power plants. Level 1 PSA/PRA plus containment performance analysis is considered in the document.

9.2 Nuclear Safety Regulation in the US

This section describes the hierarchy of the regulatory documents of the US, followed by the introduction of some of the requirements related to seismic safety of nuclear facilities.

9.2.1 Hierarchy of Regulatory Documents of the US

The hierarchy of nuclear safety regulatory documents of the US is classified as shown in Fig. 9.2. It is composed of the following structures:

- Federal Laws of the US
- Code of Federal Regulations (CFR)
- Regulatory Documents
  - Regulatory Guides (RG)
  - Standard Review Plans (SRP)
  - Generic Letters (GL)
  - NRC Bulletins (BL)
  - Codes and Standards

"The Atomic Energy Act of 1954" and "The Energy Reorganization Act of 1974" are the federal laws legislated by the federal government of the US. They provide fundamental principles regarding overall nuclear power-related activities.

"Title 10 of the Code of Federal Regulations (10 CFR)" includes the requirements prepared by the US Nuclear Regulatory Commission (USNRC) in accordance with the provisions in "The Atomic Energy Act of 1954." The USNRC has

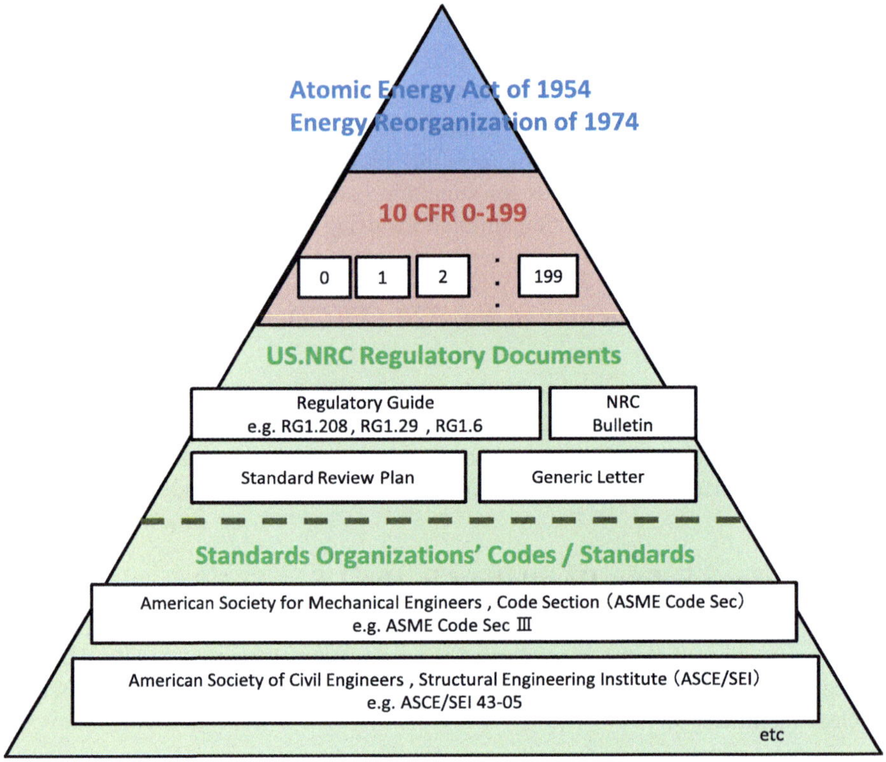

Fig. 9.2 Hierarchy of nuclear safety regulatory documents in the US [13–15, 17–19, 28]

issued various other documents as shown in Fig. 9.2. In particular, the "Regulatory Guides" provide guidance that licensees and applicants must refer to, and regulatory documents are categorized depending on their purposes.

Technical codes and standards published by standards organizations [e.g., the American Society of Mechanical Engineers (ASME) and the American Society of Civil Engineers (ASCE)] are at the bottom of the hierarchy; they provide detailed methods and techniques on the basis of the latest research findings and experiences.

In the section below, descriptions related to the following topics are summarized for the purposes of comparison with previous chapters:

- Classification of SSCs according to their importance to seismic safety,
- Determination of design earthquake ground motion,
- Determination of target performance goal, and
- Seismic probabilistic risk assessment and seismic margin analysis.

9.2.2 Seismic Design Classification

10 CFR 50.54 [13] and Appendix S to 10 CFR 50 [14] as well as 10 CFR 52 provide requirements regarding earthquake safety issues.

USNRC Regulatory Guide 1.29 (RG 1.29) [15] defines a seismic design classification for nuclear power plants. RG 1.29 utilizes a single seismic category (called "Seismic Category I"). Seismic Category I in RG 1.29 is similar to Category 1 in IAEA NS-G-1.6. SSCs necessary to ensure the following three functions are designed to withstand the effects of the safe shutdown earthquake (see Sect. 9.2.3) [14, 15]:

1. The integrity of the reactor coolant pressure boundary,
2. The capability to shutdown the reactor and maintain it in a safe shutdown condition, and
3. The capability to prevent or mitigate the consequences of accidents that could result in potential offsite exposures comparable to the guideline exposures of 10 CFR 50.34(a)(1).

Additionally, USNRC Regulatory Guide 1.143 (RG 1.143) [16] is applied to give classification of SSCs related to radioactive waste management at nuclear sites.

9.2.3 Design Earthquake Ground Motions

According to Appendix A to 10 CFR Part 100 [17], the safe shutdown earthquake (SSE) is determined based on an assessment of the maximum earthquake potential. The operating basis earthquake (OBE) is considered reasonably to occur during the plant lifetime. SSCs necessary for continued operation are designed to remain functional under the OBE.

Requirements for determination of the SSE ground motion is prescribed in 10 CFR Part 100.23 [18]. The SSE ground motion is determined considering geological, seismological, and engineering characteristics. Additionally, uncertainties inherent in such estimates must be addressed through a probabilistic seismic hazard analysis or suitable sensitivity analyses.

Formerly, design earthquake ground motions were determined by taking a deterministic approach. The design peak ground acceleration was determined on the basis of recorded historical earthquake. A standard (i.e., generic) shape of response spectra specified in USNRC Regulatory Guide 1.60 [19] has been used to define the design basis response spectra.

In 1997, USNRC Regulatory Guide 1.165 (RG 1.165) [20] introduced a probability-based approach for determination of the SSE ground motion. The SSE ground motion for the site was determined based on specified reference probability (i.e., 10^{-5}/year for the median hazard curve). In 2007, USNRC Regulatory Guide 1.208 (RG 1.208) [21] introduced a method of determining the site-specific earthquake ground motion by taking a performance-based approach, instead of a reference probability approach in RG 1.165. RG 1.165 was withdrawn in 2010. The concept of the performance-based approach in RG 1.208 is based on ASCE/SEI standards 43-05 [22], which is summarized in the next paragraph.

The results of probabilistic seismic hazard analysis are used as information for determination of those design earthquake ground motions. In RG 1.208, the utilization of the SSHAC approach [23, 24] is recommended for probabilistic seismic hazard analysis. The SSHAC approach is a standardized approach to quantify uncertainties using expert opinion elicitation.

9.2.4 Target Performance Goal for Performance-Based Approach

The ASCE Standard "*Seismic Design Criteria for Structures, Systems, and Components in Nuclear Facilities*" denoted as ASCE/SEI 43-05 [22] prescribes seismic design criteria for nuclear facilities.

Seismic Design Categories (SDC) in ASCE/SEI 43-05 are those defined in ANSI/ANS-2.26-2004 [25] on the basis of radiological consequence due to failure of SSCs. SDC 1 corresponds to conventional buildings, while SDC 5 corresponds to nuclear power plants. The seismic design criteria for SDC 3, 4, and 5 are provided in this document. SSCs categorized as SDC 1 and 2 are designed according to the procedure described in the International Building Code (IBC).

Performance goals for SDC 3, 4, and 5 are provided as the failure probability of exceeding acceptable limits. Table 9.3 summarizes the target performance goals and its related values for different SDCs.

The probability ratio R_P is defined as follows:

$$R_P = \frac{H_D}{P_F}, \tag{9.1}$$

Table 9.3 Seismic design provisions [22]

Seismic design category	Hazard exceedance probability (H_D)	Target performance goal (P_F)	Probability ratio (R_P)
SDC-1[a]	–	$(<1\times10^{-3})^b$	–
SDC-2[a]	–	$(<4\times10^{-4})^b$	–
SDC-3	4×10^{-4}/year	$\sim 1 \times 10^{-4}$	4
SDC-4	4×10^{-4}/year	$\sim 4 \times 10^{-5}$	10
SDC-5	1×10^{-4}/year	$\sim 1 \times 10^{-5}$	10

[a]Not addressed in [22]

[b]Assessment of P_F approximately achieved by building codes

where H_D denotes the exceedance probability where the uniform hazard response spectrum is evaluated, and P_F is a target performance goal.

Target performance goal P_F, probability ratio R_P, and hazard exceedance criteria H_D corresponding to SDC 5 are adopted for nuclear power plants by USNRC Regulatory Guide 1.208 (RG 1.208) [21]. The desired performance for nuclear power plants is expressed as 10^{-5}/year for mean annual probability of exceedance of the onset of significant inelastic deformation.

9.2.5 *Seismic Probabilistic Risk Assessment and Seismic Margin Analysis*

In the US, the final safety analysis report (FSAR) needs to be prepared by the licensee and be submitted to the USNRC for approval of operating a nuclear power plant. Submission of the report is required according to 10 CFR 52.47 [26].

The probabilistic risk assessment (PRA) is required for the FSAR. Only by designing each SSC against the SSE ground motion, the confidence level for a plant-level performance is not understood clearly. An overall safety of plant for beyond design basis earthquake is analyzed using a Seismic PRA as well as PRA-based seismic margin analysis.

Using a seismic margin analysis, plant-level HCLPF value is estimated for all sequences leading to core damage or containment failure. Plant-level HCLPF value for new plants needs to be greater than or equal to 1.67 times the SSE ground motion according to SECY 98-087 [27].

References

1. Fundamental Safety Principles, IAEA Safety Standard, Safety Fundamentals (SF-1), INTERNATIONAL ATOMIC ENERGY AGENCY (2006)
2. Safety of NPPs: Design, IAEA Specific Safety Requirement No. SSR-2/1, INTERNATIONAL ATOMIC ENERGY AGENCY (2012)
3. Site Evaluation for Nuclear Installations, Safety Requirements No.NS-R-3, INTERNATIONAL ATOMIC ENERGY AGENCY (2016)
4. Seismic Design and Qualification for Nuclear Power Plants, Safety Guide No NS-G-1.6, INTERNATIONAL ATOMIC ENERGY AGENCY (2003)
5. Seismic Hazards in Site Evaluation for Nuclear Installations, No.SSG-9, INTERNATIONAL ATOMIC ENERGY AGENCY (2010)
6. Seismic Evaluation of Existing Nuclear Power Plants No.28, INTERNATIONAL ATOMIC ENERGY AGENCY (2003)
7. Ground Motion Simulation Based on Fault Rupture Modelling for Seismic Hazard Assessment in Site Evaluation for Nuclear Installations, No. 85, INTERNATIONAL ATOMIC ENERGY AGENCY (2015)
8. Probabilistic safety assessment for seismic events (TECDOC-724), INTERNATIONAL ATOMIC ENERGY AGENCY (1993)

9. Safety of NPPs: Commissioning and Operation, IAEA Specific Safety Requirement No. SSR-2/2, INTERNATIONAL ATOMIC ENERGY AGENCY (2011)
10. Arrangements for Preparedness for a Nuclear or Radiological Emergency, Safety Guide No. GS-G-2.1, INTERNATIONAL ATOMIC ENERGY AGENCY (2007)
11. Evaluation of Seismic Safety for Existing Nuclear Installations, Safety Guide No. NS-G.2.13, INTERNATIONAL ATOMIC ENERGY AGENCY (2003)
12. Consideration of external events in the design of nuclear facilities other than nuclear power plants, with emphasis on earthquakes (TECDOC-1347), INTERNATIONAL ATOMIC ENERGY AGENCY (2003)
13. Nuclear Regulatory Commission Regulations Title 10 Code of Federal Regulations §50.54 Conditions of licenses (10 CFR 50.54), United States Nuclear Regulatory Commission (2015 updated)
14. Nuclear Regulatory Commission Regulations Appendix S to Title 10 Code of Federal Regulations §50—Earthquake Engineering Criteria for Nuclear Power Plants (Appendix S to 10 CFR 50), United States Nuclear Regulatory Commission (2007)
15. Nuclear Regulatory Commission Regulatory Guide 1.29 Seismic Design Classification (RG 1.29), United States Nuclear Regulatory Commission (2007)
16. Nuclear Regulatory Commission Regulatory Guide 1.143 Design Guidance for Radioactive Waste Management Systems, Structures, and Components Installed in Light-Water-Cooled Nuclear Power Plants (RG 1.143), United States Nuclear Regulatory Commission (2001)
17. Nuclear Regulatory Commission Regulations Appendix A to Part 100 - Seismic and Geologic Siting Criteria for Nuclear Power Plants (Appendix A to 10 CFR 100), United States Nuclear Regulatory Commission (2013)
18. Nuclear Regulatory Commission Regulations Title 10 Code of Federal Regulations §100.23 — Geologic and Seismic Siting Criteria (10 CFR 100.23), United States Nuclear Regulatory Commission (2007)
19. Nuclear Regulatory Commission Regulatory Guide 1.60 Design Response Spectra for Seismic Design of Nuclear Power Plants (RG 1.60), United States Nuclear Regulatory Commission (2014)
20. Nuclear Regulatory Commission Regulatory Guide 1.165 Identification and Characterization of Seismic Sources and Determination of Safe Shutdown Earthquake Ground Motion (RG 1.165), United States Nuclear Regulatory Commission (1997)
21. Nuclear Regulatory Commission Regulatory Guide 1.208 A Performance-based Approach to Define the Site-Specific Earthquake Ground Motion (RG 1.208), United States Nuclear Regulatory Commission (2007)
22. Seismic Design Criteria for Structures, Systems, and Components in Nuclear Facilities (ASCE/SEI 43-05), American Society of Civil Engineers (2005)
23. Recommendations for Probabilistic Seismic Hazard Analysis: Guidance on Uncertainty and Use of Experts, NUREG/CR-2117, United States Nuclear Regulatory Commission (1997)
24. Practical Implementation Guideline for SSHAC Level 3 and 4 Hazard Studies, NUREG-2117 Rev.1, United States Nuclear Regulatory Commission (2011)
25. Categorization of Nuclear Facility Structures, Systems, and Components for Seismic Design, ANSI/ANS-2.26-2004, American Nuclear Society (2004)
26. Nuclear Regulatory Commission Regulations Title 10 Code of Federal Regulations §52.47—Contents of Applications; Technical Information (10 CFR 52.47), United States Nuclear Regulatory Commission (2009)
27. Proposed Generic Letter 98-XX: Interim Guidance for Updated Final Safety Analysis Reports in Accordance with 10 CFR 50.71(e) (SECY 98-087) (1998)
28. Rules for Construction of Nuclear Facility Components Section III (ASME Code Sec III), The American Society of Mechanical Engineers (2015)

Further Readings

29. Earthquakes and Associated Topics in Relation to Nuclear Power Plants SitingSafety Series No. 50-SG-S1 (Rev.1), INTERNATIONAL ATOMIC ENERGY AGENCY (1991)
30. Evaluation of Seismic Hazards for Nuclear Power PlantsSafety Guide No.NS-G-3.3, INTERNATIONAL ATOMIC ENERGY AGENCY (2002)
31. External Events Excluding Earthquakes in the Design of Nuclear Power PlantsSafety Guide No NS-G-1.5, INTERNATIONAL ATOMIC ENERGY AGENCY (2003)
32. Considerations on the Application of the IAEA safety Requirements for design of NPPs (Draft TECDOC)INTERNATIONAL ATOMIC ENERGY AGENCY (2014)

Chapter 10
Planning, Construction, Operation, and Decommission of Nuclear Power Plants

Michiya Kuno and Masanori Hamada

Abstract Nuclear power plants must maintain important safety functions in the event of earthquakes and tsunamis. It is thus vital that strict quality control is implemented in all project stages including planning, construction, operation, and decommissioning to ensure that performance requirements are met. This chapter describes the planning and construction of nuclear power plants, maintenance/management activities during their operation, and decommissioning. The detailed flow of the decommissioning of a nuclear power plant is explained; the flow includes the transportation of spent fuel away from the plant, removal of radioactive substances from equipment inside the reactor and reactor building, and disposal of nuclear waste from decommissioning.

Keywords Planning · Design · Construction · Operation · Decommissioning · Inspection · Integrity evolution · Aging · Seismic reinforcement · Nuclear waste disposal

10.1 Flow of Regulatory Actions for Nuclear Power Plants

This section explains the flow of regulatory actions for nuclear power plants in the four project stages: planning, construction, operation, and decommissioning. Figure 10.1 shows the flow of regulatory actions for nuclear power plants that are taken in Japan at different times starting with the application for a reactor establishment license in the planning stage and ending with decommissioning.

M. Kuno (✉)
Chubu Electric Power Company, Nagoya, Japan
e-mail: kuno.michiya@chuden.co.jp

M. Hamada
Waseda University, Shinjuku, Tokyo, Japan

M. Hamada and M. Kuno (eds.), *Earthquake Engineering for Nuclear Facilities*,
DOI 10.1007/978-981-10-2516-7_10

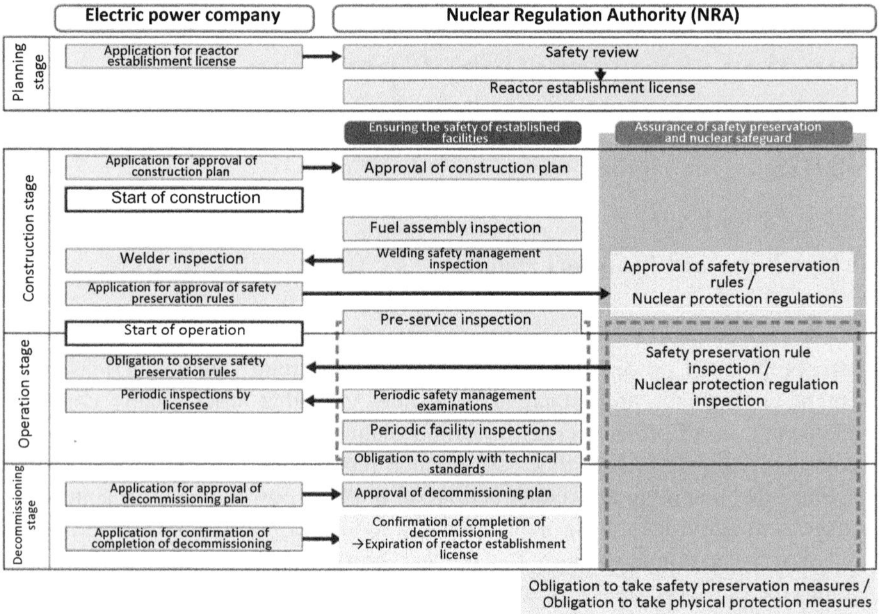

Fig. 10.1 Flow of regulatory actions for nuclear power plants [1]

10.1.1 Planning Stage

In the process of selecting a site for building a nuclear power plant, investigations and assessments are performed from diverse angles, giving attention not only to the safety of the nuclear power plant but also to effects on the lives of nearby residents and other industries. The electric power company produces three documents and submits them to the Ministry of Economy, Trade and Industry (METI) for examination: *Statement on the Consideration of Environmental Impact Assessment*, *Statement on the Method of Environmental Impact Assessment*, and *Statement on the Preparation for Environmental Impact Assessment*. In consideration of feedback from the examination process, the electric company produces an environmental impact assessment report and submits it to METI. After the contents of the environmental impact assessment report are finalized, the electric power company submits an application for a reactor establishment license to the Nuclear Regulation Authority (NRA). The NRA performs safety reviews on the application for reactor establishment and performs a safety examination to determine if the application satisfies the licensing criteria as per the Act on the Regulation of Nuclear Source Materials, Nuclear Fuel Materials and Reactors (hereinafter "Reactor Regulation Act"), and, subject to the satisfaction of these criteria, issues the reactor establishment license.

10.1.2 Construction Stage

After receiving the reactor establishment license, the electric power company applies for the approval of the construction plan to have the detailed plant design approved by the NRA before starting construction. The commercial power reactor facility constructed following the approval of the construction plan receives a pre-service inspection prior to commissioning. The pre-service inspection reviews the construction stage by means of material inspection, dimensional inspection, and appearance inspection. In addition, a welder inspection, fuel assembly inspection, and safety preservation rules are examined.

10.1.3 Operation Stage

During the operation stage after commissioning, the safe and stable operation of the nuclear power plant is ensured by periodic inspections made by the licensee (i.e., the electric power company) and statutory periodic facility inspections. In addition, an integrity assessment (i.e., an examination of equipment integrity), periodic safety reviews, and aging management activities are performed as parts of long-term facility maintenance programs.

10.1.4 Decommission Stage

The dismantling and removal of a nuclear power plant after the end of operation involves processes such as decontamination (i.e., the removal of radiological contamination by nuclear fuel materials) and the disposal of contaminated materials that must be carried out strictly in compliance with applicable laws. Before decommissioning a nuclear power plant, the licensee must produce a decommissioning plan and have it approved by the NRA. At the end of decommissioning, the licensee applies for confirmation of the completion of decommissioning by the NRA. Upon the confirmation of decommissioning by the NRA, the reactor establishment license expires and the given nuclear power plant ends its mission.

10.2 Construction of Nuclear Power Plant

This section describes the flow of construction activities from preparatory work to commissioning with explanations on particularities of different types of works involved.

10.2.1 Nuclear Power Plant Construction Works

The following are the particularities of nuclear power plant construction works:

(1) The construction works involve aggressive testing, inspection, and quality control.
(2) Construction takes a long time to complete. Notably, building work, mechanical work, and electrical work are conducted concurrently over a long period.
(3) The work volume is enormous. For example, a large site has to be developed through civil engineering work.

10.2.2 Construction Project Flow

This subsection explains the entire construction project flow after the completion of rock formation testing. A nuclear power plant construction project generally takes a little more than 4 years from the rock test under the reactor building foundation to the start of commercial operation. However, the exact period varies depending on site conditions, the reactor type (e.g., boiling water reactor or pressurized water reactor), and volume of electricity generation.

The erection of major facilities inside the reactor building is identified as a critical path that strongly affects the scheduling of the entire construction project. Because the work volume is enormous, it is important to ensure that works are conducted reliably according to schedules, and individual work processes are planned with emphasis given to concurrent execution, labor savings, rationalizing, quality control, and labor safety.

Figure 10.2 gives an overview of construction schedules for major plant buildings.

10.3 Construction of Reactor Building

This section describes the construction of the reactor building of an advanced boiling water reactor plant. The entire workflow from the start of construction to the beginning of operation starts with preparatory work, continues with construction work and mechanical/electrical work, and ends with tests and trial operation. The start of construction work is declared upon the commencement of foundation excavation work for the reactor building for example, after the approval of the construction plan. Figure 10.3 shows the workflow of the reactor building construction.

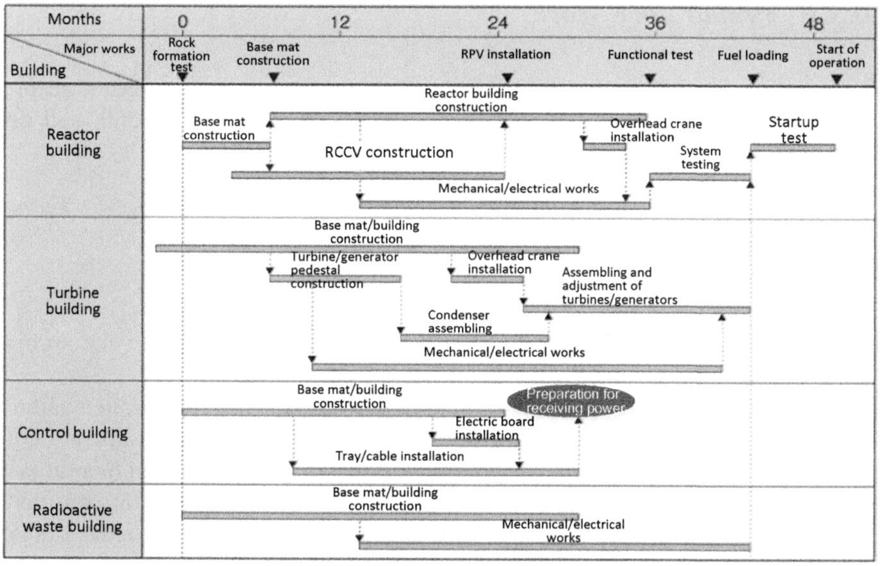

Fig. 10.2 Overview of construction schedules for major plant buildings [2]

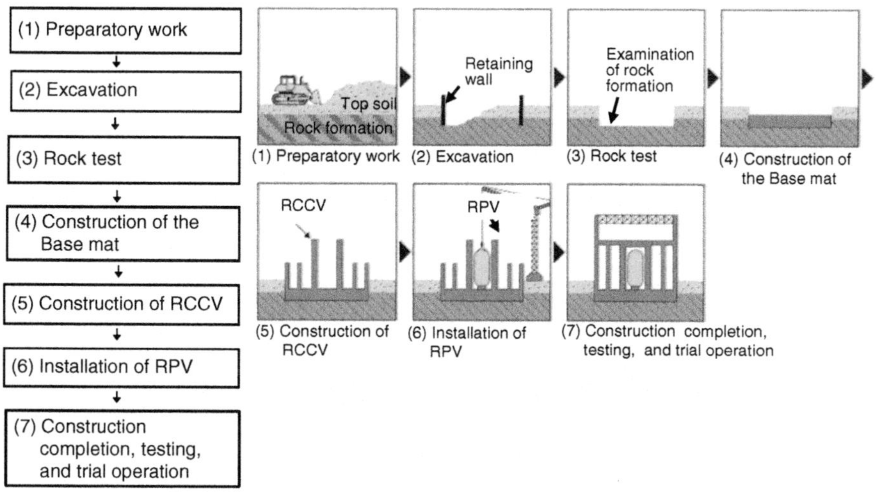

Fig. 10.3 Workflow of reactor building construction

10.3.1 Preparatory Work

The preparatory work includes the setup of temporary storage yards, shore protection work, and site development (e.g., excavation of ground landfill, soil disposal, and ground grading).

10.3.2 Excavation

The foundation excavation work begins for the reactor building, the turbine building, and water storage facilities.

To facilitate excavation work, earth-retaining walls, such as soldier pile retaining walls, soil mixing continuous walls, and underground continuous walls, are constructed. The volume of soil in a major excavation work lasting several months is in the order of 10^6 m^3. Throughout the excavation work, environmental considerations are made along with strict quality control, taking measures to reduce noise, vibration, and dust, because of the frequent traffic of large dump trucks. As a general rule, the excavated soil is reused for site development.

10.3.3 Rock Test

The rock test is performed in pursuant to the Reactor Regulation Act and other laws. The construction work starts after the completion of the rock test. Figure 10.4 shows a scene from a rock test.

The purpose of the test is to verify that the rock is strong enough to serve as the foundation of the reactor containment building. The test is performed in the

Fig. 10.4 Rock formation test (*Photograph* Chubu Electric Power Co., Inc.)

presence of NRA personnel and includes the examination of related documents. The test includes the examination of the foundation rock elevation, information about rock types/qualities and their distribution, rock test results, information about groundwater inflow and drainage facilities, and the cleanliness of the rock surface. It must be verified that the actual states of the foundation rock formation do not differ significantly from their descriptions examined in relation to the approval of the reactor establishment license and construction plan. Excavation is performed carefully for the foundation rock, and it has to finish with manually conducted finishing excavation and cleanup. Prior to the rock test, a tent is stretched over the rock surface to keep it clean and to ensure the test is unaffected by weather.

10.3.4 Construction of the Base Mat

After the completion of the rock test, the construction of the reactor building foundation begins with the setting of base mat reinforcement bars and the casting of concrete. Figure 10.5 shows the construction of base mat reinforcement bars. The base mat, 70–80 m long on each side, is composed of large-diameter reinforcement bars, laid out not only in the two horizontal rectangular directions but also radially, and a large quantity of concrete.

The construction of the reactor building involves the daily casting of a large volume of concrete (500–1,000 m^3/day). The foundation base mat of the reactor building is often of extra-large thickness (more than 4 m in thickness) and requires the use of large quantities of large-diameter reinforcement bars. From the assembling of reinforcement bars to the completion of concrete casting, the whole process

Fig. 10.5 Construction of base mat concrete reinforcement bars (*Photograph* Chubu Electric Power Co., Inc.)

takes about half a year. The strict control of concrete quality is practiced in the casting of base mat concrete. The time required for concrete casting is shortened by the use of block casting.

10.3.5 Construction of Reinforced Concrete Containment Vessel

After the production of base mat concrete, the construction of the RCCV begins. In the course of concrete works, the control of concrete quality and the management of work schedules depend largely on coordination among different works, such as the fixing of equipment supports and the installing of pipes. It is necessary to produce a work plan carefully, deciding on the division of work areas and the placement of heavy construction machinery in consideration of overlaps among construction work, mechanical work, and electrical work. Moreover, large and heavy construction machineries such as tower cranes operate in a confined construction area. To support the smooth and safe execution of reactor building construction work, it is important that the placement of construction machinery is planned carefully, appropriately considering the rotation radius of the tower crane boom and appropriately positioning work pedestals. Once the reactor building is constructed, RCCV pressure leakage testing is performed to confirm the structural integrity of the reactor containment vessel.

Figure 10.6 shows a scene from RCCV construction work. To the exterior of cylindrical steel liner plates, a reinforced concrete wall of about 2 m in thickness is constructed.

Fig. 10.6 RCCV construction (*Photograph* Chubu Electric Power Co., Inc.)

Fig. 10.7 RPV hoisting (*Photograph* Chubu Electric Power Co., Inc.)

10.3.6 Installation of Reactor Pressure Vessel

After the construction of the RCCV, the RPV is installed. Figure 10.7 shows an RPV being hoisted for installation. The RPV is hoisted carefully while avoiding the operation zones of the tower cranes for reactor building construction.

The RPV, brought to the site from the factory by sea and land transportation, is hoisted for installation using a crane specially provided for this purpose. The hoisting is done carefully to avoid the reactor containment vessel, the reactor vessel, and pre-installed piping. The installation of major equipment and piping ends with the hydraulic testing of main systems.

10.3.7 Construction Completion, Testing, and Trial Operation

After the installation of the RPV, the top story is constructed and the roof is added as the final major step in the construction of the reactor building. The construction of other buildings and structures, such as the turbine building and exhaust stack, proceeds concurrently with the construction of the reactor building and is completed around the same time, putting an end to the construction work of the nuclear power plant. After the completion of construction work, functional testing of various systems (e.g., equipment, piping, electrical installations, instruments, etc.) is performed, followed by a trial operation. After fuel loading, the reactor output is gradually increased to 100 % while monitoring the performance of various facilities at different output levels before the commencement of the commercial operation of the nuclear power plant.

10.4 Operation of Nuclear Power Plant

In relation to activities conducted at a nuclear power plan during its operation, this section explains maintenance management, aging assessment techniques, and seismic safety improvement measures.

10.4.1 Maintenance Management

The purpose of the maintenance management of a nuclear power plant is to ensure the integrity of plant facilities and to increase the assurance of safe and stable plant operation. Maintenance management activities performed at nuclear power plants either during operation or during scheduled outages (periodic inspections) include the following.

(1) Inspections are performed by the operators while the reactor is in operation, including the surveillance testing of components important to safety (approximately once a week) and daily walk-through and routine inspections.
(2) A statutory periodic facility inspection is performed while the reactor is in outage approximately once per year in pursuant to the Reactor Regulation Act. During the inspection period, the operation data of different facilities and the records from past periodic inspections are examined. In addition to this statutory inspection, there are periodic inspections performed by the electric power company according to their programs that ensure the integrity of plant facilities. Particular emphasis is placed on detecting the symptoms of aging (i.e., degradation due to long use) during periodic inspections. A plant integrity evaluation program has been introduced to collect data on the aging trend, on the basis of which repairs and replacements are performed before the performance of facilities drops below the predefined reference level,
(3) Following the amendment of the Electric Enterprise Law on October 2003, the Japan Electric Association issued the Code of Quality Assurance for Safety of Nuclear Power Plants (JEAC4111) and the Rules of Maintenance Management of Nuclear Power Plants (JEAC4209) under the Reactor Regulation Act. With guidelines on maintenance management established by these publications, the Japanese government monitors the states of maintenance management at nuclear power plants. Government-appointed nuclear safety inspectors periodically conduct examinations to determine how safety preservation rules, dictating practices to be followed to ensure safe plant operation, are observed at nuclear power plants. The inspection of certain facilities prescribed by law has to be covered by periodic inspection by the licensee. The Japan Nuclear Energy Safety Organization examines the state of implementation of such periodic inspections by licensees, and the examination result is reviewed by the Nuclear and Industrial Safety Agency of METI.

(4) In trying to maintain and improve the reliability of plant facilities, repairs and improvements need to be performed systematically, considering lessons from nuclear power plant operating experiences in Japan and abroad, and taking advantage of technological development. The aging of the RPV due to neutron irradiation embrittlement is addressed by inspections, but the replacement of the RPV in the reactor building is difficult.

(5) Every 10 years or more, the implementation of safety preservation activities and the incorporation of the latest technical knowledge during the given period are examined in a periodic safety review. Moreover, within 30 years of the commencement of commercial operation (and every 10 years thereafter), an aging technical evaluation is performed to address the aging of components and structures that are important to safety. The licensee is obliged to establish a long-term maintenance management policy based on the results of these evaluations and incorporate it into the safety preservation rules.

Figure 10.8 shows the flow of maintenance management of a power plant. The paragraphs that follow offer more information about periodic facility inspections, integrity evaluation, periodic inspections by licensees, and periodic safety reviews.

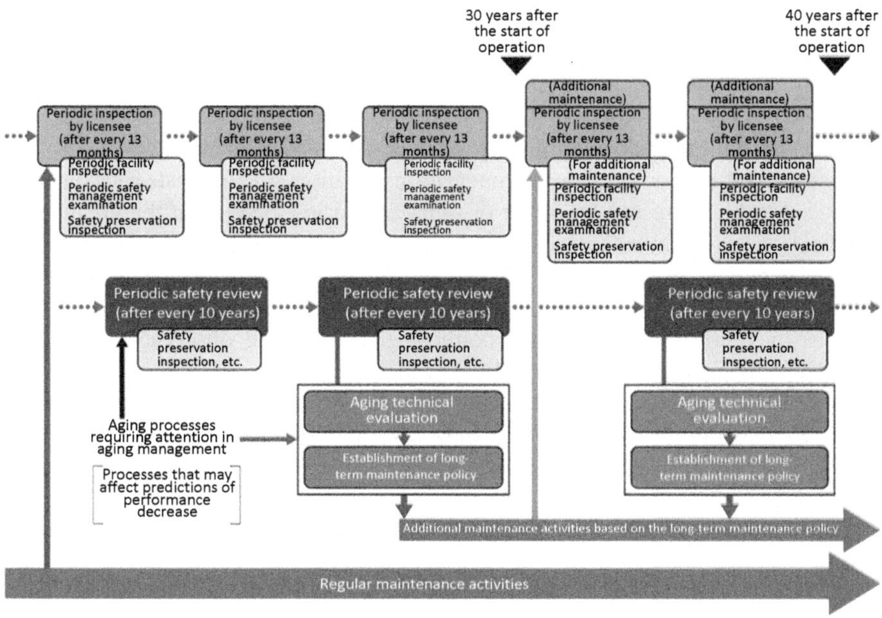

Fig. 10.8 Flow of the maintenance management of a nuclear power plant [3]

10.4.1.1 Periodic Facility Inspection

The Reactor Regulation Act requires that each commercial power reactor facility is inspected periodically every 13 months as a general rule. Conducted for the purpose of verifying the integrity of plant facilities, this inspection involves the testing of the operating performance of main facilities, facility inspection by overhaul, and leakage testing. This periodic inspection is combined with the checking and taking care of the parts of facilities that resemble or correspond to the parts that caused an accident or failed in other nuclear power plants.

The licensees (i.e., electric power companies) are responsible for conducting these periodic inspections. The inspection of important facilities is conducted in the presence of inspectors from the government. The inspection activities include the overhaul inspection of pumps and valves (e.g., the in-service inspection of containers, pipes, and support structures) and the leakage testing of the RCV and main steam isolation valve.

10.4.1.2 Integrity Evaluation Program

The integrity evaluation program gives attention to flaws discovered by periodic inspections and technically predicts their further development for evaluating future plant integrity. If a flaw such as cracking is found in the shroud inside the RPV or in a facility or component in a system that handles the reactor coolant, for example, the cause of the flaw is identified, and technical evaluation is performed to find out how far the flaw may develop during the lifecycle of the facilities.

Figure 10.9 shows the flow of integrity evaluation. In the past, each detected flaw used to be addressed by repair or replacement irrespective of the greatness of its effect on safety in reference to the design and manufacturing standards applicable to new components. In the current integrity evaluation program, however, the facilities and components in which flaws have been discovered may continue to be used as they are under enhanced monitoring or with the observation of further developments provided that they still meet their maintenance standards. According to a prediction of the gradual decrease in strength due to crack development or wearing, for example, a component may continue to be in service as long as it satisfies the safety standard. When it no longer meets the safety standard, it has to be repaired or replaced.

As standards concerning the maintenance of reactor facilities, *Codes for Nuclear Power Generation Facilities: Maintenance Standards* (JSME S NA1-2010) published by the Japan Society of Mechanical Engineers is referred to. Electric power companies are obliged to report the results of these evaluations on plant maintenance and inspection activities to the national government and preserve the records.

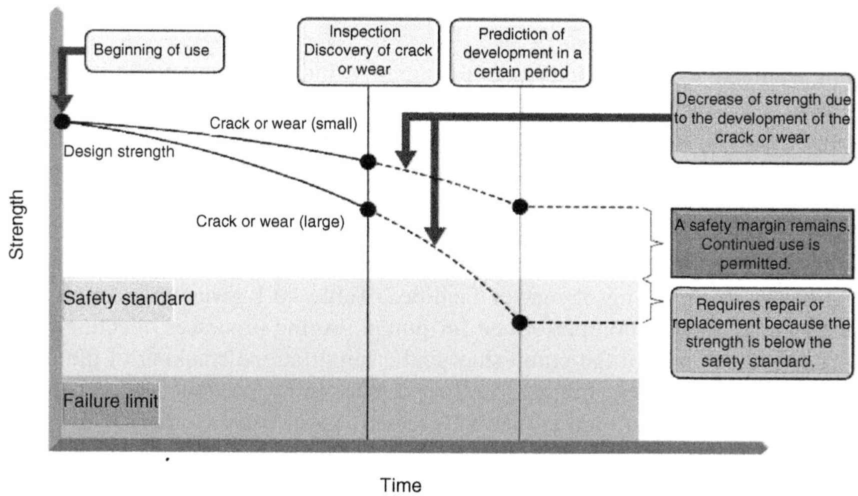

Fig. 10.9 Method for evaluating the integrity of plant facilities [3]

10.4.1.3 Periodic Inspection by Licensee

Periodic inspections by licensees were institutionalized in October 2003. Formerly, Article 54 of the Electric Enterprise Law stipulated the roughly annual execution of periodic inspections, and the licensees were also obliged to perform voluntary inspections. However, because the scope of voluntary inspections had not been clearly defined by law, Article 55 of the Electric Enterprise Law stipulated a periodic inspection by the licensee. Regarding these inspections performed by electric power companies periodically (every 13 months), the law stipulates the scope of inspection and the obligations of record keeping and reporting. Each periodic inspection by the licensee is performed together with the periodic inspection required by the Electric Enterprise Law, and their records are reviewed by the national government's periodic safety management examinations and safety preservation inspections.

10.4.1.4 Periodic Safety Review and Aging Technical Evaluation

Apart from periodic inspections, a periodic safety review is performed at intervals of less than 10 years for the implementation of safety preservation activities and the incorporation of the latest technologies during the given period, and the results are checked during safety preservation inspections. Moreover, within 30 years of the commencement of commercial operation (and every 10 years thereafter), an aging technical evaluation for extended plant operation is performed. The licensee establishes a long-term maintenance management policy based on the result of these evaluations and incorporates it into the safety preservation rules.

10.4.2 Aging Evaluation

Aging technical evaluation addresses the aging of buildings, structures, equipment, and piping systems that fulfill safety functions. The evaluation results are used in the planning of the long-term maintenance management of reactor facilities.

This subsection gives examples of aging technical evaluation performed at a nuclear power plant. Aging processes include neutron irradiation embrittlement, stress corrosion cracking, low-cycle fatigue, pipe thinning, insulation deterioration, concrete weakening, and decreasing shielding capability. Figure 10.10 shows chief contributors to the aging of reactor facilities. Table 10.1 gives an example of an aging technical evaluation performed for pumps. Aging processes affecting pumps include the corrosion of the pump shaft and casing, fatigue cracking of the casing, and fatigue cracking by fretting (i.e., repeated sliding motions at a very small amplitude between two objects that are pressed against one another).

Figure 10.11 shows an example of the parts of buildings and structures that receive attention in the evaluation of concrete weakening and decreasing shielding capability. Table 10.2 gives an example of the evaluation performed on concrete weakening. Factors contributing to concrete weakening include heat, neutron irradiation, neutralization, chloride penetration, the alkali–aggregate reaction, and mechanical vibration.

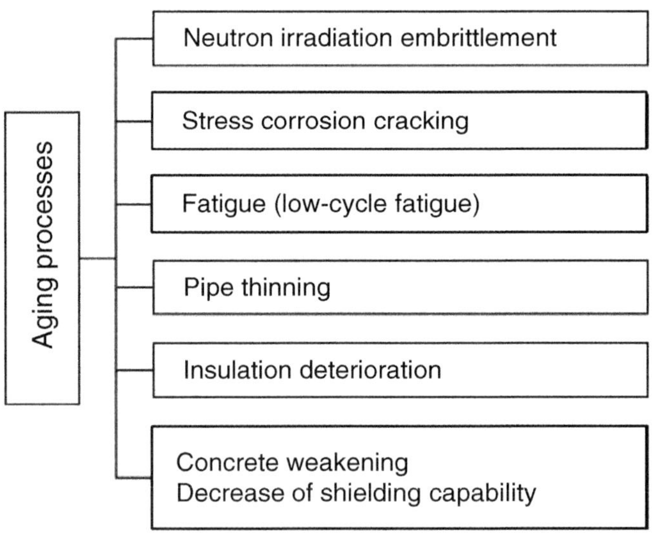

Fig. 10.10 Contributors to the aging of reactor facilities

Table 10.1 Example of an evaluation performed for processes contributing to the aging of pumps (at Hamaoka NPP) [4]

Aging process	Example of evaluation result
(a) Corrosion of shaft, casing, etc.	• Any significant corrosion can be easily found by sight. A visual inspection is performed during overhaul to find any corrosion and address it as required by replacement, etc.
(b) Fatigue cracking of casing (reactor recirculation pump)	• A fatigue life evaluation, performed in consideration of the number of thermal transient that happen due to plant startup and shutdown in the period of 60 years from the commissioning, gives an assurance that the fatigue accumulation factor will not go above the allowable limit. During the overhaul of pumps, the absence of any significant flaw is verified by the visual inspection of the casing interior, etc.
(c) Fatigue cracking due to fretting	• In the case of a pump in which the vane rotor is shrunk-fit to the shaft, the shaft is inspected after removing the vane rotor in order to confirm the absence of fatigue cracking due to fretting

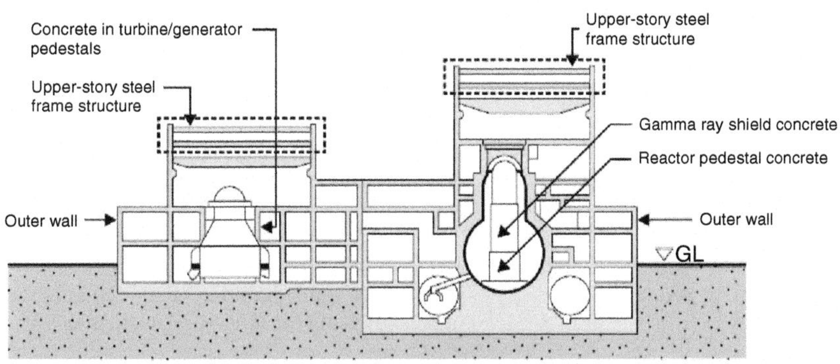

Ageing process	Factors contributing to aging		
Weakening	◎ Heat	◎ Neutron irradiation	◎ Alkali-aggregate reaction
	◎ Neutralization	◎ Chloride penetration	◎ Freezing and thawing
	◎ Mechanical vibration	◎ Chemical penetration	◎ Fatigue
	Drying shrinkage	Creep	Wear
Decrease of shielding capability	◎ Neutron irradiation (generation of heat by gamma ray)		

◎:Factors contributing to aging required by the standards for a light-water reactor

Fig. 10.11 Example of inspection targets [4]

Table 10.2 Example of an evaluation performed on concrete weakening (at Hamaoka NPP) [4]

Factors contributing to aging	Evaluation target	Example of evaluation result
Heat	Reactor pedestal	• It is confirmed that the highest temperature determined by temperature distribution analysis is below the maximum allowable temperature (65 °C generally, 90 °C locally)
Neutron irradiation	Reactor pedestal	• An analysis performed on neutron and gamma ray irradiation expected in the period of 60 years from the commissioning assures that the cumulative levels of irradiation will not be large enough to cause any significant weakening of concrete
Neutralization	Whole structure	• An estimation performed on the depth of neutralization after 60 years from the commissioning at the evaluation target points, selected in consideration of contributors such as carbon dioxide and temperature, assures that the depth of neutralization will not be large enough to initiate the corrosion of reinforcement bars
Chloride penetration	Whole structure	• An estimation performed on the thinning of reinforcement bars due to corrosion in the period of 60 years from the commissioning assures that the degree of thinning will be sufficiently lower than the level at which the cracking of the cover concrete may occur
Alkali–aggregate reaction	Whole structure	• The measurement of the total swelling ratio, performed on core samples of concrete structures in which the alkali–aggregate reaction proceeded, demonstrated that the swelling ratio was sufficiently lower in view of the requirement of "less than 0.1 % when 6-month old"
Mechanical vibration	Turbine/generator pedestals	• The average compressive strength, determined by a compressive strength testing of core samples, was sufficiently higher than the design strength

10.4.3 Seismic Reinforcement

As examples of seismic reinforcement implemented at a nuclear power plant, the following explains two projects launched by the Chubu Electric Power Company, at the Hamaoka Nuclear Power Plant: (1) the Seismic Safety Margin Improvement Project; and (2) Actions Taken in Consideration of Lessons from Damage Inflicted by the 2007 Niigata-Ken Chuetsu-Oki Earthquake on the Kashiwazaki Kariwa Nuclear Power Plant of Tokyo Electric Power Company.

10.4.3.1 Seismic Safety Margin Improvement

The Hamaoka Nuclear Power Plant exists in an area that is likely to be affected by the Tokai earthquake. To increase the seismic safety margin of the nuclear power plant, Chubu Electric Power Company launched a reinforcement project assuming that the seismic ground motion acceleration at the bedrock level may reach as high as 1,000 cm/s^2. As examples of the measures taken, the following explains exhaust stack reinforcement, ground improvement around a piping duct, and support improvement for piping and cable conduits.

(1) Exhaust stack reinforcement
During normal plant operation, the exhaust stack serves as an outlet of air from the reactor building. In an emergency, air from the secondary containment is released from the exhaust stack after passing through a charcoal filter. The exhaust stack structure must maintain its ability to support the shaft that goes through it.
The exhaust shaft is a cylindrical steel structure 100 m in height and about 8 m in diameter at the bottom. It is held by a concrete foundation that is about 20 m deep. The exhaust shaft was reinforced for seismic safety. This was done by building a steel tower around the exhaust shaft and joining this supporting steel tower with the exhaust tower via oil dampers. Figure 10.12 shows an example of exhaust shaft reinforcement.
(2) Ground improvement around a piping duct
To reduce the forces that act on a piping duct from the surroundings, the ground soil was improved by injecting a cement-based material into the ground and by mixing the material with soil. Figure 10.13 shows an example of ground improvement around the piping duct.

Fig. 10.12 Example of exhaust stack reinforcement [5]

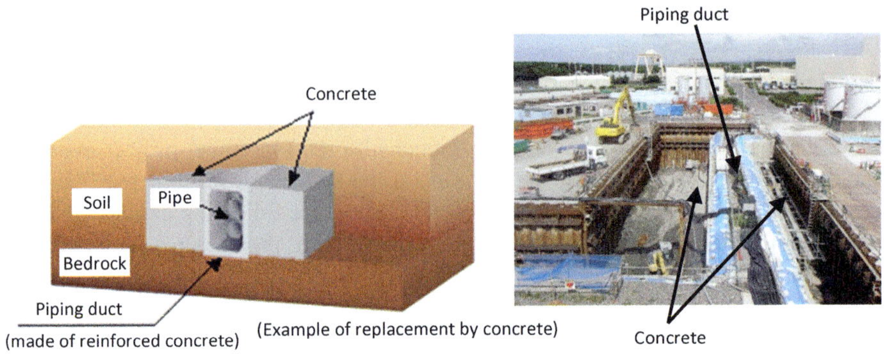

Fig. 10.13 Ground improvement around a piping duct [5]

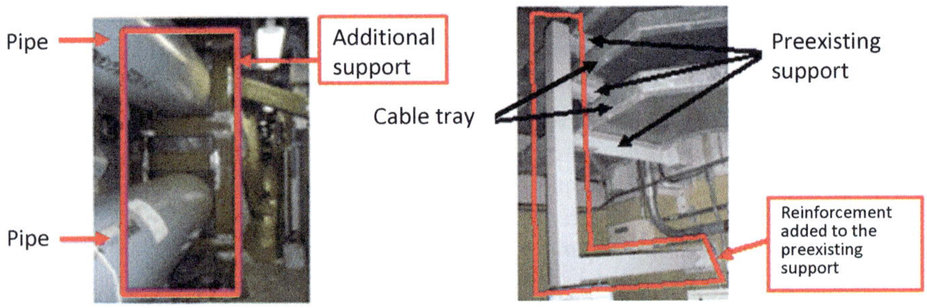

Fig. 10.14 Support improvement for piping and cable conduits [5]

(3) Support improvement for piping and cable conduits
Supports were added to reduce the horizontal and vertical displacements that can happen for piping and cable conduits inside the reactor building and some other plant buildings. The required seismic strength was determined by performing numerical analyses using models of piping and supports. After performing a seismic evaluation on piping supports at about 6,000 points in total, reinforcement was made at about 200 of the 6,000 points. In addition, cable conduit supports were reinforced at about 1,300 points. Figure 10.14 gives an overview of support improvement for piping and cable conduits.

10.4.3.2 Actions Taken in Response to the 2007 Niigata-Ken Chuetsu-Oki Earthquake

In consideration of the damaged inflicted by the 2007 Niigata-Ken Chuetsu-Oki Earthquake on the Kashiwazaki Kariwa Nuclear Power Plant of Tokyo Electric Power Company and emergency measures taken after the earthquake, Chubu

Electric Power Company, reexamined seismic safety measures for the Hamaoka Nuclear Power Plant and built a seismically isolated office building in which to facilitate initial responses to earthquake and to support the distribution of reliable information.

The seismically isolated building has four storeys above the ground with a single-storey penthouse with dimensions of about 30 × 40 m. A seismic isolation system is placed between the first storey level and the foundation to dampen earthquake motion. The seismic isolation system improves the reliance on the plant personnel's capability to make initial responses and to distribute information. Figure 10.15 shows the appearance of the seismically isolated office building.

Furthermore, in view of the need to ensure the availability of access roads used in initial responses, precast concrete plates were installed over underground constructions to prevent the collapsing of roads due to the failure of underground constructions. Figure 10.16 shows an example of access road reinforcement.

Fig. 10.15 Seismically isolated office building (*Photograph* Chubu Electric Power Company)

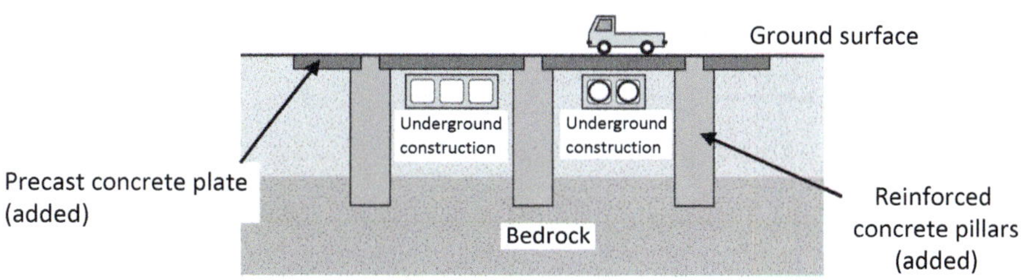

Fig. 10.16 Example of access road reinforcement

10.5 Decommissioning of Nuclear Power Plant

Nuclear power plants are dismantled and removed following the termination of their operation. This section describes the decommissioning of nuclear power plants and the disposal of the waste produced by decommissioning.

10.5.1 Decommissioning Flow

The Reactor Regulation Act stipulates "When terminating the use of a nuclear power plant, it is necessary to take measures specified by the rules of the NRA such as the dismantling of reactor facilities, the transfer of nuclear fuel materials, the removal of contamination by nuclear fuel materials, and the disposal of components contaminated by nuclear fuel materials."

In pursuant to the Reactor Regulation Act, the licensee has to prepare in advance a decommissioning plan and have it approved by the NRA. A document titled *Decommissioning of Commercial Nuclear Power Reactors* (July 15, 1985) from the Nuclear Power Committee of the Advisory Committee for Energy, the Ministry of International Trade and Industry (now METI), along with a few other documents, prescribes a standard decommissioning workflow. Figure 10.17 illustrates this standard workflow for nuclear power plant decommissioning.

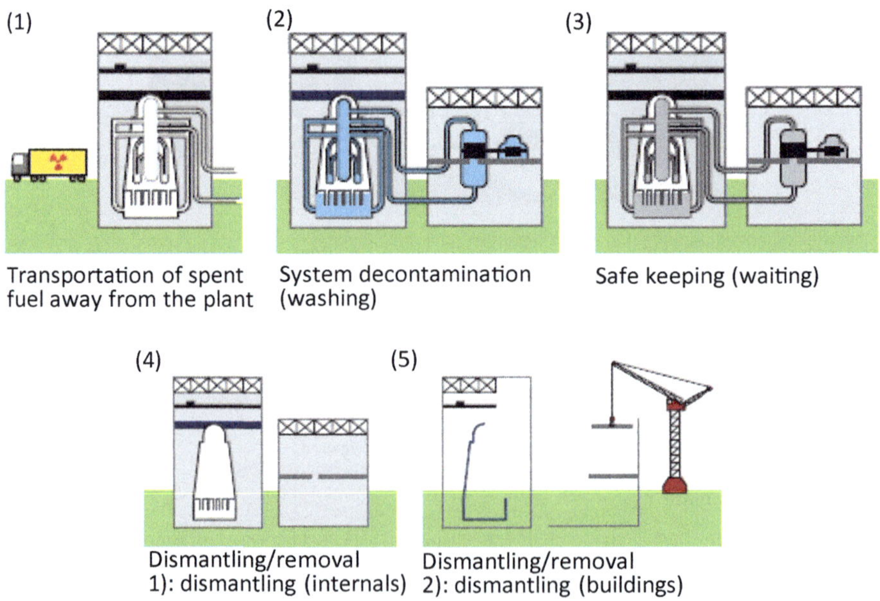

Fig. 10.17 Flow of the decommissioning of a nuclear power plant [3]

The standard workflow is as follows;

(1) Transportation of spent fuel away from the plant
(2) Removal of radioactive substances from inside the piping and containers using chemicals (i.e., system decontamination)
(3) Safekeeping for a period of 5–10 years to wait for the decrease in radioactivity by decay
(4) Dismantling and removal of piping, containers, and other components from inside the reactor and the reactor building
(5) Dismantling and removal of buildings.

10.5.1.1 Transportation of Spent Fuel from Plant

The spent fuel is transported away from the nuclear power plant. The spent fuel removed from the plant has to be properly controlled and disposed of. Before the spent fuel is removed, the nuclear power plant has to maintain safety functions that protect it from the risk of an earthquake and tsunami.

The spent fuel is put into special transportation containers (canisters) before it is carried away. Figure 10.18 shows spent fuel being transported away from a nuclear power plant. The transportation canisters must meet statutory design requirements so that they may withstand accidents such as dropping and fire without having any radiological effect on the environment. Until the transportation of spent fuel is completed, the plant facilities providing cooling and containment capabilities, as well as radioactive waste disposal facilities and radiation control facilities (e.g., radiation monitors), are periodically inspected.

Fig. 10.18 Spent fuel transported away from a nuclear power plant (*Photograph* Chubu Electric Power Company)

10.5.1.2 System Decontamination

To minimize the dose to workers who will be engaged in dismantling activities, piping, containers, and other components are washed using chemicals to clean them of radioactive substances adhering to them. During system decontamination activities, the release of radioactive substances is prevented by measures such as the restricting of work areas. Throughout the period of system decontamination, the plant facilities providing the containment capability, as well as radioactive waste disposal facilities and radiation control facilities (e.g., radiation monitors), are maintained.

10.5.1.3 Safe Keeping

The quantity of radioactive substances decreases with the passage of time. During the safekeeping period of a certain duration, determined according to the measurement and evaluation of the radioactivity of reactor components, the plant is left alone with necessary maintenance and control provided to wait until the quantity of radioactive substances decrease sufficiently to permit the easier execution of dismantling/removal activities. For the assurance of safety during the safekeeping period, the piping and containers are emptied of liquid and all valves and openings are closed at the plant to prevent the outflow of radioactive substances. Furthermore, the plant facilities providing the containment capability, as well as radioactive waste disposal facilities and radiation control facilities, are maintained.

10.5.1.4 Dismantling and Removal

To prevent the external release of radioactive substances, facilities and components inside the plant buildings are dismantled and removed first. After that, building floors and interior wall surfaces are cleaned of radioactive substances. After confirming the removal of radioactive substances from inside, the plant buildings are dismantled and removed. The work areas are covered to prevent the scattering of radioactive substances. Furthermore, the plant facilities providing the containment capability, as well as radioactive waste disposal facilities and radiation control facilities, are maintained.

10.5.2 Waste Produced by Decommissioning

In the case of a 1.1-GW-class nuclear power plant (i.e., a light water reactor plant), the waste produced by its decommissioning is expected to amount to approximately 500–550 thousand tons. Most of these wastes are nonradioactive or not radioactive enough to require treatment as radioactive waste. Figure 10.19 shows the breakdown of decommissioning waste.

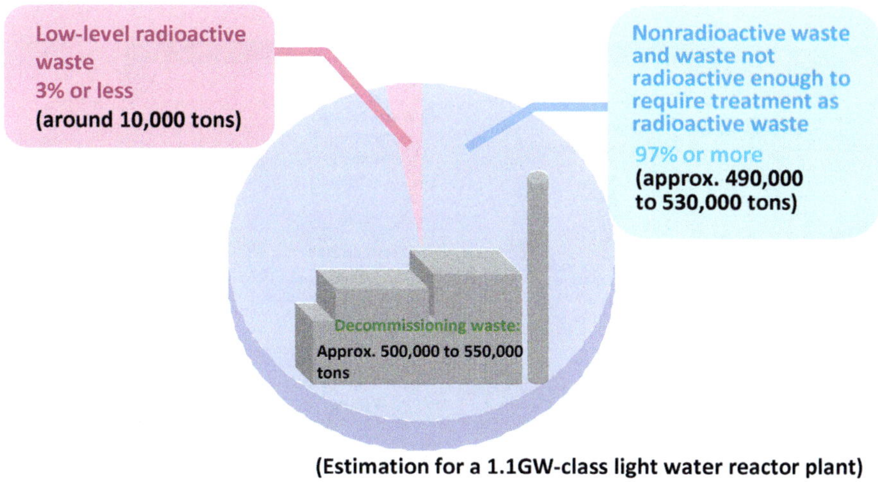

Fig. 10.19 Decommissioning waste

The nonradioactive waste and the waste not radioactive enough to require treatment as radioactive waste are recycled as resources or disposed of like typical industrial waste. Low-level radioactive waste is classified by the type of radioactive nuclides it contains and the level of radioactivity, and disposed of appropriately according to the classification.

10.5.3 Classification and Disposal of Waste from the Decommissioning of a Nuclear Power Plant

Figure 10.20 shows the classification and disposal of waste from the decommissioning of nuclear power plants. Low-level radioactive waste is classified by the types of radioactive nuclide it contains. It is then buried at a determined depth and protected by an artificial barrier such as that of a concrete pit as required.

10.6 Examples of Nuclear Power Plant Decommissioning

In Japan, four reactor units are now undergoing decommissioning: Tokai Power Station (a carbon dioxide-cooled reactor) of Japan Atomic Power, Fugen (an advanced thermal reactor) of Japan Atomic Energy Agency and Fugen Decommissioning Engineering Center, and Hamaoka Units 1 and 2 of Chubu Electric Power Company. In addition, a decision has been made to decommission Units 1–6 at Fukushima Daiichi Nuclear Power Plant of Tokyo Electric Power Company.

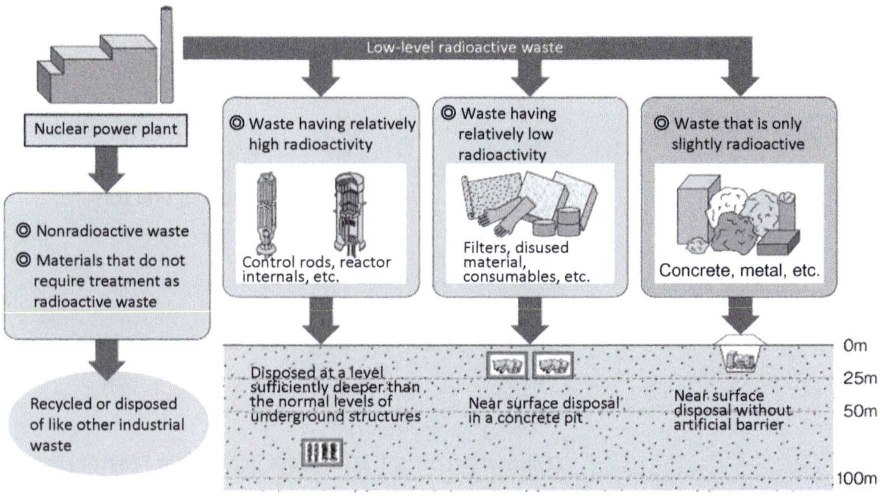

Fig. 10.20 Classification and disposal of waste from the decommissioning of a nuclear power plant

Fig. 10.21 Example of the decommissioning of a nuclear power plant in Japan [6]

Figure 10.21 shows an example of nuclear power decommissioning in Japan. The power demonstration reactor, which is called JPDR (Japan Power Demonstration Reactor) and used to be run by the Japan Atomic Energy Agency (formerly known as Japan Atomic Energy Research Institute), underwent all decommissioning processes, including dismantling, by October 2002.

As of April 2012, out of 138 commercial power reactors outside of Japan that have terminated their operation, 17 units, including Maine Yankee (0.9 GW) and Haddam Neck Nuclear Power Plant (0.6 GW) in the United States, have completed the decommissioning process.

As for the decommissioning of Hamaoka Units 1 and 2 of Chubu Electric Power Company, the first phase of the project, which continued to the end of the 2014 fiscal year, was dedicated to preparations before dismantling. Activities that have been completed or started are the carrying away of the spent fuel, the evaluation and analysis of contamination, system decontamination, and the dismantling/removal of facilities and components outside the radiation control area. The decommissioning process is scheduled to be completed in the 2036 fiscal year.

References

1. "Graphical Flip-chart of Nuclear & Energy Related Topics 2012", The Federation of Electric Power Companies of Japan, 2012
2. "*Kagenkyoukai Kouza 22; Hatsudensho no Kensetsu, Shiunntenn to Untenhoshu* (TENPES Lecture 22 Building, Commission, Operation and Maintenance of Power Plants)", Thermal and Nuclear Power Engineering Society
3. "Graphical Flip-chart of Nuclear & Energy Related Topics 2013", The Federation of Electric Power Companies of Japan
4. "*Hamaoka Genshiryoku Hatsudensho 2 gouki Koukeinenrekka Gijutsu Hyouka nado Hokokusho* (Evaluation Report of Aging Management of Hamaoka Nuclear Power Plant Unit 2)", Chubu Electric Power Co., Inc
5. Chubu Electric Power website
6. Japan Atomic Energy Agency website

Further Readings

7. "Code of Quality Assurance for Safety of Nuclear Power Plants (JEAC4111)", The Japan Electric Association
8. "Rules of Maintenance Management of Nuclear Power Plants (JEAC4209)", The Japan Electric Association
9. "JSME Codes for Power Generation Facilities; Maintenance Codes", The Japan Society of Mechanical Engineers

Chapter 11
Radioactive Waste Treatment and Disposal Technique

Michiya Kuno and Masanori Hamada

Abstract Collaboration with various related industries is important for the running of nuclear power plants. Particularly important is the collaboration with industries that are involved in the fabrication and reprocessing of nuclear fuel and the disposal of radioactive waste. This chapter first gives an overview of the nuclear fuel cycle and its facilities, and describes the reprocessing of spent fuel and the treatment and disposal of radioactive waste. Furthermore, the earthquake-resistant design of facilities used for the nuclear fuel cycle is explained. In particular, for the disposal of slightly radioactive waste, the concept of clearance and the clearance level are explained.

Keywords Nuclear waste cycle · Spent fuel treatment · Reprocessing of spent fuel · Spent fuel disposal · Clearance

11.1 Nuclear Fuel Cycle Facilities

This section describes the spent fuel reprocessing plant and other facilities that concern the nuclear fuel cycle. An overview of the spent fuel reprocessing plant is presented, and the earthquake-resistant design is discussed.

M. Kuno (✉)
Chubu Electric Power Company, Nagoya, Japan
e-mail: kuno.michiya@chuden.co.jp

M. Hamada
Waseda University, Shinjuku, Tokyo, Japan

M. Hamada and M. Kuno (eds.), *Earthquake Engineering for Nuclear Facilities*,
DOI 10.1007/978-981-10-2516-7_11

11.1.1 Overview of Nuclear Fuel Cycle Facilities

Natural uranium ore is transformed into fuel assemblies for use at nuclear power plants by a series of processes comprising refinement, conversion, enrichment, reconversion, and fabrication. The spent fuel contains unspent uranium and plutonium produced by nuclear fissions. The recycling of plutonium through the reprocessing of spent fuel ensures the continued availability of atomic energy for a long time. Also with the contribution of facilities for the management, treatment, and disposal of radioactive waste, the nuclear fuel cycle is established. Figure 11.1 illustrates the nuclear fuel cycle.

Currently, Japan Nuclear Fuel Limited (JNFL), a company funded by electric power companies in Japan, is running three facilities: (1) uranium enrichment plant, (2) high-level radioactive waste storage management center, and (3) low-level radioactive waste burial center. The construction of (5) mixed oxide fuel fabrication plant is under way, anticipating the startup of (4) reprocessing plant that plays a central role in the nuclear fuel cycle. The completion of these facilities allows the recycling of nuclear fuel, thus ensuring the continued availability of atomic energy in Japan.

As shown in Fig. 11.2, these nuclear fuel cycle facilities are constructed in the village of Rokkasho in the Shimokita Peninsula, Aomori Prefecture.

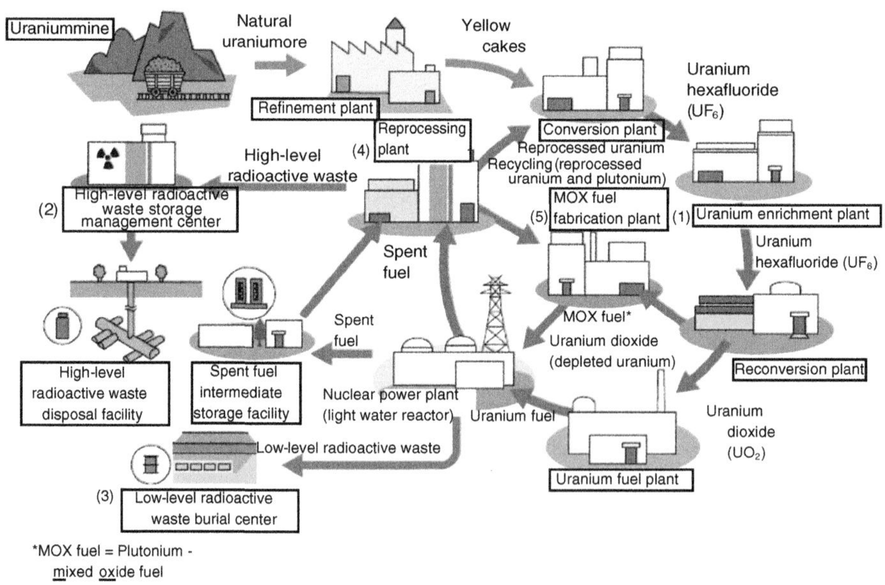

Fig. 11.1 Overview of nuclear fuel cycle facilities [1]

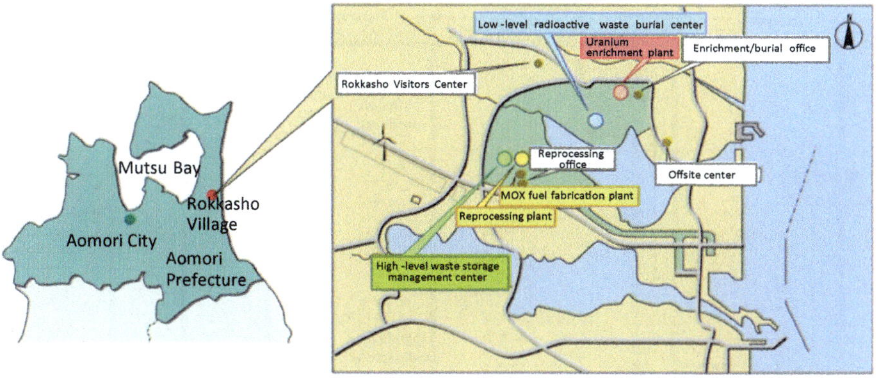

(Source: JNFL brochure)

Fig. 11.2 Locations of nuclear fuel cycle facility [1]

11.1.2 Overall Workflow of Spent Fuel Reprocessing

The reprocessing plant, playing a central role in the nuclear fuel cycle, takes in spent fuel from nuclear power plants and retrieves from it unspent uranium along with plutonium, which can be reused as nuclear fuel.

As shown in Fig. 11.3, the spent fuel from reactors is stored in a pool before being shredded and dissolved using nitric acid. The solution is separated into fission products, uranium and plutonium. The refinement process increases the purity of uranium and plutonium. Fission products are separated in the form of high-level liquid waste, and are subsequently turned into vitrified radioactive waste by being fused with melted glass material. As will be described in Sect. 11.4, the radioactive waste eventually undergoes deep underground disposal.

11.1.3 Reprocessing Plant and Nuclear Power Plant

Table 11.1 shows that the reprocessing plant differs from the nuclear power plant in terms of process characteristics and safety functions.

Most prominent is a difference in building layout. At a nuclear power plant, major facilities are concentrated in the reactor building. At a reprocessing plant, major facilities are distributed in several buildings. As another difference, a chain nuclear reaction (fission) takes place in nuclear power plants but not in the reprocessing plant. The main safety functions at nuclear power plants are *shut down, cooldown,* and *confine*. At the reprocessing plant, the chief safety function is the removal of decay heat from high-level radioactive waste, while *decay heat removal, confinement* and *criticality prevention* are also important.

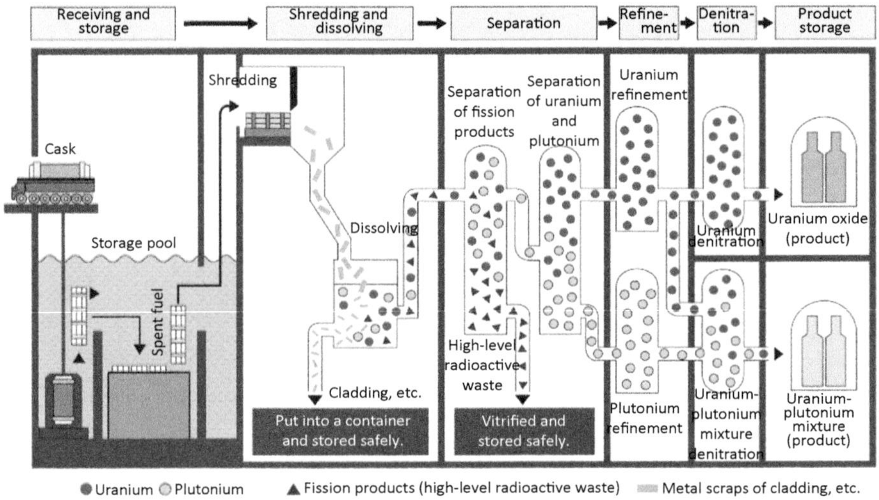

Fig. 11.3 Reprocessing workflow [1]

11.1.4 Earthquake-Resistant Design of a Reprocessing Plant

Like nuclear power plants, the reprocessing plant is designed with the consideration of the seismic classification of facilities. Table 11.2 shows the seismic classification of different facilities at a reprocessing plant. Facilities that handle highly concentrated radioactive materials, as well as facilities that provide the capabilities of *decay heat removal*, *confinement*, and *criticality prevention*, fall into the Class-S facilities.

11.2 Classification and Disposal of Radioactive Waste

This section describes different types of high-level and low-level radioactive wastes and how they are disposed off.

11.2.1 Types of Radioactive Waste

As shown in Fig. 11.4, radioactive waste is produced at nuclear power plants and other nuclear-power-related facilities, and also at hospitals and research institutions.

In Japan, radioactive waste is grouped into two categories according to the level of radioactivity: high-level radioactive waste and low-level radioactive waste.

Table 11.1 Reprocessing plant vs. nuclear power plant [2]

	Reprocessing plant	Nuclear power plant
Major facilities	Distributed presence of buildings containing radioactive materials: Spent fuel acceptance/storage facility, pre-processing building, separation building, refinement building, denitration building, etc.	Concentration of major facilities in the reactor building
Process characteristics	A number of independent chemical processes take place. - Mechanical operation (shredding) - Chemical operation (dissolving, separation, and extraction) - Product handling (powder handling and storage) - Waste treatment (melting, vitrifying, and storage) Individual processes take place in multiple concrete cells.	Power generation processes: - Nuclear reaction (criticality) - Heat removal (cooling) - Generating power using turbine-driven generator
Temperature and pressure ranges	Normal temperature and normal pressure (or slight negative pressure) processes: Dissolver: 100–110 °C, −0.007 atm Separation process: 20–50 °C, −0.007 atm Evaporator: 100–110 °C,−0.007 atm Acid removal evaporator: 70 °C, −0.85 atm Uranium denitrator: 300 °C, −0.005 atm Glass melting furnace: 1,100 °C, −0.01 atm	High temperature and high pressure processes: PWR: 323 °C, 156 atm BWR: 286 °C, 70 atm
Safety functions	Speed of hazard development after functional failure is slower compared with reactors: - Decay heat removal - Containment of radioactive materials - Prevention of criticality	Basic safety functions: - Reactor shut down - Reactor core cool down - Confine of radioactive materials

Table 11.2 Seismic category [2]

Seismic category	Reprocessing plant facilities
S	• Fuel storage pool • Dissolver • High-level radioactive waste evaporator • Systems containing plutonium solution • Safety coolant system • Main exhaust stack, etc.
B	• Low-level radioactive liquid waste treatment facility • Analysis facility
C	• General-use coolant system • Power receiving switchgears

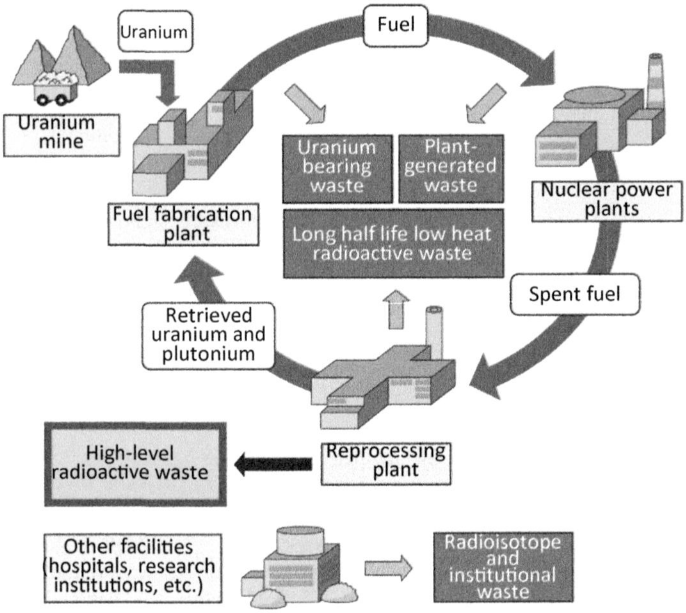

Fig. 11.4 Sources and types of radioactive waste [3]

High-level radioactive waste comprises fission products left after the separation of uranium and plutonium. As mentioned in Sect. 11.1.2, a reprocessing plant produces high-level radioactive waste during the reprocessing of spent fuel.

Other waste, namely the waste produced at nuclear power plants, medical institutions, and research institutions, falls into the category of low-level radioactive waste. Radioactive waste is grouped by source as plant-generated waste, transuranium waste (long-half-life low-heat radioactive waste), uranium-bearing waste, and radioisotope and institutional waste.

11.2.2 Radioactive Waste Disposal Method

Radioactive waste undergoes underground disposal at a depth determined by the radioactivity level. Figure 11.5 shows disposal methods for different types of radioactive waste.

High-level radioactive waste from a reprocessing plant undergoes geological disposal; i.e., disposal in a deep stable geological layer. Low-level radioactive waste is disposed off near the surface or at intermediate depth, considering the radioactivity. Low-level radioactive waste containing long-life transuranium nuclides at levels beyond what is allowable has to be isolated for a long time from the

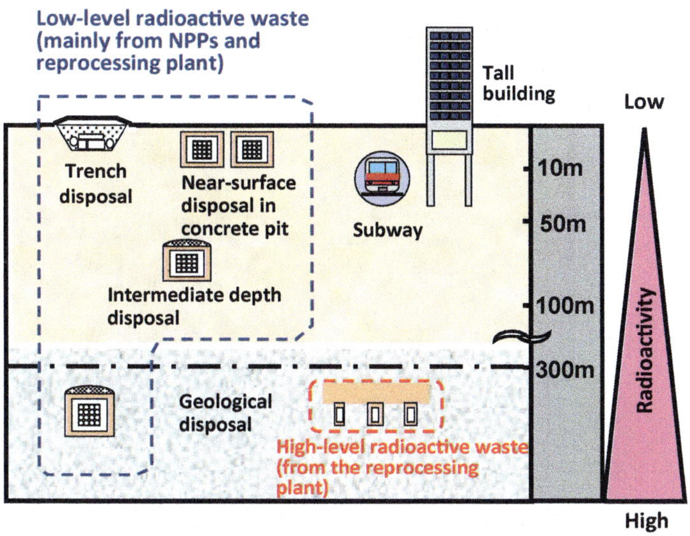

Fig. 11.5 Disposal methods for different types of radioactive waste

biosphere, and therefore undergoes geological disposal like high-level radioactive waste.

11.3 Low-Level Radioactive Waste

Low-level radioactive waste is produced in many forms and is disposed off in a manner that corresponds to its type. This section describes different methods of low-level radioactive waste disposal and discusses the earthquake-resistant design of disposal facilities.

11.3.1 Waste from Nuclear Power Plants

Nuclear power plants produce radioactive waste in the forms of solid, liquid, and gas. Liquid waste and gaseous waste are cleaned of radioactive substances at nuclear power plants and released to the external environment while the radiation is monitored to ensure that it is at a level that will not affect the environment.

Figure 11.6 shows solid waste disposal methods. Waste with a very low concentration of radioactivity such as concrete and metal waste is disposed off near the surface in a trench without an artificial barrier. This is referred to as trench disposal. Low-level radioactive waste such as ash from the incineration of paper and cloth, filter sludge, and used ion exchange resin is disposed off near the surface with a

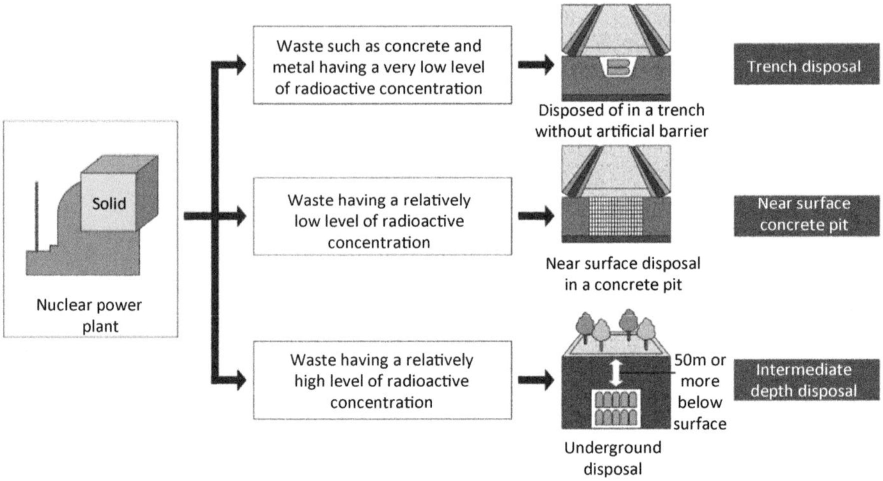

Fig. 11.6 Solid waste disposal methods

concrete barrier structure. This is referred to as near-surface concrete-pit disposal. Waste having a relatively high radioactivity concentration, such as control rods, is disposed off underground at a depth of 50 m or more. This is referred to as intermediate depth disposal, where the term "intermediate depth" means a depth sufficiently greater than the depth of subways and underground building stories.

11.3.2 Trench Disposal

Figure 11.7 shows the trench disposal of very-low-level radioactive waste of concrete and metal produced by the decommissioning of JPDR, the Japan Power Demonstration Reactor that used to be run by the Japan Atomic Energy Agency. This is so far the only example of trench disposal in Japan. Because large quantities of waste destined for trench disposal will arise in the future from the dismantling of reactor facilities, disposal facilities need to be developed in a well-planned manner.

11.3.3 Near-Surface Concrete-Pit Disposal

Figure 11.8 shows an example of near-surface concrete-pit disposal. At this facility in the village of Rokkasho, Aomori Prefecture, drums containing low-level radioactive waste from nuclear power plants all over Japan undergo near-surface disposal. The drums stacked in rows are placed inside concrete structures, and the space between the drums and concrete structures is filled with cement-based

Fig. 11.7 Trench disposal (at the burial disposal demonstration site of the Japan Atomic Energy Agency) [4]

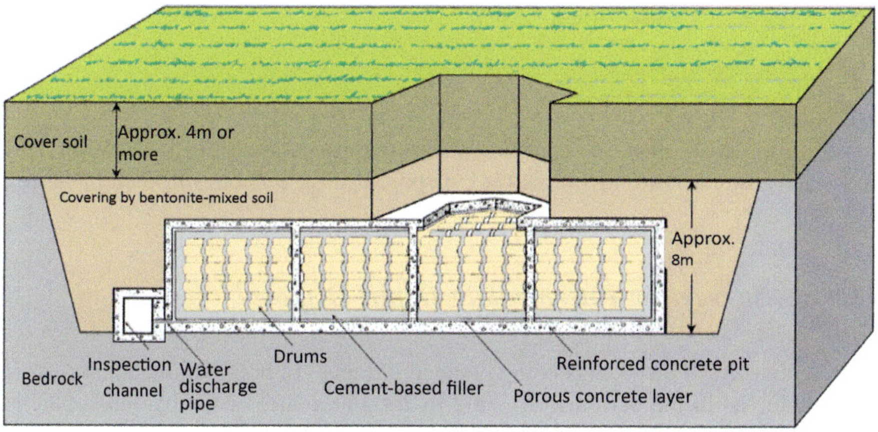

Fig. 11.8 Concept of near-surface concrete-pit disposal [5]

material. The maximum storage capacity is 400,000 drums. At the end of November 2013, the facility stored 260,000 drums.

11.3.4 Intermediate Depth Disposal

Figure 11.9 illustrates the intermediate depth disposal method. Metal components that have been neutron-irradiated while being used in the reactor, such as control rods, channel boxes (of boiling-water reactor fuel assemblies), and burnable poison (in pressurized water control rods), undergo intermediate depth disposal. Since such radioactive waste has relatively high radioactivity, it is disposed off at a depth that is

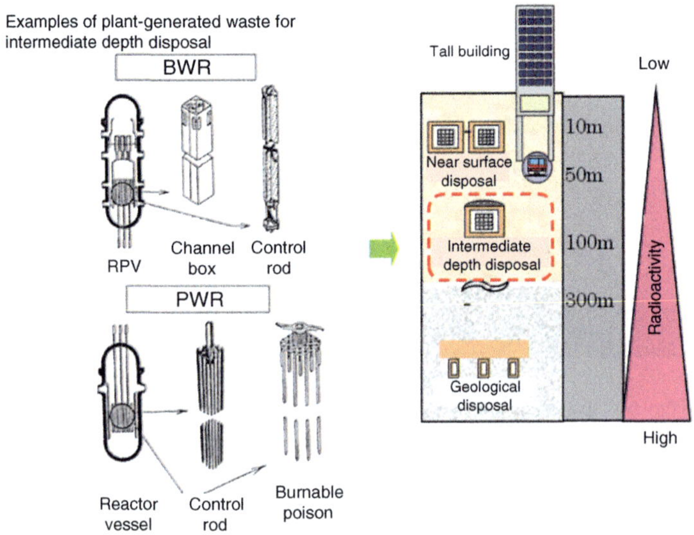

Fig. 11.9 Concept of intermediate depth disposal

sufficiently greater than the depth of underground constructions such as subways and underground building stories (i.e., disposed off at a depth of 50 m or more).

11.3.5 Survey for Intermediate Depth Disposal

Intermediate depth disposal is being planned in Japan. In preparation, a basic survey was conducted in the period from 2002 to 2006 at a site owned by Japan Nuclear Fuel Limited in Rokkasho, Aomori Prefecture.

At the site, located on a plateau (at an elevation of 30–40 m above sea level), the survey was conducted in a tunnel about 100 m below the ground surface. According to the survey results, the horizontal layout and depth of intermediate depth disposal were determined. To evaluate the stability of the test repository, geological, geotechnical, and groundwater surveys were conducted. Figure 11.10 illustrates the survey for intermediate depth disposal.

11.3.6 Earthquake-Resistant Design of Low-Level Radioactive Waste Disposal Facilities

Low-level radioactive waste disposal facilities require an earthquake-resistant design like other nuclear-power-related facilities. Facilities whose contribution to

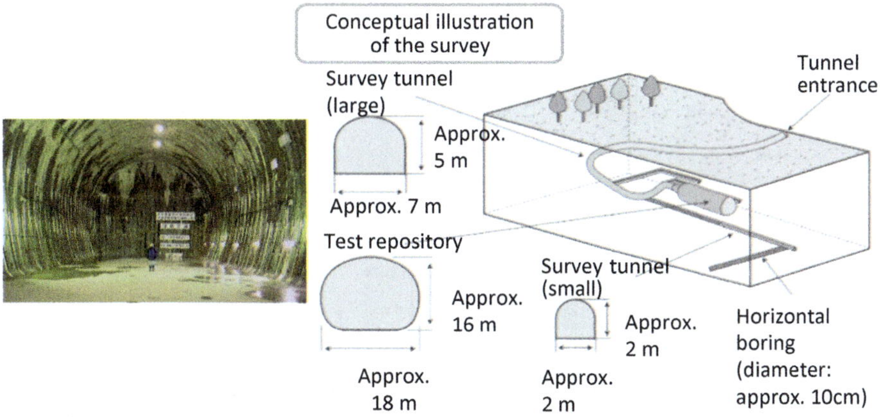

Fig. 11.10 Survey for intermediate depth disposal [6]

the risk of radiological impacts on the public is sufficiently insignificant fall in the Class-C under the seismic classification. Waste burial facilities and waste management buildings (where waste is handled) are Class-C facilities.

11.4 High-Level Radioactive Waste

This section describes the method of geologically disposing of high-level radioactive waste, the geological conditions required for the disposal site, and the seismic safety of geological disposal. In addition, this section gives examples of initiatives relating to geological disposal from Japan and abroad.

11.4.1 Treatment and Disposal of High-Level Radioactive Waste

As shown in Fig. 11.3, spent fuel is shredded and then dissolved using nitric acid. From the solution thus produced, uranium and plutonium are separated as recyclable products. In this process, high-level radioactive waste is produced. It is mainly composed of fission fragments from uranium and plutonium.

High-level radioactive waste is produced in the form of liquid and may therefore splatter easily. It is thus fused with glass material melted at a high temperature and then cooled so that it solidifies in a stainless steel container (Fig. 11.11). Each container is about 1.3 m in height, 150 L in capacity, and 500 kg in weight.

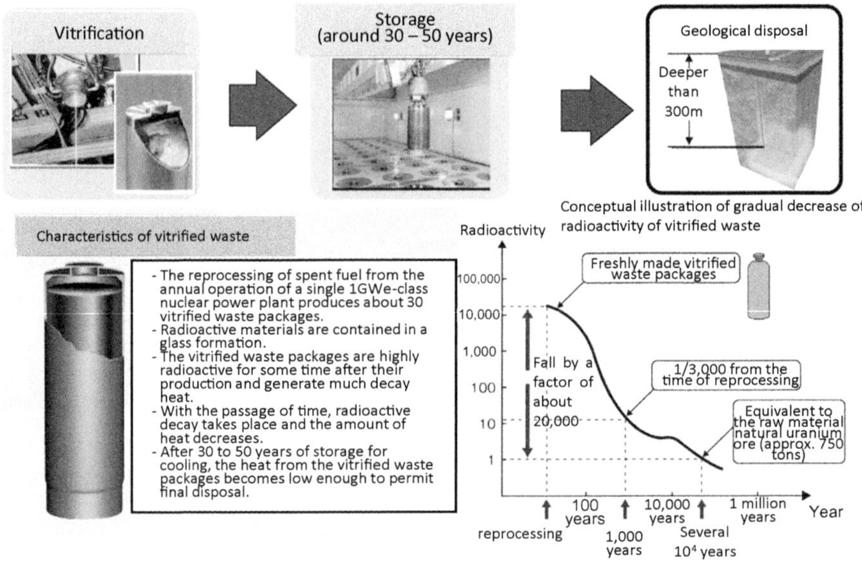

Fig. 11.11 Treatment and disposal of high-level radioactive waste

The vitrified waste takes several tens of thousands years (several 10^4 years) before its level of radioactivity falls to that of naturally occurring radioactive substances. Figure 11.11 shows the level of radioactivity of vitrified waste.

As a way to keep high-level radioactive waste away from the reach of humans for an enormously long period of time, methods such as disposal in space were proposed in the past. However, disposal in a stable geological layer at a great depth has received support internationally as the most realistic and reliable method. In Japan, it is mandatory under law to dispose off high-level radioactive waste geologically at a depth of 300 m or more.

11.4.2 Concept of Geological Disposal

The concept of geological disposal is based on containment by the combination of an engineered barrier and natural barrier (i.e., natural rock formation). The combination of engineered and natural barriers provides a multiple-barrier system. Figure 11.12 illustrates the concept of geological disposal.

The artificial barrier consists of three types of barrier. The first is the vitrified waste itself. Since glass is hardly soluble, it prevents radioactive substances from dissolving into groundwater. The second is a steel container around vitrified waste;

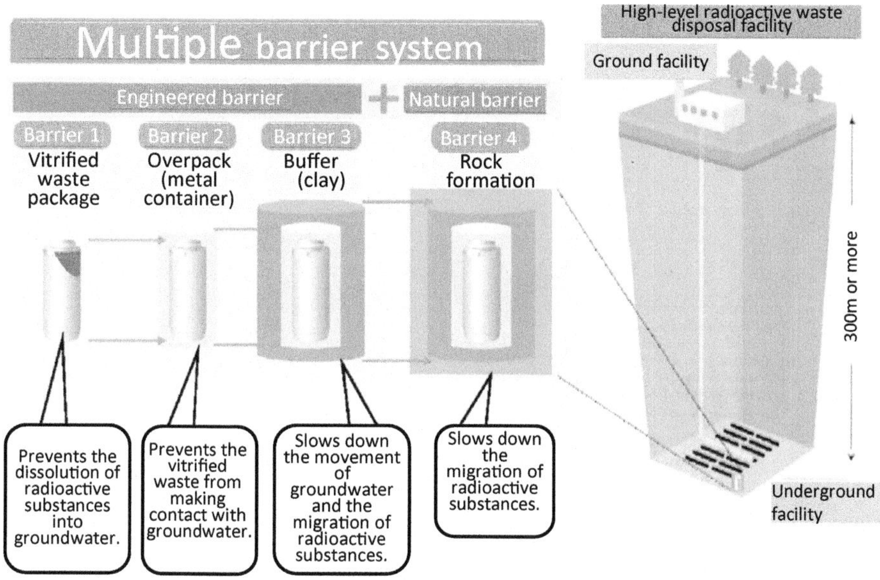

Fig. 11.12 Concept of geological disposal [7]

it is called the overpack and has a wall thickness of about 20 cm. This barrier is capable of preventing the vitrified waste from making contact with groundwater for a period of at least 1,000 years. The third barrier is formed by a buffer layer (of clay), several tens of centimeters in thickness, around the steel containers, which prevents the movement of groundwater and the migration of radioactive substances. A natural rock formation exists around this clay layer. The rock formation provides an environment characterized by slow groundwater movement and the scarcity of oxygen, which are ideal for the stable containment of substances. This forms the natural barrier.

Figure 11.13 shows an example of naturally produced multiple barriers that are functioning effectively. The figure shows a uranium deposit that remained in a geological layer for a period of about 1.3 billion years at a depth of about 450 m in Cigar Lake in Northern Canada. A clay layer formed around the uranium served as a natural barrier, which continued to prevent the uranium from making contact with groundwater. This clay layer, like the buffer layer (also made of clay) of the artificial barrier, was surrounded by a rock formation, forming a multiple-barrier system. This analog demonstrates the feasibility of containing radioactive substances for a long time by means of multiple barriers formed by the combination of an appropriate geological environment and artificial barriers.

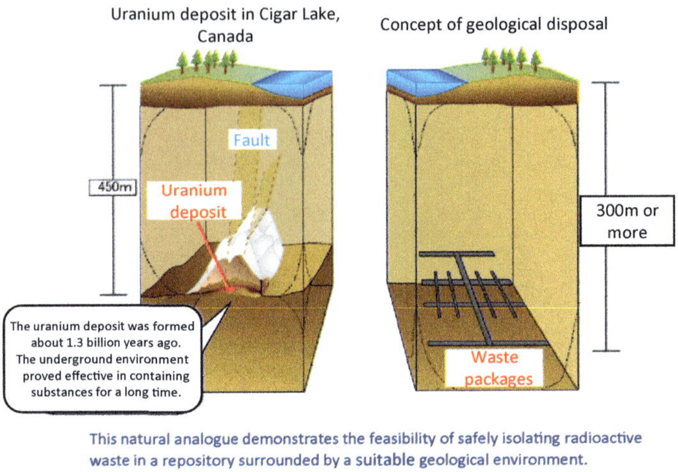

Fig. 11.13 Example of containment by multiple barriers [8]

11.4.3 Geological Conditions Suitable for the Disposal of High-Level Radioactive Waste

A repository for geological disposal is built inside a rock formation at a depth of 300 m or more under the ground surface. Seismic safety is of no concern because the ground motion is less at this depth than at the ground surface and the rock formation is stable.

Issues requiring attention when building a repository for the disposal of high-level radioactive waste are fault and volcanic activities. If an active fault exists across the waste packages on the underground repository, a faulting may break the packages, resulting in the leakage of radioactive substances. Moreover, the activity of a fault near the repository may produce a fracture zone, causing the groundwater to reach the ground surface. This reduces the ability of the natural barrier (i.e., rock formation) to contain radioactive substances.

If a volcano exists near the repository, its activity may destroy the repository. Magma may heat the underground rock formation, reducing the ability of artificial and natural barriers to contain radioactive substances. Considering these risks, a candidate site for repository construction should be at least 15 km from quaternary volcanos and there should be no active fault within the site area.

Figure 11.14 illustrates the disposal site selection process. The process begins with a literature search, which is followed by a geological survey conducted through boring exploration. In the final stage of the site selection process, a test repository (an underground facility for geological survey) is constructed to verity the availability of a suitable geological environment. Japan still does not have any statutory guidelines concerning the earthquake-resistant design of a repository for

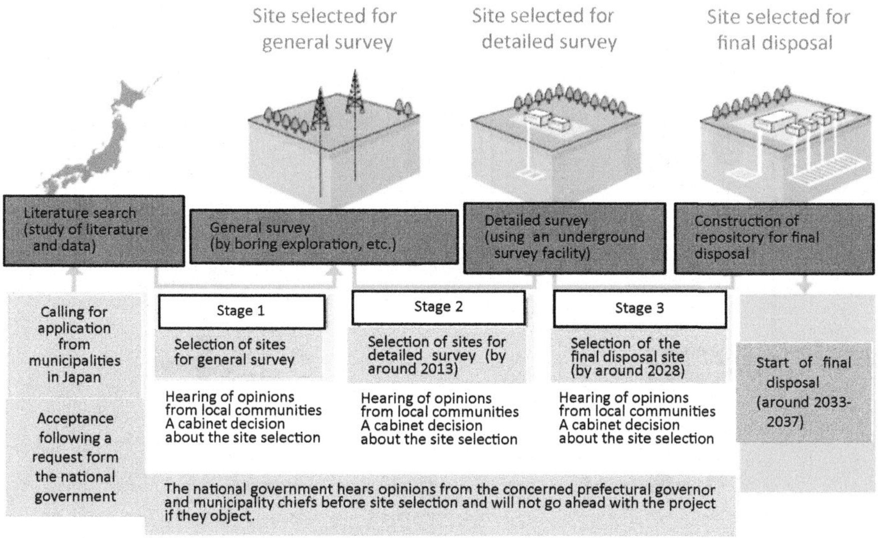

Fig. 11.14 Disposal site selection process [1]

the disposal of high-level radioactive waste. However, since the handling of high-level radioactive waste takes place, the ground facilities for waste handling and temporary storage require Class-S earthquake-resistant design. In general, earthquake-resistant design is deemed unnecessary for underground disposal facilities because earthquakes are unlikely to cause the release of radioactive substances from inside the waste packages.

11.4.4 History of Progress in Japan Toward the Construction of a Repository for the Disposal of High-Level Radioactive Waste and Situations in Other Countries

In 2000, Japan established a law concerning the disposal of high-level radioactive waste in the hope of realizing the disposal of high-level radioactive waste. On the basis of this law, a private organization for project implementation, called the Nuclear Waste Management Organization of Japan (NUMO), was established. In 2002, NUMO called for applications from municipalities in Japan that are interesting in hosting a repository. Toyo-cho, Kochi Prefecture, responded to the call, but subsequently canceled their application after the reelection of the mayor. From then until today (November 2014), no municipality has responded to the call. Thus, selecting the site for repository construction is difficult in Japan.

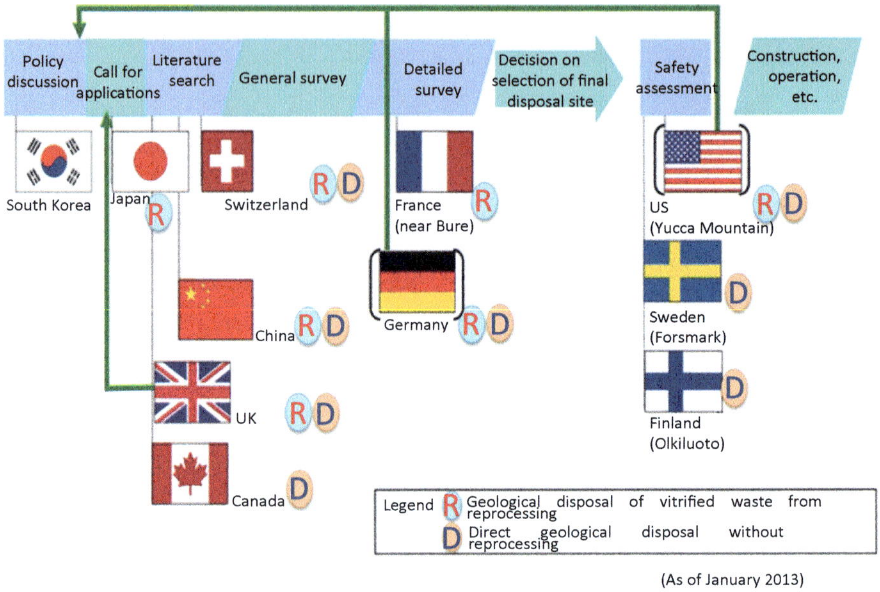

Fig. 11.15 State of progress in other countries [7]

Other countries share a similar difficulty in siting a repository for the disposal of high-level radioactive waste. Figure 11.15 shows the state of progress in other countries.

In Finland, a site for repository construction has been decided, and a survey tunnel, which later will be a part of the repository, is being excavated. In the United States, a site in Yucca Mountain, Nevada, had been selected for repository construction, but the United States government changed policy and canceled the project in 2009. In Sweden, a site in Östhammar has been selected, and an application for a construction license has been made. In France, a detailed survey is being conducted at a site near Bure, Meuse. In Switzerland, a project implementing organization has announced three candidate sites. In Great Britain, which calls for applications from municipalities like in Japan, two municipalities have responded to the call but both have canceled their applications owing to objection from the local assembly.

11.5 Clearance

This section describes the approach taken for the disposal of waste having radioactivity that is so minor that the waste may be treated as typical industrial waste rather than radioactive waste, and outlines the process involved in the clearance of such waste.

11.5.1 Clearance and Clearance Level

The radioactive waste from nuclear power plants includes a significant quantity of waste that is only slightly radioactive. It is considered acceptable to treat this slightly radioactive waste like typical industrial waste.

A material having such a low concentration of radioactive substances that its impact on the health of people can be ignored is not treated as radioactive material. This exclusion is referred to as clearance. The radiation level that serves as the criterion for clearance is referred to as the clearance level. Internationally, a level of 0.01 mSv/y, which is less than 1/200 of the background radiation level (2.4 mSv/y), is accepted as the clearance level.

Waste having a radioactivity concentration below the clearance level is treated similarly to typical industrial waste. That is to say, it can be disposed off at normal waste disposal facilities or otherwise recycled. Clearance requires the approval of and verification by the national government. Figure 11.16 illustrates the clearance process and the roles of the national government and waste producers.

11.5.2 Example of the Quantity of Decommissioning Waste that May Pass Clearance

The decommissioning of a nuclear power plant produces a very large quantity of radioactive waste. According to an example of an actual decommissioning project, the following gives an idea of how much may pass clearance.

Figure 11.17 shows the quantity of waste from the decommissioning of the Tokai Power Station of Japan Atomic Power Company that has passed clearance. The total quantity of decommissioning waste amounted to about 200,000 tons. About 21 % of this quantity passed clearance because of its very low radioactivity. Additionally, a major proportion (approximately 67 %) turned out to be nonradioactive. Such waste is free from the adhesion of radioactive substances or secondary contamination by the penetration of radioactive liquid. The waste of these

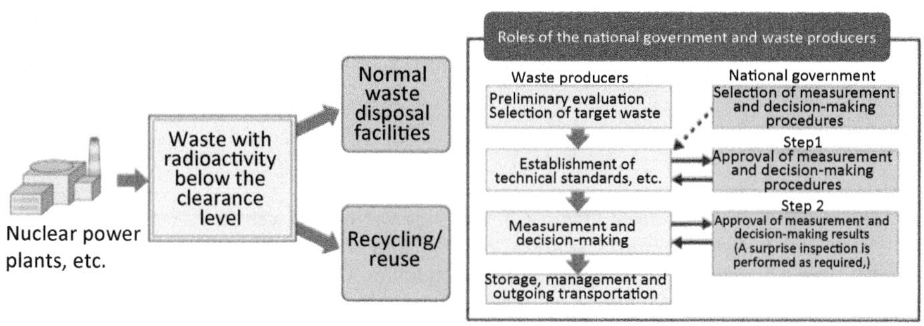

Fig. 11.16 Clearance process and the roles of the national government and waste producers [5]

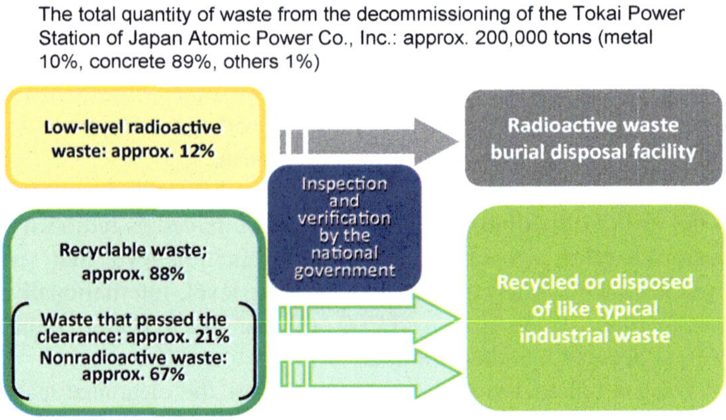

Fig. 11.17 Example of the quantity of decommissioning waste that may pass clearance

types can be recycled or disposed at normal waste disposal facilities like typical industrial waste. The disposal of cleared waste is controlled by the national government by means of verification and inspection processes.

References

1. "Graphical Flip-chart of Nuclear & Energy Related Topics 2013", The Federation of Electric Power Companies of Japan, 2013
2. "*Jishin, Jishindou Hyouka Iinkai oyobi Shisetsu Kenzensei Hyouka Iinkai Dai 13 kai Working Group 4* (The 13th Working Group 4 of Earthquake and Ground Motion Evaluation Committee and Facility Soundness Evaluation Committee)", (Material No. WG4-13-1, WG4-13-2), Nuclear Safety Commission, December 10, 2009
3. Nuclear Regulation Agency website
4. Japan Atomic Energy Agency website
5. "Graphical Flip-chart of Nuclear & Energy Related Topics 2008", The Federation of Electric Power Companies of Japan, 2008
6. Japan Nuclear Fuel Limited website
7. "*Genshiryoku Seisaku wo meguru Saikin no Doukou* (Resent Status of Nuclear Policies)", The second meeting of the Coordination Subcommittee of the Advisory Committee for Natural Resources and Energy (April 23, 2013), Material 4, 2013
8. Website of Agency for Natural Resources and Energy, Ministry of Economy, Trade and Industry

Further Readings

9. Nuclear Waste Management Organization of Japan (NUMO) website
10. "*Dainishu Haikibutsu Maisetsushisetsu no Ichi, Kouzou oyobi Setsubi no Kijun ni kansuru Kisoku no Kaishaku* (Interpretation of Rules regarding Location, Structure and Equipment of LLW and MLW Disposal Facilities)", Decided on November 27, 2013, Nuclear/Control/Waste No. 1311277, NRA's Decision

Chapter 12
Future Technology for the Seismic Safety of Nuclear Power Facilities

Satsuya Soda, Masanori Hamada and Michiya Kuno

Abstract Because earthquakes occur frequently in Japan, attention has been given to seismic safety in the design and construction of Japanese nuclear power facilities. A serious accident at the Fukushima Dai-ichi Nuclear Power Plant of the Tokyo Electric Power Company that was caused by the 2011 Great East Japan Earthquake prompted us to review measures to ensure structural safety against future earthquakes and tsunamis. In recent years, progress has been made in investigations and research on the earthquake-resistant design of nuclear power plants. This has contributed to the development of various new technologies based on the new findings. As examples of future technologies for the seismic safety of nuclear power facilities, this chapter describes structure control technologies for nuclear power facilities and technologies to allow for greater diversity in the siting of nuclear power plants.

Keywords Strong earthquake motion · Base isolation · Damper · Siting of nuclear power plant · Quaternary layer siting · Underground nuclear power plant · Offshore nuclear power plant

12.1 Structure Control Technologies for Nuclear Power Facility

This section outlines the principles and characteristics of the seismic isolation and seismic response control techniques that have begun to be applied to nuclear power facilities to improve seismic safety.

S. Soda (✉) · M. Hamada
Waseda University, Shinjuku, Tokyo, Japan
e-mail: satsuya.soda@waseda.jp

M. Kuno
Chubu Electric Power Company, Nagoya, Japan

M. Hamada and M. Kuno (eds.), *Earthquake Engineering for Nuclear Facilities*,
DOI 10.1007/978-981-10-2516-7_12

12.1.1 Improvement in Protection Against Large-Amplitude Earthquakes

Table 12.1 shows the seismic ground motion properties of maximum acceleration and velocity observed in various parts of Japan over 15 years since the 1995 Kobe Earthquake. According to the Building Standard Act of Japan, specific buildings, such as buildings taller than 60 m, need to have their seismic safety verified by response analysis. A rule established by the Ministry of Land, Infrastructure, Transport, and Tourism requires an assumption of 50 cm/s in the evaluation of ultimate safety as the maximum velocity of "earthquake ground motion that can appear in very rare cases." Table 12.1 shows that the maximum ground motion velocity exceeded 100 cm/s in a number of cases.

To resist such large-amplitude ground motions that have been observed in recent years, it is not impossible to try to maintain the structural integrity of buildings by increasing the structural strength and/or deformability. However, any attempt to increase the structural strength may result in an increase in construction costs and an acceleration response. This may lead to negative consequences such as a higher stress to the substructure and a higher risk of secondary hazards from the displacement, toppling, and falling of equipment attached to the building and objects inside the building. When trying to improve the seismic safety of typical existing buildings, modifications that result in a higher stress to the substructure are usually unfeasible because of space restrictions and unfavorable in terms of cost. Added excessive deformability may trigger unexpected behaviors of structural and non-structural building components. Buildings at nuclear power facilities should be prevented from making excessive dynamic responses to ensure the safety of facilities that exist inside them. To achieve general improvements in seismic safety of conventional buildings and nuclear power facilities, it is advisable to implement seismic isolation and seismic response control as new structural design concepts.

The following are some basic findings from earthquake resistance engineering and vibration theories:

(1) In ground motion from each earthquake, a certain vibration period is dominant, even though this period may vary depending on the seismogenic mechanism.
(2) A proximity between the natural period of structures and the dominant period in the ground motion results in significant vibration of the structures.
(3) The vibration of structures is damped by the absorption of input energy by the structures themselves.
(4) The force acted on structures as a result of an earthquake is determined by multiplying the structural acceleration by the structure mass. An excessively high acceleration is unfavorable because it increases the stress acted on the structures, and may cause falling and toppling of equipments attached inside buildings.

Table 12.1 Large-amplitude seismic ground motion observed in various parts of Japan (from 1995 to 2011)

Year	Earthquakes	Monitoring point	Direction	Max. acceleration (cm/s^2)	Max. velocity (cm/s)
2011	Tohoku Earthquake	K-NET Tsukidate	NS	2,700.0	116.6
		K-NET Furukawa	EW	571.5	87.8
		K-NET Sendai	NS	1,517.0	84.3
		K-NET Haga	EW	1,197.0	77.8
2008	Iwate/Miyagi Inland Earthquake	K-NET Naruko	NS	440.2	78.6
2007	Niigata-ken Chuetsu-oki Earthquake	K-NET Kashiwazaki	EW	513.6	79.9
		K-NET Kashiwazaki	NS	667.0	129.0
2007	Noto Peninsula Earthquake	K-NET Anamizu	EW	781.7	99.0
		JMA Wajima	EW	438.8	76.6
		JMA Wajima	NS	463.6	98.2
2004	Niigata-ken Chuetsu Earthquake	JMA Kawaguchi	EW	1,676.0	120.0
		K-NET Ojiya	EW	1,308.0	130.0
		K-NET Ojiya	NS	1,147.0	96.0
		JMA Takezawa	EW	721.8	88.7
		JMA Takezawa	NS	538.4	101.0
		K-NET Nagaoka Branch	EW	705.9	120.0
		K-NET Nagaoka Branch	NS	870.4	111.0
2003	Tokachi-oki Earthquake	K-NET Chokubetsu	EW	785.0	112.0
1995	Kobe Earthquake	JMA Kobe	EW	619.2	72.3
		JMA Kobe	NS	820.6	82.8
		JR Takatori station	EW	666.2	128.0
		JR Takatori station	NS	641.7	134.0
		Osaka Gas Fukiai	NS	810.1	125.9

K-NET strong quake monitoring network (National Research Institute for Earthquake Science and Disaster Prevention); *JMA* Japan Meteorological Agency's strong quake monitoring point; *JR* Japan Railway

As shown by the acceleration response spectrum in Fig. 14.5 (Chap. 14, Part II), the dominant period in earthquake ground motion is generally shorter than and 1.0 s in most cases. Low- and medium rise buildings have a natural period that corresponds approximately to this range, and therefore, they are shaken excessively. Skyscrapers are not shaken as hard by the same ground motion because their natural period range from 2 to 5 s usually does not match the dominant period in ground motion.

However, strong earthquake ground motions contain many long-period components, which may match the natural period of skyscrapers. This may result in their large vibration and excessive deformation. It is possible to reduce the dynamic response by increasing the damping property of the building. Low- and medium rise buildings have a short natural period. After inserting a seismic isolation device (material) between the ground and its foundation, which is markedly less rigid than the superstructure, the natural oscillation period of the entire building increases and the dynamic response of the superstructure is reduced. However, to prevent the seismic isolation layer deformation from becoming excessively large, various energy absorbing devices (dampers) are used as a basic rule to this approach.

12.1.2 Seismic Isolation

A typical seismic isolation technique can be applied so successfully that it may reduce the response acceleration at the buildings' lowest story level by a factor of three to five compared with the acceleration at ground level. This is possible without causing excessive deformation to the seismic isolation device by prolonging the natural period of the buildings and adding a damping capacity to the buildings.

By using laminated rubber bearings, such as the one shown in Fig. 12.1, it is possible to make the buildings above it (termed the "superstructure") virtually immune to short period earthquake ground motion, by prolonging its natural period to $\sim$4 s. As shown in Fig. 12.1b, laminated natural rubber bearings show an almost linear load–displacement relationship, and have no energy-absorbing capacity. Therefore, they are often used with energy-absorbing dampers such as those in Figs. 12.2 and 12.3. Besides dampers of these types, oil dampers that are used extensively in other industries can also be applied to buildings. In recent years, such a device as that in Fig. 12.4 has been characterized by the bearing of mat slab, which serves as a foundation of the superstructure, and exists on top of a concrete pile with connection via a friction material.

Seismic isolation is an excellent technique for improving seismic safety, and it can be applied to existing structures as demonstrated by its use in the Tokyo railway station building. However, this technique is not suitable for use in all applications. For example, it may not be suitable for use in structures built on soft ground where dominant periods in earthquake ground motion tend to be longer. Furthermore, such ground motion can produce a large overturning moment that may threaten the safe application of the seismic isolation device because of the tensile force.

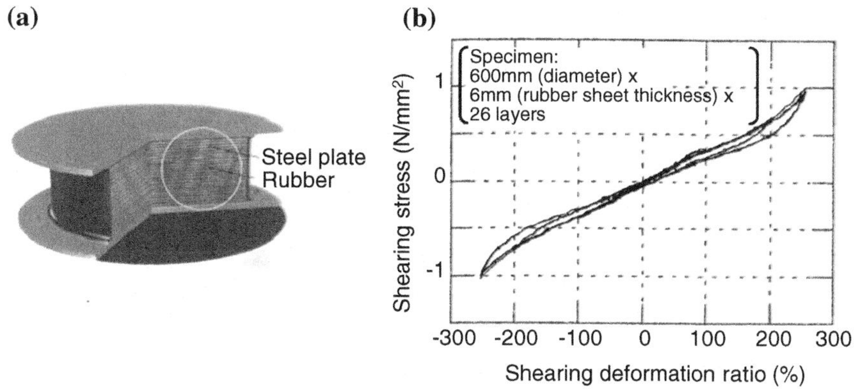

Fig. 12.1 Laminated natural rubber bearing and its load–deformation relationship. **a** Configuration of laminated natural rubber bearing. **b** Load–displacement curve of laminated natural rubber bearing [1, 2]

Fig. 12.2 Lead damper [2]

In most cases, the seismic isolation technique is applied to address horizontal ground motion. However, a three-dimensional isolation system, which is capable of addressing ground motion in the horizontal and vertical directions, has also been developed and applied.

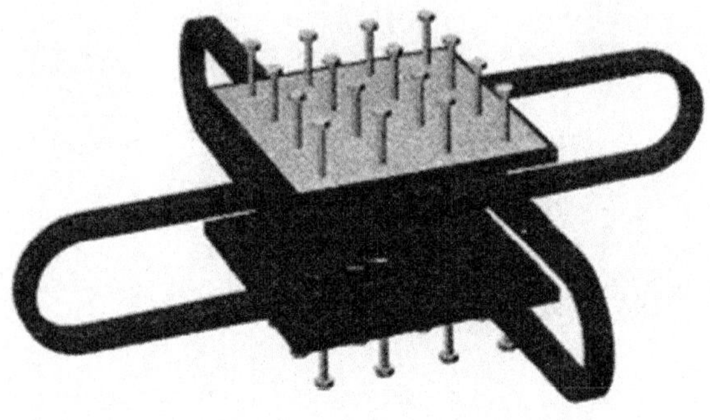

Fig. 12.3 Steel bar damper [2]

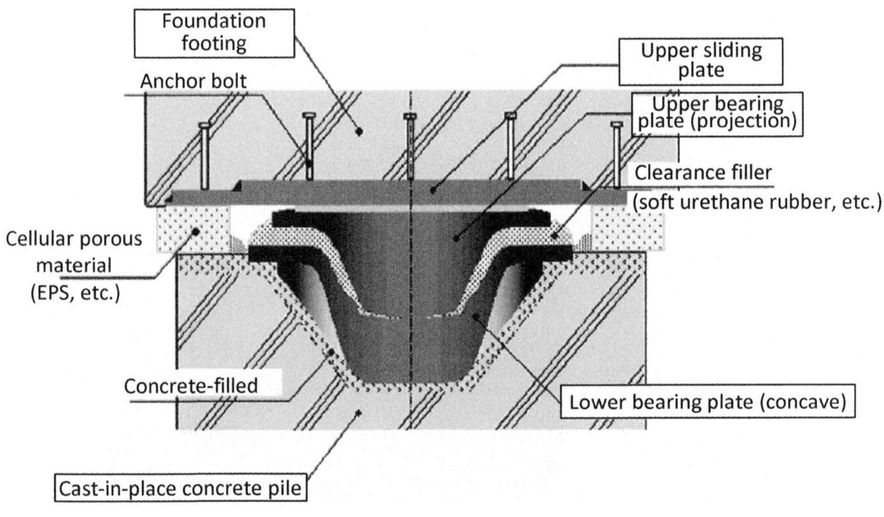

Fig. 12.4 Rotary slide bearing (BSL) [3]

12.1.3 Seismic Vibration Control

In the conventional approach to structural design, it is often difficult to design a building with high seismic safety by reducing the deformation and acceleration of the structure simultaneously; an attempt to reduce the response deformation by increasing the rigidity through expanding the dimensions of the structural members results in an increase in the response acceleration. Therefore, a new technique, termed seismic damping, has been developed. In this technique, the target structure is provided with structural components with a particularly high hysteresis capacity (i.e., dampers) apart from the main structural members that bear the vertical loads.

Dampers absorb input energy from the ground motion by their damping capability to reduce swaying of the structure; this makes the main structural components less prone to damage. A structure that uses this technique is termed a "(seismic) damper protected structure."

Seismic dampers exist as three types: active type (with an active control mechanism), passive type (without any active control mechanism), and semi-active type (with a simple active control mechanism). Active dampers require an external energy supply for active control, whereas passive dampers do not. Dampers are classified into the following types according to the style of installation: additional mass type, inter-story type, inter-building type, and distributed installation type. Figure 12.5 shows the damping system types.

An additional mass-type damping system includes a "mass" that is usually placed on top of the target structure, and decreases the dynamic response of the structure using a force that acts between the mass and the structure. The active-tuned mass damper (TMD) is an example of this type of damping system. An active mass driver amplifies the motion of the mass in the TMD to increase its effectiveness. Figure 12.6 illustrates the concept of an active TMD. The photograph on the left shows a scene from shaking table testing of a model that represents TMD installation in the attic of a residential house.

Inter-story dampers are installed between building stories so that they absorb energy as they deform with the building in response to ground motion, and thereby protect the building from damage. Specific examples of this type of damping system include curved steel dampers and oil dampers.

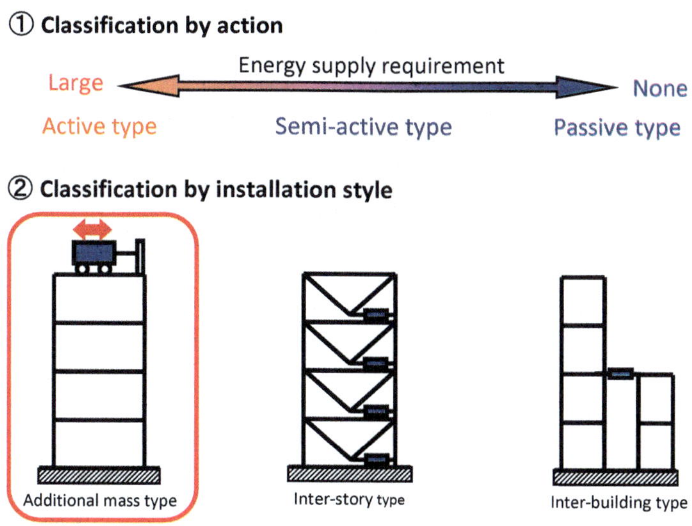

Fig. 12.5 Damping system types

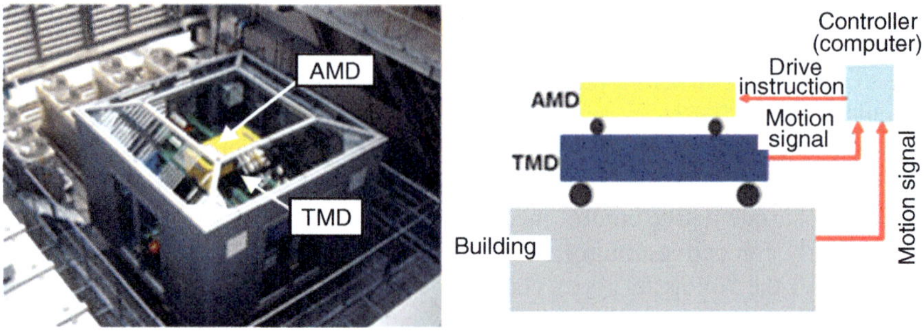

Fig. 12.6 Concept of active-tuned mass damper [4]

Fig. 12.7 Example of curved steel damper installation [4]

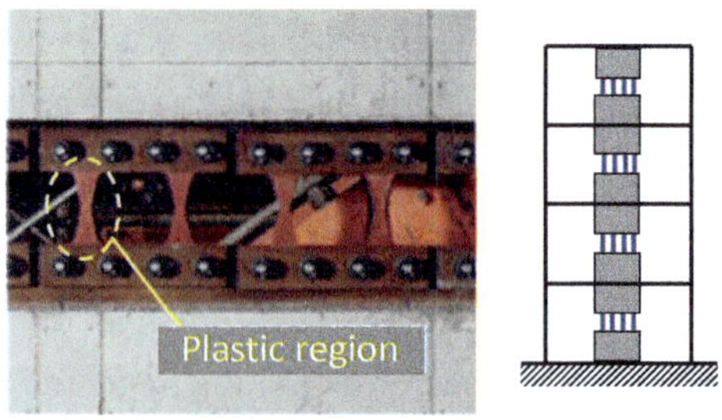

Fig. 12.8 Example of oil damper installation [4]

Curved steel dampers have elastoplastic characteristic that are designed to yield faster than main structural components at a smaller deformation. Figure 12.7 shows an example of how curved steel dampers may be installed.

The oil damper is designed to damp the dynamic response as its piston follows the inter-story deformation with a certain degree of resistance produced by the viscosity of oil that is pushed by the piston through the damping adjustment valve inside the damper. Figure 12.8 provides an example of how an oil damper may be installed.

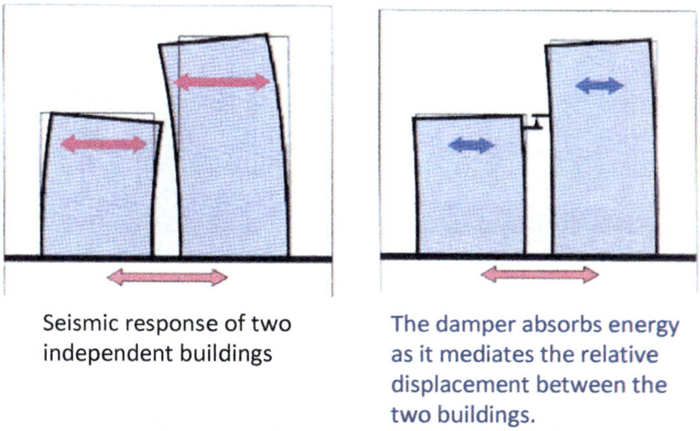

Fig. 12.9 Concept of inter-building damping system

An inter-building damper is installed between two buildings and structures. The damper mediates the relative displacement between the two buildings and structures, which arises from a difference in the natural period of vibration of the buildings to reduce the dynamic response. Figure 12.9 illustrates this concept of the inter-building damping system.

12.1.4 Examples of Application to Nuclear Power Facility

The seismic isolation technique has been used in the Koeberg Nuclear Power Plant in South Africa and the Cruas Nuclear Power Plant in France. Figure 12.10a shows the construction of the Koeberg Nuclear Power Plant, while Fig. 12.10b shows the construction of the Cruas Nuclear Power Plant.

The Koeberg Nuclear Power Plant (Fig. 12.10a) uses a seismic isolation system to reduce earthquake motion. The two reactor buildings and the auxiliary building are placed on the same foundation and are supported with laminated rubber bearings and slide bearings. The Cruas Nuclear Power Plant (Fig. 12.10b) uses seismic isolation for the reactor building. The entire reactor building structure is supported with laminated rubber bearings. Both plants use the seismic isolation technique to enable the building of a plant according to the standard plant model of Électricité de France without modification but with the assurance of seismic safety in a region of high seismic activity.

An example of using the seismic damping technique at a nuclear power plant is presented by a reinforcement of exhaust stacks at a nuclear power plant, as mentioned in Chap. 10 (Fig. 10.12). Moreover, oil dampers and mechanical dampers are used to decrease the dynamic response of equipment and piping.

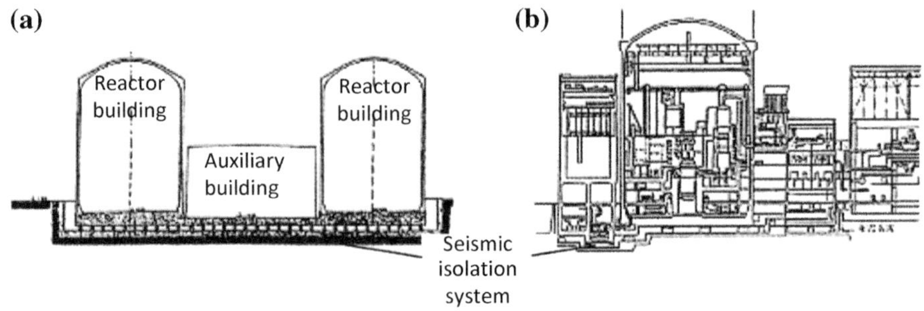

Fig. 12.10 Koeberg Nuclear Power Plant and Cruas Nuclear Power Plant [5]

12.1.5 Application to Next-Generation Light Water Reactor

Based on the accident that occurred at the Fukushima Dai-ichi Nuclear Power Plant of the Tokyo Electric Power Company in 2011, the Japanese government formed a view that "nuclear power should be promoted while ensuring completeness in the management of its risks so that its adverse impacts on the life of people may be kept sufficiently small." The Japanese government therefore demanded a concentration of effort in the development of basic or core technologies and science that may contribute to the robustness of nuclear power facilities against natural disasters and the enhancement of severe accident management measures, including an improvement of their reliability.

As of fiscal 2013, research and development initiatives described below regarding seismic isolation techniques are underway based on outputs from the next-generation light water reactor development program and aimed at advancements in plant safety measures:

- Assuming wide usage of lead–rubber bearings for seismic isolation nuclear power facilities, efforts should be made to develop a technique to design seismic isolation systems for nuclear power facilities, which should allow for the simultaneous consideration of input ground motion in the horizontal and vertical directions; a technique to determine the safety margin provided by a seismic isolation system (in terms of the maximum bearable load); and a technique to evaluate the seismic safety of pipes, including high-temperature and high-pressure pipes that run from building to building.
- The impact of long-lasting long-period motions, such as that produced by the 2011 Great East Japan Earthquake, should be considered in the development of the seismic isolation system. From the viewpoint of environmental protection, the feasibility of substituting lead–rubber bearings with tin–rubber bearings should be studied.

12.2 Techniques for Allowing Greater Diversity in the Siting of Nuclear Power Plants

For a site to be deemed appropriate for the construction of a nuclear power plant, it must satisfy three conditions: the availability of a vast space, the availability of good geotechnical conditions, and the availability of large quantities of water for cooling. When designing a nuclear power plant, particular attention must be given to the fact that the plant is so heavy that stress applied to foundation ground may amount to several N/mm^2. A nuclear power plant, therefore, has to be built at an available site where the foundation ground has sufficient strength and rigidity. Because of such restrictions, previous nuclear power plants have been constructed mainly in coastal areas where large quantities of cooling water and strong and rigid rock formations are available. However, because of the importance of allowing for a greater diversity in the siting of nuclear power plants, studies regarding the feasibility of siting a nuclear power plant differently (such as underground, inland, and offshore) have been initiated and continue today. Table 12.2 lists alternative siting options for nuclear power plants.

Siting on a Quaternary layer (1) implies siting on a geological layer formed in the Quaternary period, that is to say, a geological period that began after the Tertiary period and continues today.

Table 12.2 Overview of alternative siting options for nuclear power plants [6]

Siting style	Type	Overview
(1) Siting on a quaternary layer	–	Siting on a geological layer formed in the quaternary period (from about 1.8 million years ago up to today) Japan: two examples of siting a research reactor on a quaternary layer World: many examples in Europe and the United States
(2) Underground siting	Vertical type • Partial underground construction • Full underground construction Horizontal Type • Partial underground construction • Full underground construction	Excavating into a hard rock formation, an underground space is created for the installation of major plant facilities Japan: no example World: six examples of siting a research or commercial reactor underground (Five out of the six have been disused and decommissioned.)
(3) Offshore siting	Man-made island type • Reclamation • Caisson-type Floating type: • Buoy type • Submerged type	Offshore siting on a man-made island or a floating body Japan: no example World: under construction in Russia

Since the Examination Guide for Seismic Design of Nuclear Power Reactor Facilities (1981) stated that “important buildings and structures should be supported by a rock formation,” it has been a practice in Japan to construct reactor facilities on a rigid rock formation that dates to the Tertiary period or earlier (Appendix 12.1). The 2006 revised edition of the Examination Guide, however, states that it should be possible to ensure seismic safety by building facilities on ground with a load-bearing capacity that is sufficiently large compared with the design load level. The new regulation standards issued by the Nuclear Regulation Authority (NRA) in July 2013 state that the ground on which reactor facilities are constructed should have a sufficiently large load-bearing capacity.

Underground siting (2) means excavating into a hard rock formation to create an underground space for the installation of major plant facilities (Appendix 12.2).

Offshore siting (3) means the siting of a nuclear power plant either on a man-made island built on the seabed or on a floating body (Appendix 12.3).

Each siting option has its advantages and disadvantages. To realize these alternative siting options, risks associated with each option and measures should be studied in detail, and efforts should be made to reduce the risks.

Table 12.3 summarizes the advantages and disadvantages of the alternative siting options.

Table 12.3 Advantages and disadvantages of alternative siting options [6]

Siting style	Advantages	Disadvantages
(1) Siting on a quaternary layer	• Proximity to the electricity demand area • Construction cost and construction period similar to conventional siting	• Siting difficulty due to societal restrictions • Requirement of huge foundation structure • Unavailability of technical guidelines concerning the load-bearing capacity and settlement of the foundation ground
(2) Underground siting	• Preservation of the original landscape • Reduction of seismic force • Possible simplification of engineered structure by reliance on the strength of rock formation around the underground space	• Higher construction cost and longer construction period • Unavailability of technical guidelines concerning the stability of underground space and the evaluation of groundwater flow • Many restrictions concerning building/facility layout, construction and maintenance
(3) Offshore siting	• Scarcity of societal restrictions • Proximity to the electricity demand area • Smallness of impact on land use	• Higher construction cost and longer construction period • Measures against sea wave and tsunami • Measures against marine traffic • Requirement of R&D for the establishment of a new system of safety assessment and performance evaluation • Risk of hazard escalation in the case of accident

Siting on a Quaternary layer has the advantage of facilitating the choice of a site located close to the electricity demand area. Siting close to a densely populated demand area; however, is often difficult because of safety and societal restrictions. Since the Quaternary layer is relatively poor in strength and rigidity, the foundation has to be placed deeply into firm ground.

The advantages of underground siting include the possibility of preserving the original landscape by building most facilities underground and the possibility of reducing the seismic load acting on structures *t*. Disadvantages include high construction costs and operational restrictions.

Offshore siting has advantages of space utilization and disadvantages include its high construction cost, the need for sea wave and tsunami protection, and access difficulty in an emergency.

Appendix 12.1: Siting on Quaternary Layer

Figure 12.11 shows an example of a geological structure. The Quaternary layer is composed of the alluvium and diluvium layers. Alluvium is the youngest geological layer formed by sedimentation from 10,000 years ago to date. It is a relatively weak layer that is composed of sand and silt layers, and is spread widely over plain areas of Japan. The diluvium layer is a geological layer that is formed between 1.8 million and 10,000 years ago. Composed of sand and clay layers, it often forms a tightly compacted, rigid ground. For siting on a Quarterly layer, a site covered by sedimentary soft rock, hard cohesive soil or gravel should be chosen.

There are many examples of commercial and demonstration reactors built on a Quaternary layer in the world. In the United States, almost half of the nuclear power plants are built on a Quaternary layer. Many similar examples exist in Europe, the Middle East, and South Africa. In Japan, no reactor is built on a Quaternary layer. One of the reasons for this is the need in Japan to ensure safety against a large seismic force.

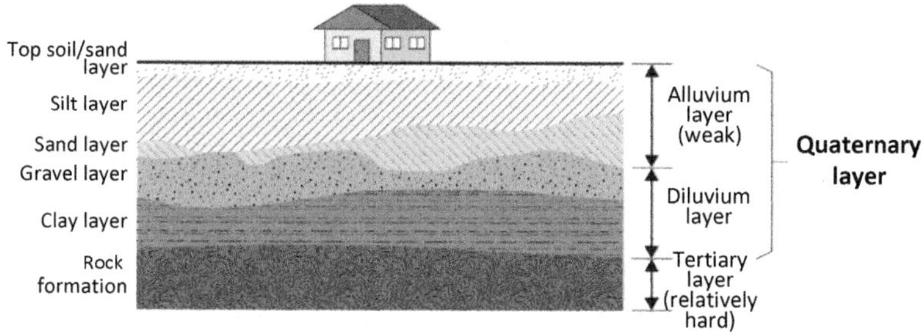

Fig. 12.11 Example of geological structure

Appendix 12.2: Underground Siting

By excavating into a hard rock formation, an underground space is created for the installation of major nuclear plant facilities. There are several types depending on underground space geometry (vertical or horizontal type) and construction style (partially underground or completely underground). The most appropriate design and style are chosen while considering site characteristics. If the local rock formation has enough strength and rigidity, a horizontal design is preferred; if not, a vertical design is preferred. If the site exits on steep terrain, full underground construction is preferred; if the site exists on flat or nearly flat ground, partially underground construction is preferred.

Irrespective of these choices in style and construction type, underground siting has the advantage of facilitating the preservation of the original landscape at ground level. In the world, six reactors in total, including research and commercial reactors, have been constructed underground. Five of the six reactors, however, have been disused and decommissioned. Figure 12.12 shows the type of underground siting of nuclear power plants. Figure 12.13 shows an example of the underground nuclear power plant layout.

Appendix 12.3: Offshore Siting

Offshore siting means the construction of a nuclear power plant in a sea area. The nuclear power plant is constructed either on a man-made island built on the seabed or on a floating body. Figure 12.14 shows different types for the offshore siting. There are two types of construction of man-made islands: reclamation and caisson-type man-made island. Figure 12.15 shows the concept of a reclaimed man-made island, whereas Figure 12.16 shows the concept of a floating body.

Japan has no experience of offshore nuclear power plants. However, abundant experience exists in Japan in the construction and operation of non-nuclear offshore facilities such as fossil-fired power plants on man-made islands and liquefied natural gas terminals on reclaimed grounds in bays. Because offshore facilities in Japan can be subjected to large tsunamis, the feasibility of offshore siting has to be determined after evaluating the risk of tsunamis at the proposed site.

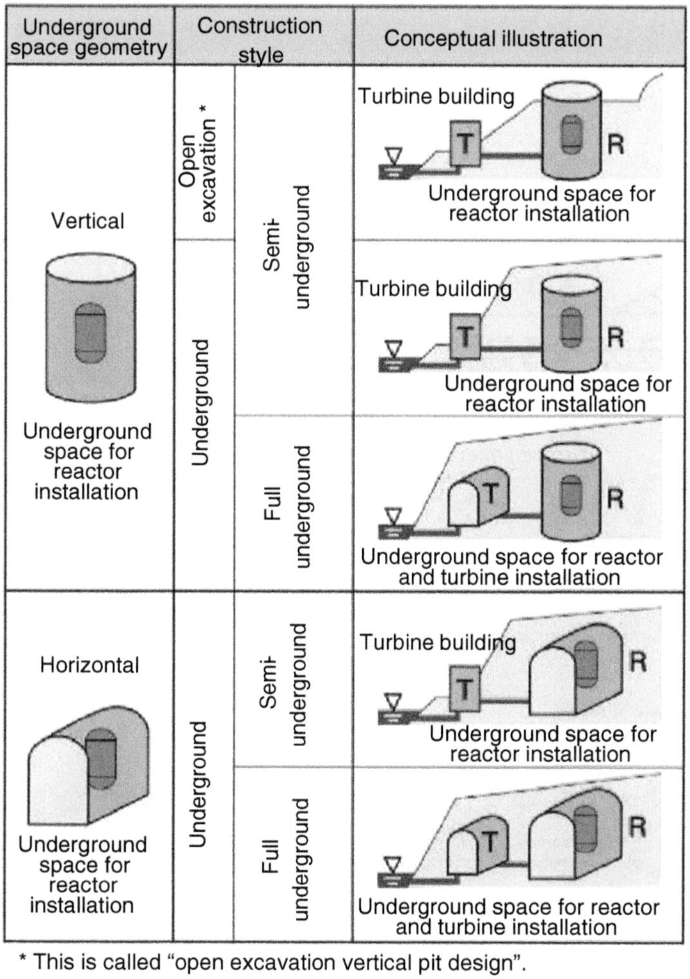

Fig. 12.12 Classification of underground siting techniques [6]

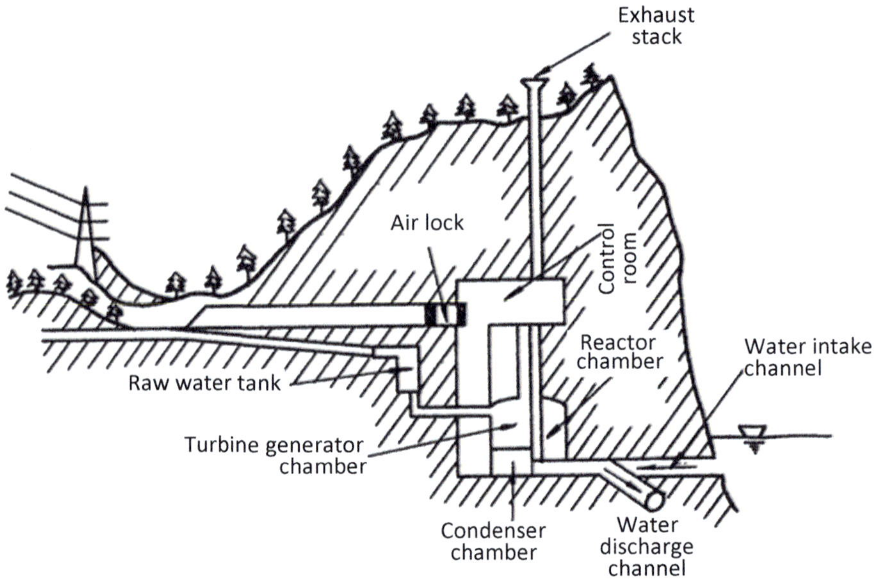

Fig. 12.13 Example of underground nuclear power plant layout [7]

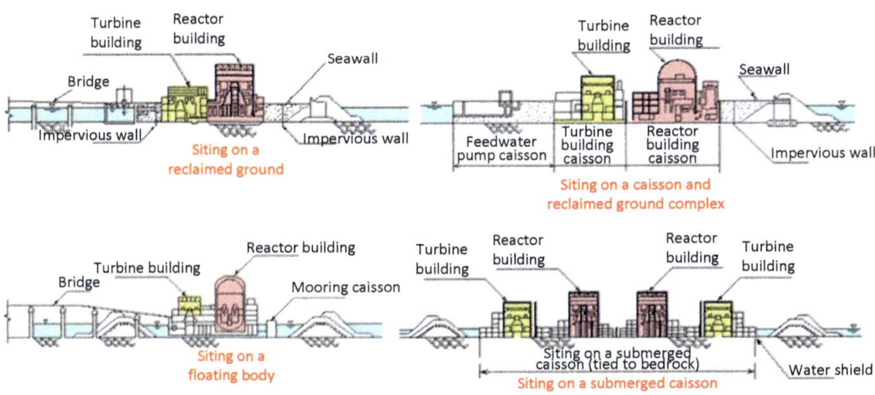

Fig. 12.14 Different techniques for the offshore siting of nuclear power plants [6]

Fig. 12.15 Concept of offshore siting using a man-made island [6]

Fig. 12.16 Classification of offshore siting using a floating body [8]

References

1. The Japan Society of Seismic Isolation website
2. "Design Recommendations for Seismically Isolated Buildings", Architectural Institute of Japan, 2013
3. Dynamic Design Inc. website
4. Kajima Corporation website
5. Material for Earthquake and Ground Motion WG, Subcommittee for the Regulatory Guide for Reviewing Seismic Design, Nuclear Safety Commission
6. "*Genshiryoku Hatsudensho no Ricchi Tayouka Gijutsu* (Diversification Technique for Nuclear Power Plant Siting)", Japan Society of Civil Engineers, 1996
7. "Annual Report on the Promotion of Science and Technology 1975", Ministry of Education, Culture, Sports, Science and Technology, 1975
8. "Feasibility Study on Floating Nuclear Power Plant, 1; Conceptual Design Study of FNPP", Noriaki Yabuuchi et al., JAERI-Research 2000-063, February 2001

Further Readings

9. The Institute of Applied Energy website

Part II
Fundamentals of Seismology and Earthquake Engineering

Chapter 13
Earthquake, Ground Motion, and Tsunamis

Masanori Hamada and Michiya Kuno

Abstract This chapter explains the roles of seismology and earthquake engineering for earthquake-resistant design of nuclear facilities. It also describes the various sources of earthquakes and the types of active faults. The propagation characteristics of S-waves among four kinds of seismic waves, and the reflection and transmission of S-waves at soil layer boundaries with different soil properties are explained. The amplification of S-waves through surface ground propagation, and the methods used to estimate the dominant ground motion periods are discussed. Tsunami sources and their propagation on the sea surface as well as on land, the external force on structures from tsunamis, and examples of tsunami-damaged structures are introduced.

Keywords Seismology · Earthquake engineering · S-wave · P-wave · Surface wave · Earthquake motion amplification · Dominant period · Reflection and transmission of seismic waves · Tsunami forces · Tsunami damage · Tsunami propagation

13.1 Seismology and Earthquake Engineering

The purpose of seismology is to study the scientific characteristics of earthquake sources, such as active faults and plate boundary movements, and seismic wave propagation in the crust, as shown in Fig. 13.1. Its major goal is to predict the time and the location of future earthquakes. However, failure to predict the 2011 Great East Japan earthquake of magnitude 9.0 reiterated that predicting earthquakes 1 day to several weeks before the occurrence is actually impossible.

M. Hamada (✉)
Waseda University, Shinjuku, Tokyo, Japan
e-mail: hamada@waseda.jp

M. Kuno
Chubu Electric Power Company, Nagoya, Japan

M. Hamada and M. Kuno (eds.), *Earthquake Engineering for Nuclear Facilities*,
DOI 10.1007/978-981-10-2516-7_13

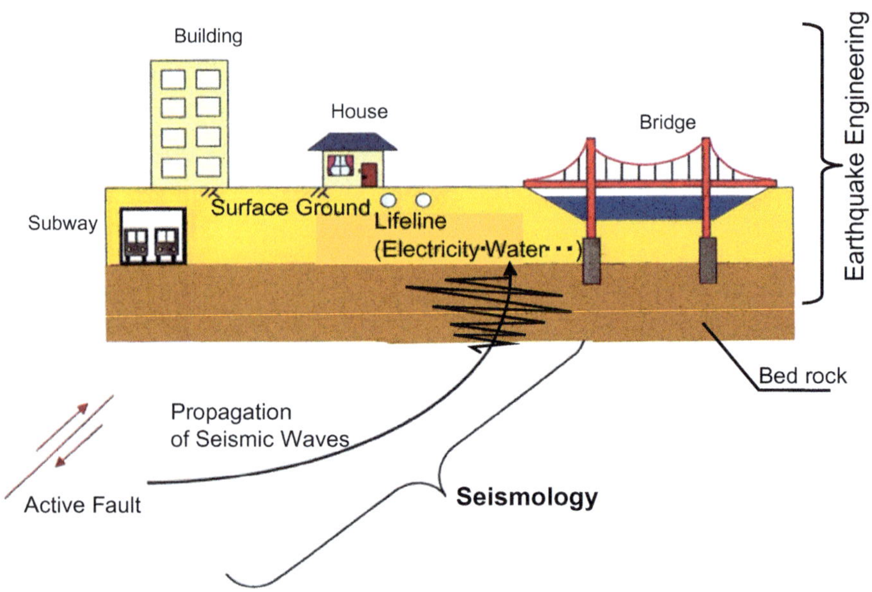

Fig. 13.1 Earthquake engineering and seismology

The purpose of earthquake engineering is to analyze the response of structures, such as buildings, bridges, and surface ground against seismic motion and to ensure the safety of structures and ground. The ultimate goal of earthquake engineering is to create safer and more secure societies, based on the design and construction of structures.

In seismology, researchers are asked to make predictions of earthquake probability and severity. In earthquake engineering, researchers are tasked with preventing the loss of life and property by ensuring the safety of structures and ground against future earthquakes and tsunamis. If both seismology science and earthquake engineering execute their individual obligations and collaborative roles, a safe and risk-free society in the face of earthquakes and tsunamis is possible. Unfortunately, the two fields have never been fully cooperative, and this was one of the principal reasons that the 2011 earthquake and tsunami were so devastating.

13.2 Earthquakes and Active Faults

13.2.1 Origins of Earthquakes

Earthquakes are categorized into the following three groups depending on their origins (Fig. 13.2).

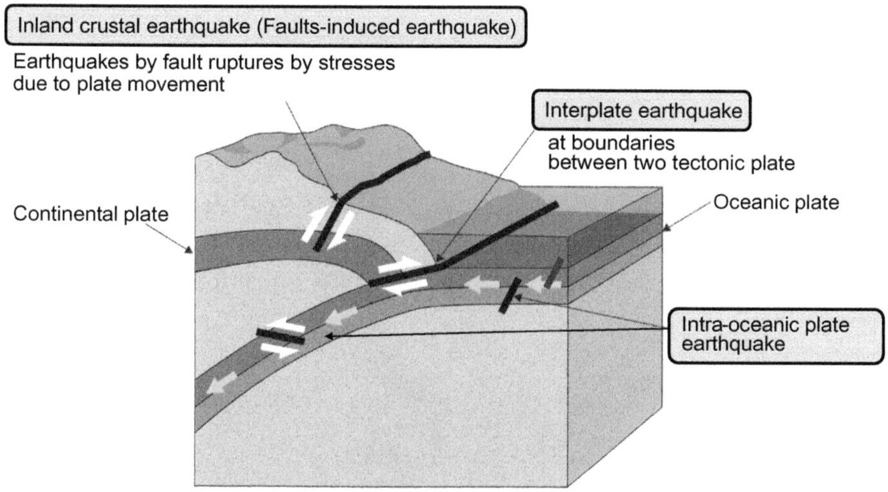

Fig. 13.2 Mechanism earthquake occurrence in Japan

(1) Interplate earthquake

This type of earthquake is caused by sliding at the boundary between two tectonic plates. Most of these earthquakes have return periods of several hundred years, and in most cases the source areas can be identified based on past earthquake records.

The 2011 Tohoku earthquake was caused by slippage along a length of approximately 500 km and a width of approximately 200 km between the Pacific and Northern American plates.

(2) Intra-oceanic plate earthquake

This type of earthquake is caused by the destruction of oceanic intra-oceanic plate, the source areas of which are difficult to define their locations, meaning that they cannot be predicted beforehand.

(3) Inland crustal earthquake

This type of earthquake happens when active faults rupture. The return period is generally over a thousand years, and this type of earthquake causes serious damage when they occur directly beneath highly urbanized areas.

As shown in Fig. 13.3, there are four plates, i.e., the Pacific, Eurasian, North American, and Philippine Sea plates, around the Japanese Archipelago. The Pacific plate is moving west at a speed of approximately 80 mm per year, while the Philippine Sea plate is moving north at a speed of 30–50 mm per year. Therefore, compressive stresses are dominant in the Japanese islands on the American and Eurasian plates, and many adverse faults have caused earthquakes.

In particular, the four abovementioned plates meet in the large Tokyo metropolitan area, which is Japan's is most earthquake-prone area.

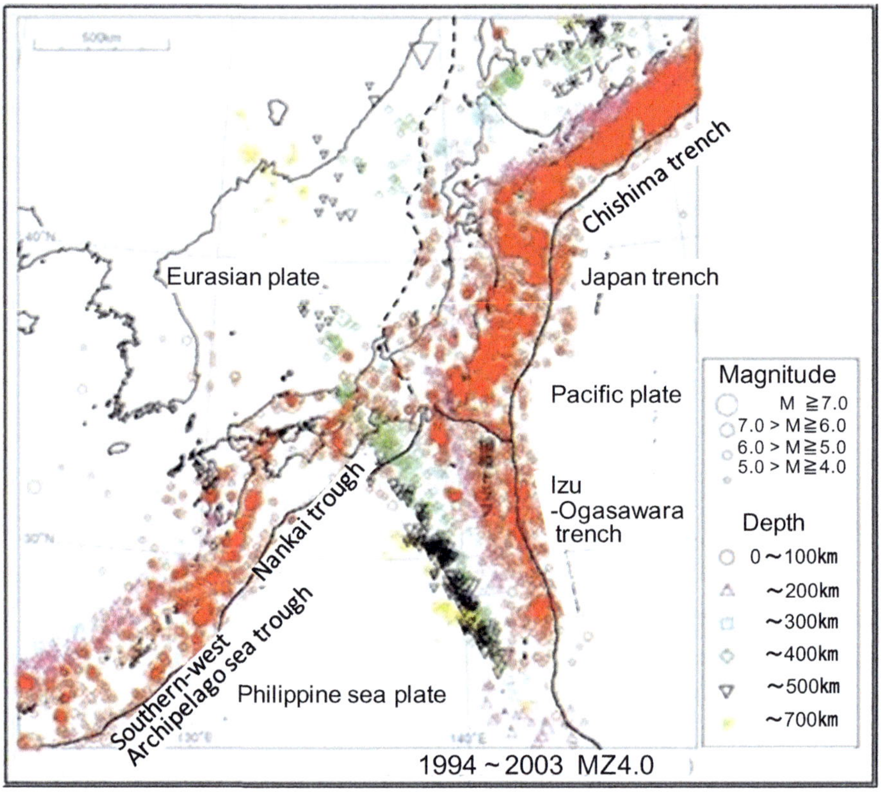

Fig. 13.3 Tectonic structure around the Japanese Archipelago [1]

13.2.2 Active Faults

Active faults are defined as those that have had repeated activities during the recent geological age, and have any possibility of movement in the future. According to this definition, there are over two thousand active faults in the Japanese Archipelago. The Japanese government is continuously surveying 98 active faults and fault zones with especially high probabilities of activity.

For the earthquake-resistant design of nuclear facilities, active faults, i.e., that have been active since the late Pleistocene epoch (~120,000–130,000 years ago), should be taken into consideration. If necessary, fault activity should be surveyed back to middle Pleistocene epoch (~400,000 years ago) (Part I, Chap. 2. Appendix 2.1, Fig. 2.15).

Active faults that cause earthquakes are referred as earthquake source faults, Active faults the ruptures of which appear at the ground surface are either referred as surface earthquake faults or earthquake faults. As shown in Fig. 13.4, the point of the fault rupture initiation is the hypocenter and the point on the ground surface above the hypocenter is the epicenter. The hypocentral and epicentral distances are the distances from the hypocenter and the epicenter, respectively, to a specific point on the surface.

As shown in Fig. 13.5, active faults are classified as strike-slip and vertical-slip faults. Strike-slip faults are classified as right and left slip faults, while vertical slip faults are classified as normal and reverse faults. Reverse faults occur under compressive stress conditions of the earth's crust, while normal faults occur under tensile conditions. The ground motions on the upper rock of the reverse fault observed during past earthquakes were much larger than those on the lower rock, because the upper rock runs up onto the lower rock.

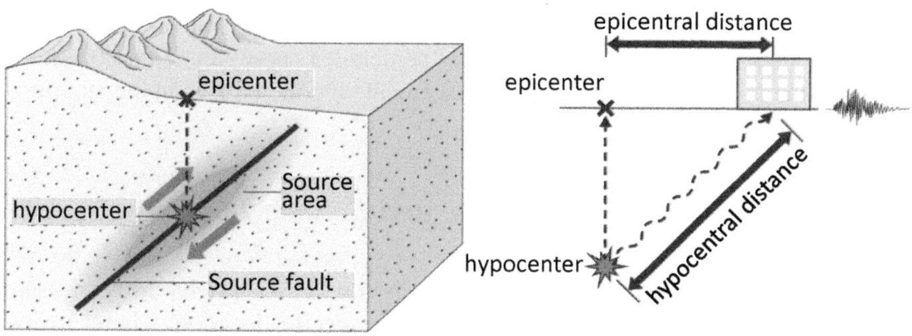

Fig. 13.4 Epicenter and hypocenter

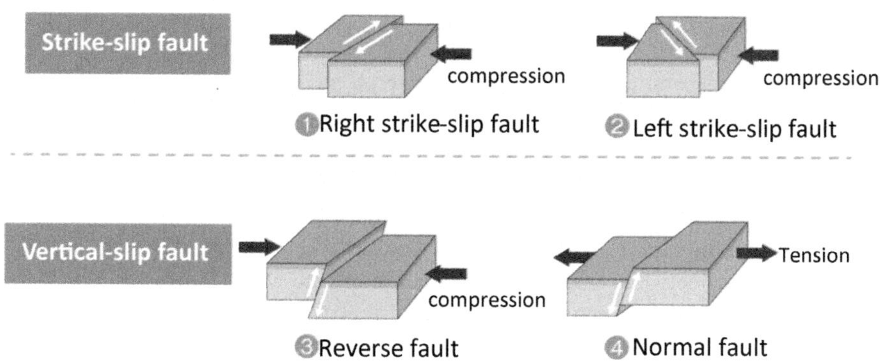

Fig. 13.5 Active faults

For the design and construction of nuclear facilities, whether or not active faults exist at the plant site, particularly under the reactor building and other important structure foundations is investigated by tectonic, geological, and geophysical surveys (Part I, Chap. 2, Sect. 2.2). When an active fault is found, it is impossible to design and construct nuclear facilities according to Japan's present regulations. If an active fault is found after construction, plant operation is required to be ceased immediately and the plant must be demolished.

However, the degree of damage to nuclear facilities depends on the magnitude of the differential displacement of faults. For example, if the fault displacement is several centimeters or less, a reactor building with a concrete foundation approximately 5 m thick may not sustain serious damage and the facilities inside the building can maintain important safety-related functions, e.g., *shut down*, *cool down*, and *confine*. Furthermore, the underground ducts for the cooling water pipes may absorb the fault displacement by using flexible joints.

Therefore, the functions of important facilities, piping systems, buildings and components should be examined based on the detailed survey or of the location of the faults, the magnitude of the fault displacement, and the effects on the structures.

13.2.3 Magnitude of Earthquake and Seismic Intensity

Magnitude measures the scale of earthquakes, while the seismic intensity measures the intensity of earthquake ground motion at a specific location. There are two scales of magnitude, these are the moment magnitude (M_w) and the Japan Meteorological Agency (JMA) magnitude (M_J). The moment magnitude is estimated as a product of the area of the fault rupture, fault displacement, and rock stiffness, which gives the total energy of the fault movement. M_J is estimated based on the displacements recorded on the surface at several locations. The effects of epicentral distance and hypocentral depth are taken into consideration when estimating M_J.

The modified Mercalli intensity scale (MMI) is used worldwide, while in Japan Meteorological Agency intensity scale (JMAI) is used. The JMAI has eight ranks, 0–7, and seismic intensities of 5 and 6 are more precisely divided into two levels, e.g., 5+ and 5−. JMAI was estimated based on the sensible vibration of the ground motion and damage to structures, but at the present time it is estimated by the intensity, frequency, and duration of the recorded accelerations. Table 13.1 shows the relationship between maximum acceleration and JMA intensity when the dominant period of acceleration is assumed to be 0.5 s. Table 13.2 also shows the relationship between JMAI and MMI when the dominant period is 0.5 s.

Table 13.1 JMAI and maximum acceleration on the ground surface (The dominant period of ground motion is assumed as 0.5 s)

JMAI	Maximum ground acceleration (cm/s^2)
1	~5
2	5~10
3	10~30
4	30~100
5−	100~160
5+	160~270
6−	270~500
6+	500~850
7	850~

Table 13.2 JMAI and MMI (The dominant period of ground motion is assumed as 0.5 s)

JMAI	MMI
0~1	I
1~3	II~III
3	IV
4	V
5−	VI
5+	VII
6−	VIII
6+~7	IX
7	X~

13.3 Seismic Waves and Propagation

13.3.1 Seismic Waves

As shown in Fig. 13.6, there are four kinds of seismic waves; P (primary) and S (secondary) waves, and Rayleigh and Love waves. P- and S-waves are to referred as body waves, which are directly generated by a fault movement and a plate-boundary slippage, and propagate in the crust to a specific point. P- and S-waves are amplified through the propagation of subsurface ground that is tens of meters thick. The dominant period of P- and S-wave vibrations is usually <1 s, and may strongly vibrate ordinary buildings and structures, which causes a resonant vibration with the seismic waves because the dominant period of the seismic motion is close to the natural periods of buildings and structures coincide.

The Rayleigh and Love waves are referred to as surface waves, which are generated at the free ground surface and propagate horizontally along the ground surface. Rayleigh and Love waves are amplified by surface ground structures of several kilometers thickness. The dominant periods of Rayleigh and Love waves vibrations are usually much longer than those of P- and S-waves, several seconds to several ten seconds. Therefore, the long period components of Rayleigh and Love waves cause resonant vibrations in skyscrapers and tall towers with natural periods

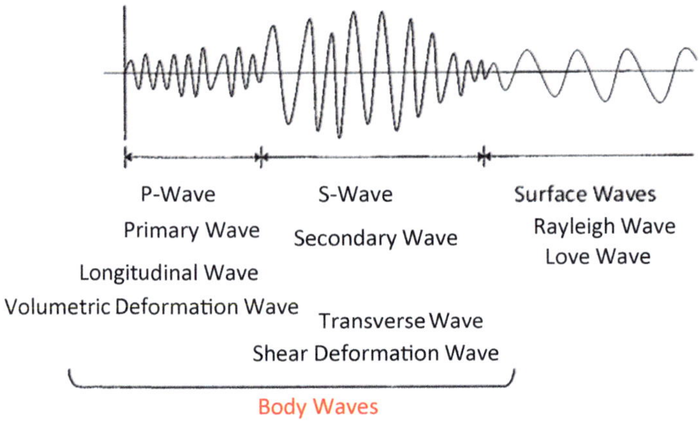

Fig. 13.6 Seismic waves

of several seconds, and the sloshing vibration in oil contained in large-diameter tanks.

For earthquake-resistant design, S-wave vibrations have a serious impact on the safety and stability of reactor buildings and other facilities. Therefore, the intensity, frequency characteristics, and ground motion duration should be carefully assessed based on various geological and soil condition surveys.

13.3.2 S-Wave Propagation

When an S-wave is propagating through homogeneous surface ground in a vertical direction (as shown in Fig. 13.7). The equilibrium equation on a soil segment at depth z can be written as follows:

$$\left(\tau + \frac{\partial \tau}{\partial z}\mathrm{d}z\right)A - \tau A - \rho A\mathrm{d}z\frac{\partial^2 U}{\partial t^2} = 0, \tag{13.1}$$

where $U(t, z)$ is ground displacement in the horizontal direction induced by S-wave propagation, and is a function of time t and the vertical coordinate z. A and dz are area and thickness of the soil segment, and ρ is soil density. τ is the shearing stress on the segment, and is expressed below using shearing strain $\partial U/\partial z$.

$$\tau = G\frac{\partial U}{\partial z}, \tag{13.2}$$

where G is the elastic shear modulus of the soil.

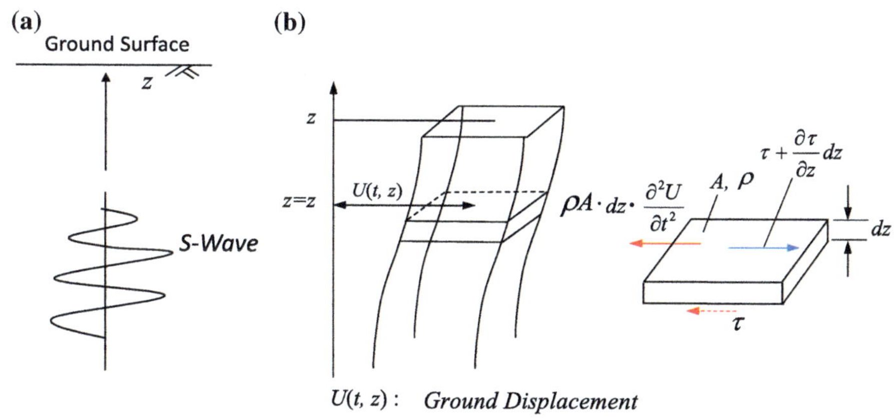

Fig. 13.7 Propagation of S-wave and wave equation. **a** Propagation of S-wave. **b** Force equilibrium of a soil segment

Assuming that G is constant, the following is obtained:

$$\frac{\partial^2 U}{\partial t^2} - v_s^2 \frac{\partial^2 U}{\partial z^2} = 0, \tag{13.3}$$

where v_s is the S-wave velocity (phase velocity), obtained as follows:

$$v_s = \sqrt{\frac{G}{\rho}}. \tag{13.4}$$

General solutions satisfying Eq. 13.3 are obtained as follows:

$$U(t, z) = F\left(t - \frac{z}{v_s}\right) \tag{13.5}$$

$$U(t, z) = D\left(t + \frac{z}{v_s}\right). \tag{13.6}$$

These two solutions show the waves propagating in plus or minus directions on the z axis, respectively, as shown in Fig. 13.8. The wave represented by Eq. (13.5) is propagating in the plus direction and is called the forward wave. The wave represented by Eq. (13.6) is propagating in the minus direction, and is called the backward wave.

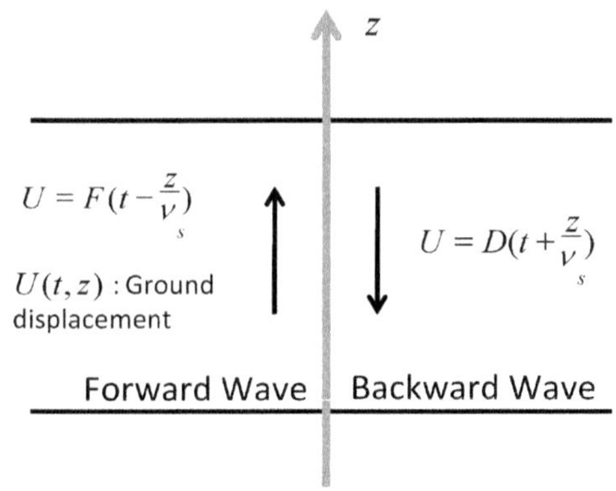

Fig. 13.8 Forward and backward waves

13.3.3 Reflection and Transmission of S-Waves at Soil Layer Boundary

Figure 13.9 depicts an S-wave propagating upwards (z+ direction) and reaching the boundary between medium (soil) I and II, which have different S- wave velocities and densities. The S-wave propagating upwards in the medium I is written as follows:

$$F_{\mathrm{I}}\left(t - \frac{z}{v_{S1}}\right), \tag{13.7}$$

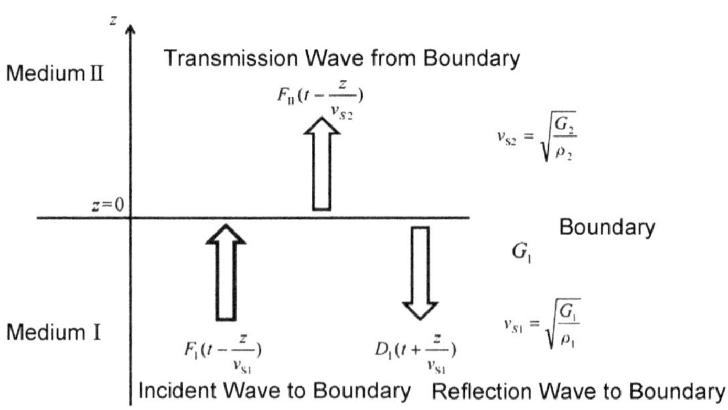

Fig. 13.9 Reflection and transmission of seismic waves at the boundary of soil layers

where v_{s_1} is the shear wave velocity in the medium I. The S-wave that is transmitted through the boundary and is propagated upwards in the medium II is written as follows:

$$F_{\mathrm{II}}\left(t-\frac{z}{v_{s_2}}\right). \tag{13.8}$$

The S-wave that reflects at the boundary surface and propagates downwards in medium I is written as follows:

$$D_{\mathrm{I}}\left(t+\frac{z}{v_{s_1}}\right). \tag{13.9}$$

At the boundary (z = 0) the displacement of ground I ($F_{\mathrm{I}}(t)+D_{\mathrm{I}}(t)$) is equal to that in the medium II $F_{\mathrm{II}}(t)$, while the shearing stresse in the medium I at the boundary is equal to that in the medium II:

$$G_1\left(\frac{\mathrm{d}F_{\mathrm{I}}}{\mathrm{d}z}+\frac{\mathrm{d}D_{\mathrm{I}}}{\mathrm{d}z}\right)_{z=0}=G_2\left(\frac{\mathrm{d}F_{\mathrm{II}}}{\mathrm{d}z}\right)_{z=0}. \tag{13.10}$$

Therefore, the transmitting $F_{\mathrm{II}}(t)$ and the reflecting $D_{\mathrm{I}}(t)$ waves are obtained as follows:

$$F_{\mathrm{II}}(t)=\frac{2}{1+\kappa}\cdot F_{\mathrm{I}}(t), \tag{13.11}$$

$$D_{\mathrm{I}}(t)=\frac{1-\kappa}{1+\kappa}\cdot F_{\mathrm{I}}(t), \tag{13.12}$$

$$\kappa=\frac{G_2 v_{s_1}}{G_1 v_{s_2}}. \tag{13.13}$$

where κ is referred as the impedance ratio at the boundary, and $2/(1+\kappa)$, $(1-\kappa)/(1+\kappa)$ are the transmission and reflection coefficients at the boundary, respectively.

13.3.4 Multi-reflection Theory (One Dimensional Wave Theory)

The surface ground consists of multiple soil layers with different elastic shear moduli and densities (different s-wave velocities), as shown in Fig. 13.10. The S-waves entering upward from the bedrock repeat the reflection and transmission at the boundaries of multi-soil layer boundaries. By accumulating reflected and

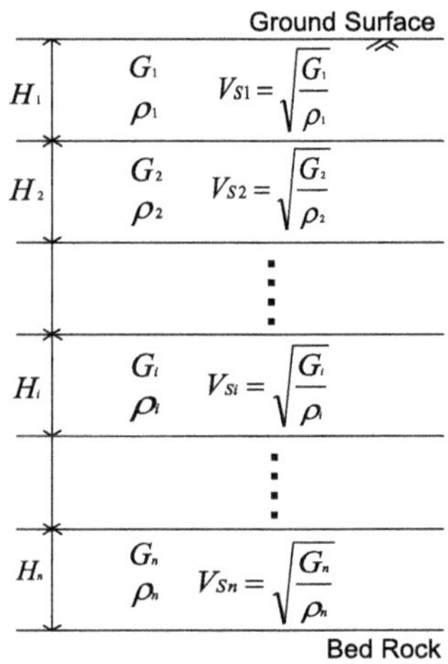

Fig. 13.10 Subsurface ground with multi-soil layers

transmitted waves in each soil layer, the dynamic response (displacement, acceleration, stress, and strain) to the input earthquake motion (upwards wave in the bedrock) can be obtained. This method is called multi-reflection theory (one dimensional wave theory).

If the bedrock appears at the surface (outcrop), as shown in Fig. 13.11. The reflected wave from the outcrop $D_1(t)$ is equal to $F_1(t)$, because the impedance ratio κ (Eq. 13.13) is 0.

$$D_1(t) = F_1(t). \tag{13.14}$$

Therefore, displacement $u_s(t)$ at the surface at the outcrop is obtained as follows:

$$\begin{aligned} u_s(t) &= D_1(t) + F_1(t) \\ &= 2F(t). \end{aligned} \tag{13.15}$$

Equation (13.15) shows that the magnitude of the ground motion at the outcrop is twice that of the incident wave propagating in the bedrock upwards.

Two kinds of input ground motion at the base rock are defined for the earthquake-resistant design of nuclear facilities. The first input ground motion is defined at the outcrop, where the effect of surface ground and structures above the bedrock on the earthquake motion is eliminated. This ground motion is called earthquake motion at the virtually free surface of bedrock, which is assumed to have a wide, and flat surface. When the shear wave propagating upwards in the bedrock is $E(t)$, the ground motion on the outcrop is $2E(t)$ as explained by Eq. (13.15). The ground motion at the bedrock $E + D$. D is the S-wave propagating downwards in

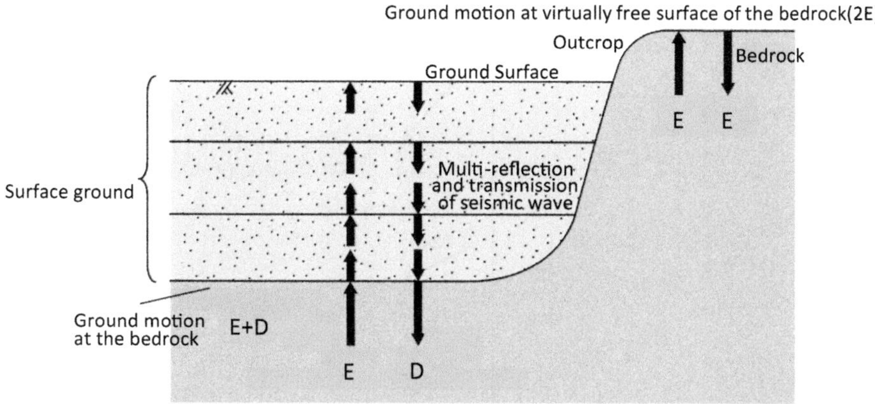

Fig. 13.11 Earthquake ground motion on the bedrock and at the virtually free surface

the bedrock, which can be calculated by the multi-reflection and transmission of S-wave in the surface ground (multi-reflection theory). The basic earthquake ground motion Ss for the design of nuclear power plants is defined as $2E$ on the virtually free surface of the bedrock.

13.4 Earthquake Motion Amplification and Dominant Period at the Ground Surface

Figure 13.12 is an example of horizontal accelerations observed in artificial land reclaimed from Tokyo Bay during the 2011 Tohoku earthquake and their Fourier spectra. As depicted in Fig. 13.12a, the ground at the observation point is composed of filled soil to −13 m depth and alternating layers of sand and clay, including gravel, to −42 m. Layers deeper than −42 m are either hard gravel or sand layers with N values exceeding 50, which can be regarded as the bedrock. S-wave velocities of the ground above the bedrock are 160–240 m/s. Earthquake ground motion was observed at the ground surface, at −22 and −53 m from the ground surface and on the hard gravel layer. Figure 13.12b shows that acceleration at the ground surface was much greater than that in the ground, and it is understood that the earthquake motion was amplified by the surface ground. Fourier spectra of the acceleration in Fig. 13.12c shows that the earthquake motion was amplified in the frequency range from 0.8 to 1.6 Hz. Amplification of the earthquake motion at 1.2 Hz (1/s) was especially significant; this frequency is recognized as the natural frequency of the surface ground. The above findings revealed that earthquake motion at the ground surface had two characteristics: amplification; and dominant vibration at natural frequencies (natural period).

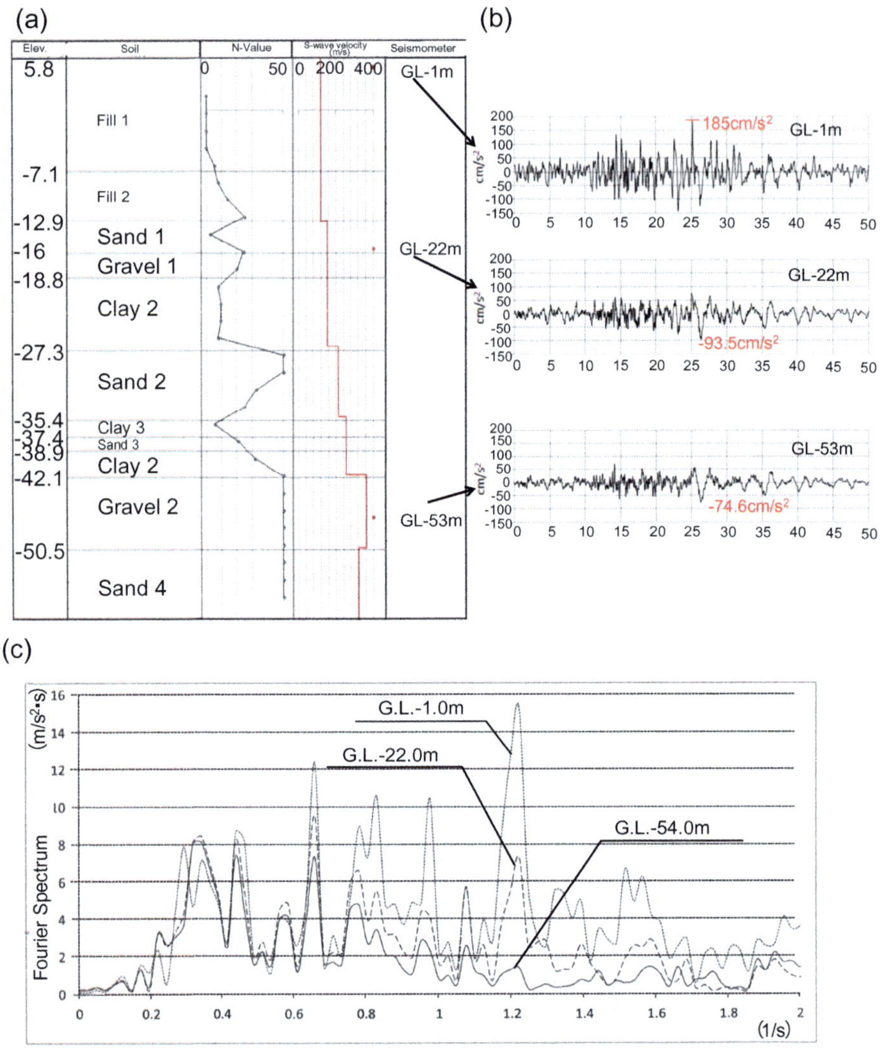

Fig. 13.12 Earthquake ground motion observed in artificial land reclaimed from Tokyo Bay **a** Soil condition, **b** Observed ground motion, **c** Fourier spectrum of observed ground acceleration

If the surface ground is composed of multiple soil layers of different shear moduli and densities (Fig. 13.10), the following two simplified methods are proposed to calculate the natural period of the surface ground T:

$$T = \frac{4\sum_{i=1}^{n} H_i}{\bar{v}_s}, \tag{13.16}$$

where $\bar{v}_s$ can be obtained as a mean of S-wave velocities, weighted by the thickness of each soil layer.

$$\bar{v}_s = \frac{\sum_{i=1}^{n} H_i v_{si}}{\sum_{i=1}^{n} H_i}, \tag{13.17}$$

where H_i is thickness of the ith layer, and v_{si} is S-wave velocity of the ith soil layer and is obtained by:

$$v_{si} = \sqrt{\frac{G_i}{\rho_i}}, \tag{13.18}$$

where G_i and ρ_i are elastic shear modulus and density of the ith soil layer, respectively.

The following equation is also a simple method for determining the natural period of the surface ground T:

$$T = \sum_{i=1}^{n} \frac{4H_i}{v_{si}}. \tag{13.19}$$

In Specification for Road Bridges and Explanations, *V* Earthquake Resistant Design [3], a method to classify the subsurface ground type by a characteristic value T_G representing the natural period of the surface ground is proposed. The ground condition is classified into the following groups:

Soil condition I: $T_G < 0.2\,s$,
Soil condition II: $0.2\,\text{s} < T_G < 0.6\,\text{s}$,
Soil condition III: $0.6\,s < T_G$.

13.5 Tsunamis

13.5.1 Tsunami Mechanism and Propagation

When a fault movement in the oceanic crust emerges on the surface of the seabed, the seabed either rises up or sinks down. Tsunamis are caused by the rise and sink of the sea surface, which are induced by the vertical movement of the seabed. Landslides in the sea and on land as well as meteorites can also cause tsunamis. The change in vertical elevation on the sea surface propagates in all directions, as shown in Fig. 13.13. The length of a tsunami wave is much larger than the sea depth; tsunami propagation is governed by the long wave theory. The propagation velocity of a tsunami v is calculated as follows:

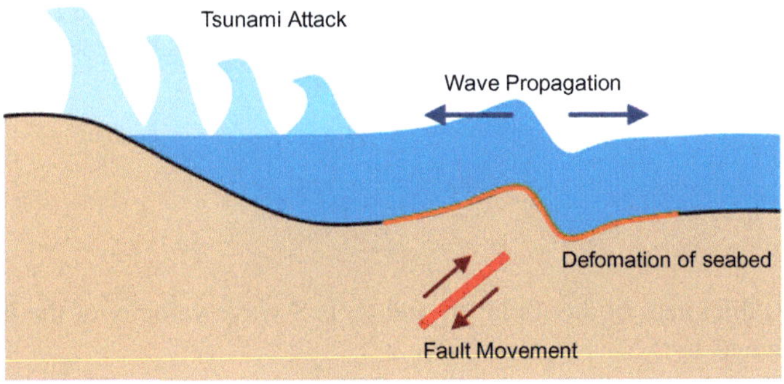

Fig. 13.13 Mechanism of occurrence of tsunami [2]

$$v = \sqrt{gh}, \tag{13.20}$$

where h and g are the depth (m) of the sea and the gravitational acceleration (9.8 m/s^2), respectively.

As shown in Eq. 13.20, the propagation velocity increases proportionally with the square root of sea depth. For example, with depths of 100 and 5000 m, the propagation velocities are approximately 220 m/s and 30 m/s, respectively. In inshore areas where the water is 10 m, the propagation velocity would be approximately 10 m/s, making it difficult for a human to escape a tsunami by running.

In contrast, in the ocean where depths are far greater the wave length increases with increasing propagation velocity. Therefore, sometimes people on ships do not realize that a tsunami is propagating. The Japanese word *tsu* means harbor/seashore, where a tsunami is very obvious because of the short wave length. The Japanese word *nami* means wave.

In recent years, GPS systems developed to measure vertical elevation changes in the sea surface have been stationed around the Japanese islands. The highly accurate information on sea level changes from GPSs is used for the tsunami warning system. Numerical simulation methods have been developed to analyze tsunami propagation. If the sea level elevation change is defined as the initial condition, the time of tsunami arrival at the sea shore and the elevation of sea water along the shore line can be calculated.

Tsunami wave height varies considerably depending on local sea surface topography. As shown in Fig. 13.14a, tsunami height is generally amplified around the top of a cape. This is caused by a concentration of the tsunami around the cape, because the tsunami is refracted as a result of the change in sea depth. Figure 13.14b illustrates tsunami wave height amplification at the end of a V-shaped bay as result of the wave being concentrated. This is the reason that the ria coastline in the Tohoku region of Japan has been repeatedly attacked by high tsunamis.

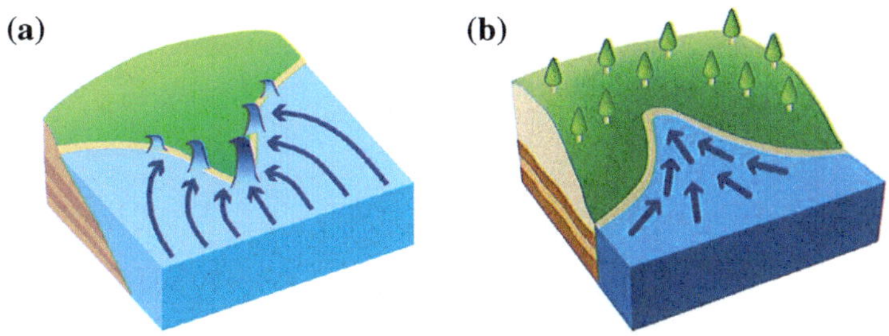

Fig. 13.14 Amplification of tsunami height by topographical conditions. **a** Amplification at top of cape. **b** Amplification at end of V-shaped bay [2]

13.5.2 Characteristics of Tsunami Waves

Tsunami wave height is determined by four measurements. The first measure is sea surface elevation from normal tide level along the coast. The second is run-up height. The landed tsunami runs towards a higher elevation on land, and the run-up height is the elevation difference between the highest point of the tsunami run-up on land and normal sea level. The tsunami trace height is the elevation difference between the highest trace of tsunami flood and normal sea level. The flood height is the elevation difference between the highest flood level and the normal sea level (Fig. 13.15).

The motion of water particles in a tsunami differs from that in a normal sea wave. As shown in Fig. 13.16a, in the case of a normal sea wave, the water particles near the sea surface are moving, while in the case of a tsunami all of the water particles from the sea surface to the seabed are moving as shown in Fig. 13.16b. Large forces are exerted on structures, such as breakwaters, as a result of the movement of water particles in a tsunami.

Figure 13.17 shows the collapse of a breakwater at the gate of Kamaishi Bay as result of the 2011 Great East Japan Earthquake-caused tsunami. The height of the

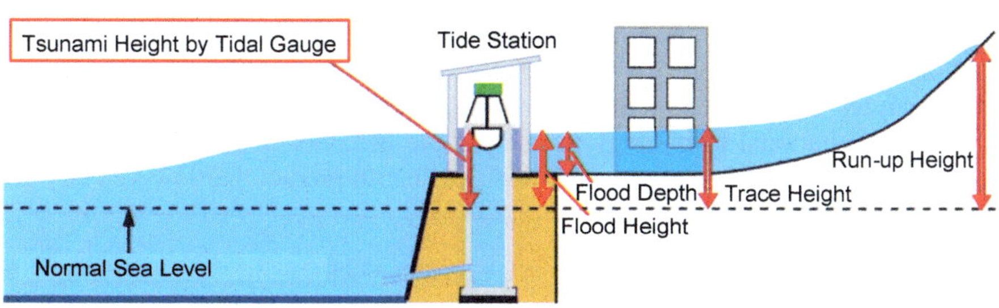

Fig. 13.15 Definitions of tsunami height [2]

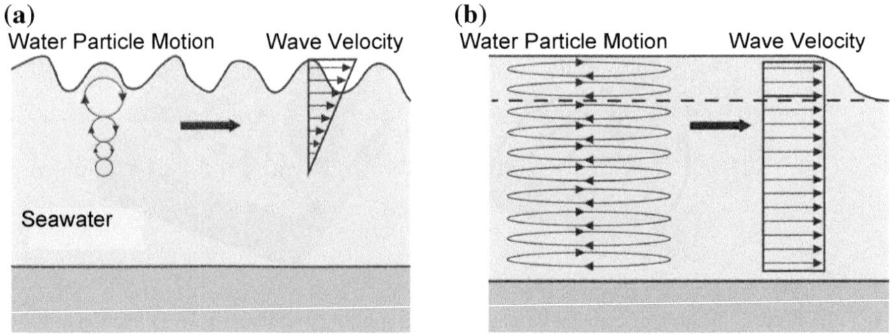

Fig. 13.16 Water particle motion. **a** In a normal sea wave. **b** In a tsunami [2, 3]

breakwater from the seabed was approximately 60 m. The lower part of the breakwater was a 30 m high gravel mound and the upper part was the concrete caissons with gravel infills to resist the force of a tsunami with its weight. This breakwater was designed to resist a tsunami 5.6 m in height, but the tsunami height largely exceeded the design height, causing the gravel mound under the concrete caissons to slide, as shown in Fig. 13.18. However, the Port, Harbor, and Airport Research Institute of Japan reported that the breakwater delayed the tsunami arrival time by approximately 6 min in the residential areas in Kamaishi city, and reduced the original tsunami height of 13 m to ~7–9 m, even though it collapsed. However, the tsunami broke the breakwater and hit residential areas washing away numerous houses, buildings, and various types of infrastructures. After the 2011 tsunami disaster the Ministry of Land, Transport, and Tourism recommended the construction of a resilient tsunami breakwater structure that will not collapse even if it is badly damaged.

Studies on the characteristics of tsunami force on structures have been carried out since the 1960s in Japan. Most of these studies were done using model tests in water tanks under centrifugal and gravitational conditions, and the characteristics of the tsunami forces have been clarified.

During the 2011 tsunami, many buildings were destroyed and washed away, but some of them survived with considerable damage, as shown in Fig. 13.19. Based on the results of back analysis of structures partially damaged by the tsunami, the actual force exerted on the structures can be calculated. A reliable and rational method for the evaluation of tsunami forces should be developed based on these back analyses and previous knowledge from model tests.

The computational technology to analyze the behavior of landed tsunamis has advanced remarkably in recent years. It is possible to predict the flow velocity and

Fig. 13.17 Collapse of the breakwater at Kamaishi Bay [4]

the height of a landed tsunami, as well as the external forces that tsunamis exert on structures. By employing these computational techniques, effective locations of evacuation shelters and anti-tsunami embankments can be determined to save lives and create tsunami-resilient living spaces.

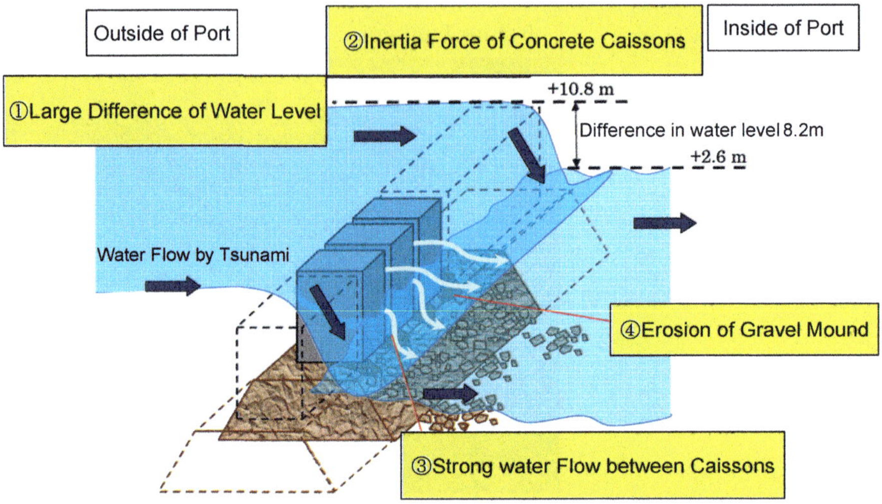

Fig. 13.18 Structure of the collapsed breakwater at Kamaishi Bay [5]

Fig. 13.19 Damage to a building during the 2011 Great East Japan earthquake-caused tsunami

References

1. "Jishin no Hassei Mekanizumu wo Saguru (Study of Earthquake Generation Mechanism)" Ministry of Education, Culture, Sports, Science and Technology, 2004
2. Japan Meteorological Agency website
3. "Electrical Review" June 2005, Electrical Review Company
4. Website of Kamaishi Port Office, Ministry of Land, Infrastructure, Transport and Tourism
5. "Tohoku Chihou Taiheiyou Oki Jishin Tsunami ni yoru Kanmaishi-kou Tsunami Bouhatei no Hisai Youin no Kentou Besshi 3(H23.4.1)(Appendix 3, Study of Damage Factors of Tsunami Breakwater in the Kamaishi Port by 2011 Tohoku Earthquake Tsunami (April 1, 2011))", Port and Airport Research Institute, 2011

Chapter 14
Dynamic Response Analysis

Masanori Hamada

Abstract The dynamic response characteristics of a one-mass-spring-damper system are outlined. The resonant vibration, damping, and natural periods of the dynamic system are explained. Multi-masses-springs-dampers systems for the dynamic response analysis of buildings and other structures, and three methods of solving the differential equations governing the motions of the system: the direct integration in the time domain, the integration in frequency domain (Fourier transformation), and the modal analysis method are explained. Furthermore, the basic concept of the Finite Element Method (FEM), which is frequently used for the dynamic response analysis of reactor buildings/facilities, and foundation ground/surrounding slopes is introduced.

Keywords One-mass-spring-damper system · Resonant vibration · Free vibration · Forced vibration · Modal analysis · Direct integration method · Fourier transformation · Natural period (frequency) · Natural vibration mode

14.1 Dynamic Response of One-Mass-Spring-Damper Model

14.1.1 One-Mass-Spring-Damper Model

Surface ground with a uniform elastic shear modulus and density, and one-story buildings can be modeled by a one-mass-spring-damper model, as shown in Fig. 14.1. In the case of a uniform surface ground, the mass is calculated as 1/2 $HA\rho$, where H, A, and ρ are the thickness, area, and density of the surface ground, respectively. The spring constant k is obtained as GA/H, where G is the elastic shear modulus. In the case of a one-story building, the mass of the model m is obtained

M. Hamada (✉)
Waseda University, Shinjuku, Tokyo, Japan
e-mail: hamada@waseda.jp

M. Hamada and M. Kuno (eds.), *Earthquake Engineering for Nuclear Facilities*,
DOI 10.1007/978-981-10-2516-7_14

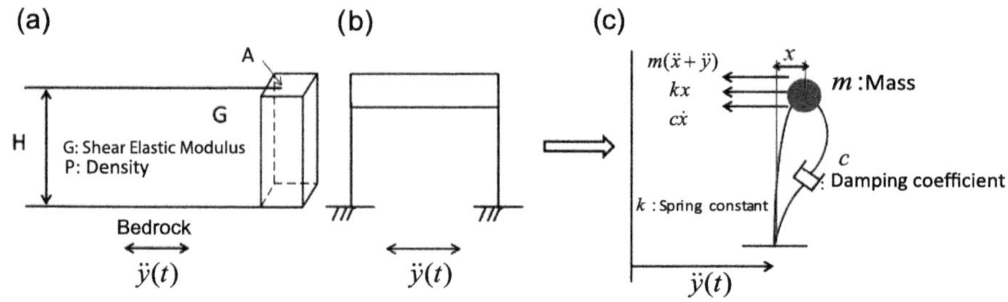

Fig. 14.1 One-mass-spring-damper model, **a** Surface ground, **b** Building, **c** One-mass-spring-damper model (A:unit area)

from the masses of the roof, columns, and walls, while the spring constant k is evaluated based on the stiffness of the walls and columns.

All ground and structures have a damping. Both ground and structure vibration will cease after external excitations have stopped. The mechanism of damping has not yet been clearly resolved. The effect of damping can be demonstrated by a damper, as shown in Fig. 14.1. The constant c of the damper is the damping coefficient, and the damper exerts a force on the one-mass-spring-damper system, which is negatively proportional to the relative velocities between the mass and the base of the systems. Based on the equilibrium condition of the system, the following equation can be obtained:

$$m\ddot{x}(t) + c\dot{x}(t) + kx(t) = -m\ddot{y}(t), \tag{14.1}$$

where $m\ddot{x}, c\dot{x}$, and kx are the inertia force, damping force, and restoring force, and $\ddot{y}(t)$ is the input acceleration at the base of the system. Equation 14.1 can be rewritten as follows:

$$\ddot{x}(t) + 2\omega_0 h\dot{x}(t) + \omega_0^2 x(t) = -\ddot{y}(t), \tag{14.2}$$

where ω_0 and h are natural circular frequency and the damping constant, respectively. These are calculated as follows:

$$\omega_0 = \sqrt{\frac{k}{m}}, \tag{14.3}$$

$$h = \frac{1}{2}\frac{c}{\sqrt{mk}}. \tag{14.4}$$

14.1.2 Free Vibration of a One-mass-spring-damper Model

When $\ddot{y}(t)$ is zero, the free vibration of the system $x(t)$ is obtained as follows:

$$x(t) = e^{-h\omega_0 t}(A_1 \cos\sqrt{1-h^2}\omega_0 t + B_1 \sin\sqrt{1-h^2}\omega_0 t), \tag{14.5}$$

where A_1 and B_1 can be determined by the initial conditions. For example, when the initial displacement and velocity at $t = 0$,

$$x(0) = x_0, \quad \dot{x}(0) = 0, \tag{14.6}$$

the free vibration of the system $x(t)$ is:

$$x(t) = x_0 e^{-h\omega_0 t}\left(\cos\sqrt{1-h^2}\omega_0 t + \frac{h}{\sqrt{1-h^2}}\sin\sqrt{1-h^2}\omega_0 t\right). \tag{14.7}$$

As shown in Fig. 14.2, the displacement $x(t)$ gradually decreases along the exponential function.

14.1.3 Stationary Vibration of One-mass-spring-damper Model

When the base acceleration $\ddot{y}(t)$ is a harmonic wave with the circular frequency ω and the amplitude α_0

$$\ddot{y}(t) = a_0 \cdot e^{i\omega t}, \tag{14.8}$$

$x(t)$, and $\ddot{x}(t)$ are obtained as follows:

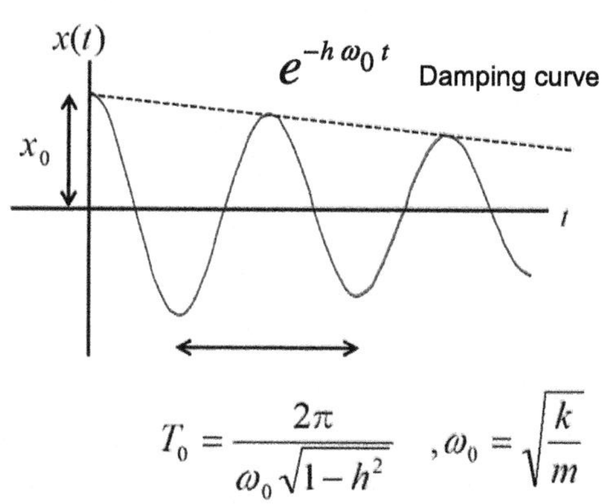

Fig. 14.2 Free vibration of a one-mass-spring-damper model

$$x(t) = C \cdot e^{i\omega t}, \quad C = \frac{a_0}{\omega^2 - \omega_0^2 - 2\omega\omega_0 hi} \tag{14.9}$$

$$x(t) = \frac{a_0}{\omega^2 - \omega_0^2 - 2\omega\omega_0 hi} e^{i\omega t} \tag{14.10}$$

$$\ddot{x}(t) = \frac{-\omega^2}{\omega^2 - \omega_0^2 - 2\omega\omega_0 hi} a_0 e^{i\omega t} \tag{14.11}$$

Equation 14.11 can be rewritten as follows:

$$\ddot{x}(t) = \frac{\left(\frac{\omega}{\omega_0}\right)^2}{\sqrt{\left\{1 - \left(\frac{\omega}{\omega_0}\right)^2\right\}^2 + 4\left(\frac{\omega}{\omega_0}\right)^2 h^2}} a_0 e^{i(\omega t - \phi)}, \tag{14.12}$$

where φ is the phase angle obtained as follows:

$$\tan\varphi = \frac{2\left(\frac{\omega}{\omega_0}\right)h}{1 - \left(\frac{\omega}{\omega_0}\right)^2}. \tag{14.13}$$

The absolute amplitude ratio $|x(t)/a_0|$ and the phase angle φ are expressed by the function ω/ω_0, as shown in Fig. 14.3. Fig 14.3a is the resonant curve, where the amplitude of the response acceleration rapidly increases when the circular frequency of the one-mass-spring-damper system ω_0 closes to the circular frequency of the input sinusoidal motion ω. The amplification depends on the constant h. If, h is 0 and ω equals ω_0, the phase angle becomes 90°.

Based on the excitation experiment of structures and equipment using a shaking table and/or vibrator, the resonant curve and the phase angle curve can be obtained, and from those curves, the natural period and the damping constants of the structures and equipment are calculated.

14.1.4 Response Spectra

The solution of Eq. (14.2); relative displacement $x(t)$, relative velocity $\dot{x}(t)$, and absolute acceleration $\ddot{x}(t) + \ddot{y}(t)$ of one-mass-spring-damper model can be obtained as follows:

$$x(t) = -\frac{1}{\omega_0\sqrt{1-h^2}} \int_0^t e^{-\omega_0 h(t-\tau)} \cdot \sin\omega_0\sqrt{1-h^2}(t-\tau) \cdot \ddot{y}(\tau)d\tau \tag{14.14}$$

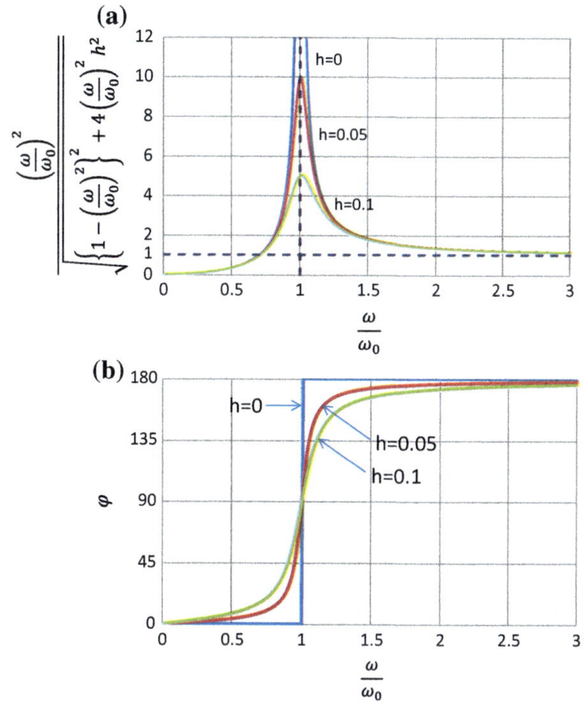

Fig. 14.3 Resonant curve and phase angle **a** resonant curve, **b** phase angle

$$\dot{x}(t) = -\int_0^t \mathrm{e}^{-\omega_0 h(t-\tau)} \cdot \left\{ \cos \omega_0 \sqrt{1-h^2}(t-\tau) - \frac{h}{\sqrt{1-h^2}} \sin \omega_0 \sqrt{1-h^2}(t-\tau) \right\} \ddot{y}(\tau) d\tau \tag{14.15}$$

$$\ddot{x}(t) + \ddot{y}(t) = \omega_0 \frac{1-2h^2}{\sqrt{1-h^2}} \int_0^t \mathrm{e}^{-\omega_0 h(t-\tau)} \cdot \sin \omega_0 \sqrt{1-h^2}(t-\tau) \cdot \ddot{y}(\tau) d\tau + 2\omega_0 h \int_0^t \mathrm{e}^{-\omega_0 h(t-\tau)} \cos \omega_0 \sqrt{1-h^2}(t-\tau) \cdot \ddot{y}(\tau) d\tau + \ddot{y}(t) \tag{14.16}$$

By taking the maximum value irrespective of time t of the relative displacement, relative velocity, and absolute acceleration, the response spectra S_D, S_V, and S_a are as obtained follows:

$$S_D = |x(t)|_{\max} \tag{14.17}$$

$$S_V = |\dot{x}(t)|_{\max} \tag{14.18}$$

$$S_a = |\ddot{x}(t) + \ddot{y}(t)|_{\max} \tag{14.19}$$

Generally, h is much smaller than 1.0 for ordinary structures, and hence, the response spectra S_D and S_V can be rewritten as follows:

$$S_D = \frac{1}{\omega_0}\left|\int_0^t e^{-\omega_0 h(t-\tau)} \cdot \sin\omega_0(t-\tau)\cdot\ddot{y}(\tau)d\tau\right|_{\max} \tag{14.20}$$

$$S_V = \left|\int_0^t e^{-\omega_0 h(t-\tau)} \cos\omega_0(t-\tau)\cdot\dot{y}(\tau)d\tau\right|_{\max} \tag{14.21}$$

The input acceleration $\ddot{y}(t)$ is generally much smaller than the response acceleration $\ddot{x}(t)$; therefore, the acceleration response spectrum can be obtained as follows:

$$S_a \fallingdotseq \omega_0\left|\int_0^t e^{-\omega_0 h(t-\tau)} \cdot \sin\omega_0(t-\tau)\cdot\ddot{y}(\tau)d\tau\right|_{\max} \tag{14.22}$$

The maximum response values of displacement S_D, velocity S_V, and acceleration S_A are calculated at each natural circular frequency ω and critical damping constant h. Figure 14.4a shows the concept of the response spectrum and Fig. 14.4b is a schematic figure of a displacement response spectrum, where the horizontal axis is

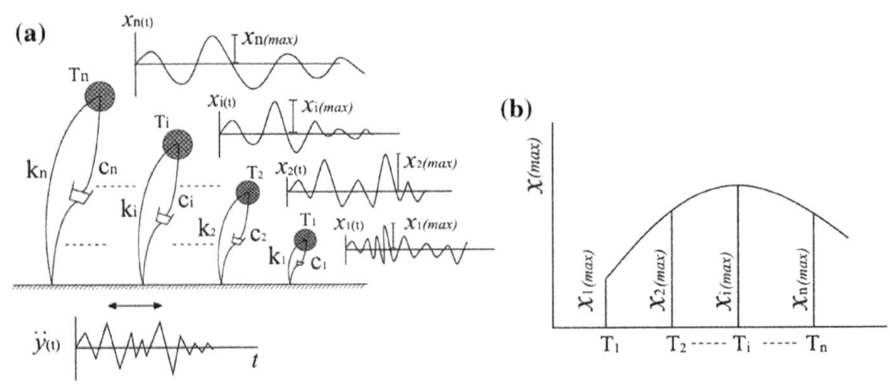

Fig. 14.4 Response spectra, **a** maximum response of one-mass-spring-damper system, **b** displacement response spectrum

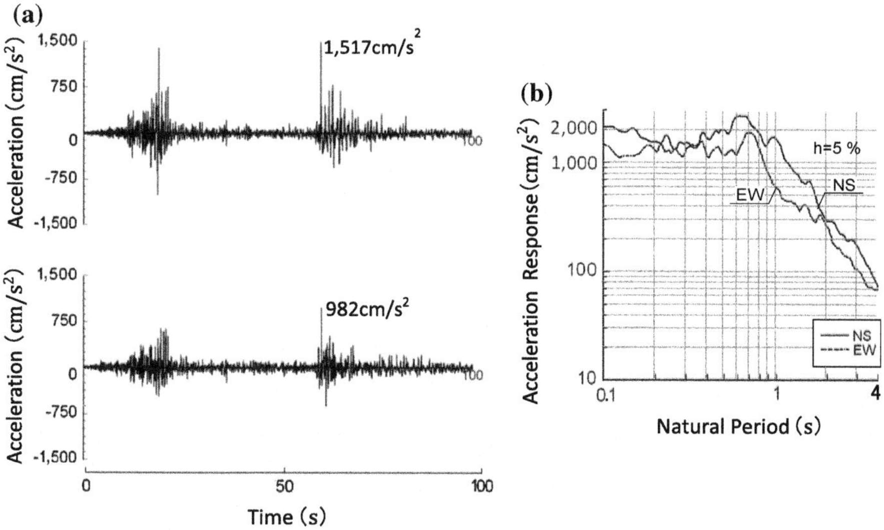

Fig. 14.5 Ground acceleration and acceleration response spectra (Great East Japan Earthquake at Sendai K-Net). **a** Accelerations observed at the ground surface. **b** Acceleration response spectra [1]

the natural period ($T_1 \sim T_n$) instead of the natural circular frequency ω_0 under a constant critical damping constant h.

Figure 14.5 shows an example of horizontal accelerations at the ground surface and their acceleration response spectra at K-NET [1] in Sendai during the 2011 Great East Japan earthquake. The maximum accelerations were 1,517 and 982 cm/s^2 in the north–south and east–west directions, respectively. The response acceleration in the north–south direction exceeded 2,000 cm/s^2 between 0.5 and 0.8 s, as shown in Fig. 14.5b. This demonstrates that the earthquake ground motion was dominant in these periods, and that dynamic responses of structures with these natural periods are amplified.

If the duration of the earthquake ground motion $y(\tau)$ in Eqs. (14.20), (14.21), and (14.22) are sufficiently long, the influence of the difference obtained by multiplying it by either $\sin \omega_0(t-\tau)$ or $\cos \omega_0(t-\tau)$ becomes negligible for the evaluation of the maximum response, and the following equations are obtained:

$$S_D \fallingdotseq \frac{1}{\omega_0} \cdot S_V \tag{14.23}$$

$$S_a \fallingdotseq \omega_0 \cdot S_V \tag{14.24}$$

Equations (14.23) and (14.24) mean that if one of the acceleration, velocity, or displacement spectrum is defined, the other two response spectra can be obtained. The relationship among the three kinds of response spectra can be displayed as one figure with four axes, as illustrated in Fig. 14.6.

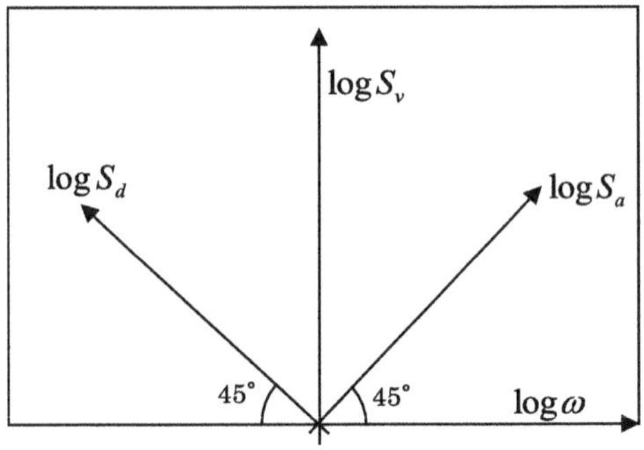

Fig. 14.6 Response spectra displayed as one figure with four axes

14.2 Dynamic Analysis of Surface Ground and Structure

14.2.1 Multi-masses-Springs-Dampers Model

Computing capabilities have continued to advance since the 1970s, facilitating the development of dynamic response analysis methods of ground and structures for use in earthquake-resistant design. Figure 14.7 shows multi-masses-springs-dampers model of surface ground composed by multiple soil layers and a multi-story building.

The equilibrium equation at the mass point i as shown in Fig. 14.8 is obtained as follows:

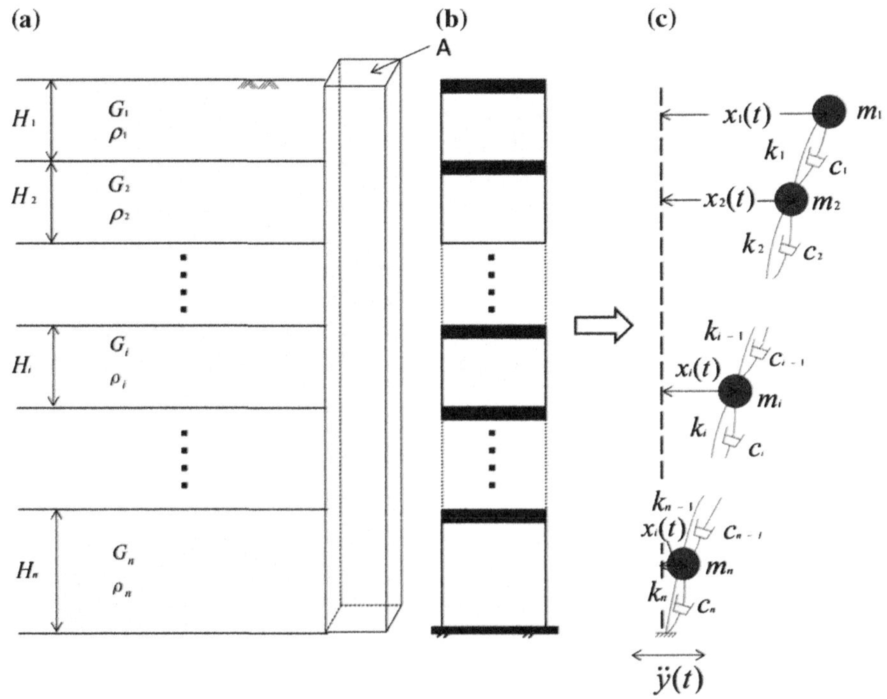

Fig. 14.7 Multi-masses-springs-dampers model, **a** surface ground, **b** building, **c** analytical model

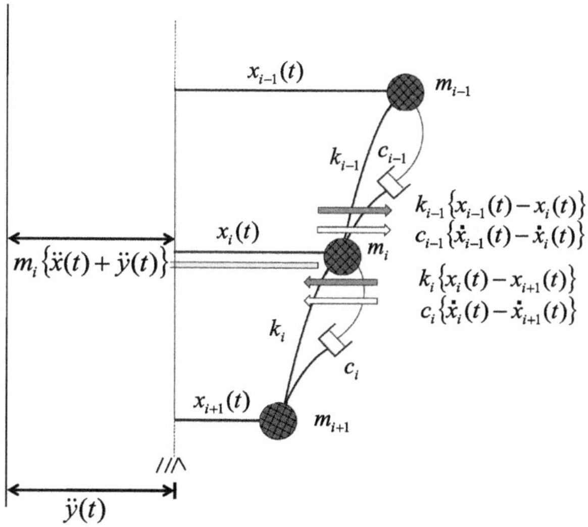

Fig. 14.8 Equilibrium of forces at point *i*-mass

$$\begin{aligned} &- m_i\{\ddot{x}_i(t) + \ddot{y}(t)\} + k_{i-1}\{x_{i-1}(t) - x_i(t)\} - k_i\{x_i(t) - x_{i+1}(t)\} \\ &+ c_{i-1}\{\dot{x}_{i-1}(t) - x_i(t)\} - c_i\{\dot{x}_i(t) - \dot{x}_{i+1}(t)\} = 0 \end{aligned} \tag{14.25}$$

where $x_i(t)$, $\dot{x}_i(t)$, and $\ddot{x}_i(t)$ are relative displacement, relative velocity, and relative acceleration between the *i*th mass and the bedrock, and $\ddot{y}(t)$ represents the input acceleration at the bedrock. m_i is mass of the *i*th mass point, and c_i and k_i are the viscous damping coefficient and spring constant that link the mass point *i* and mass point *i* + 1, respectively. The constants m_i and k_i can be obtained in the case of surface ground with multi-soil layers as follows:

$$m_i = \frac{1}{2}(\rho_{i-1}H_i + \rho_i H_i) \cdot A \tag{14.26}$$

$$k_i = \frac{G_i A}{H_i} \tag{14.27}$$

where A is the area of the soil column and usually takes as unit, G_i and H_i are the elastic shear modulus and the thickness of *i*th layer. In the case of multi-story buildings, m_i is calculated from the masses on each floor, while k_i is obtained from the structural stiffness of the columns, walls, and beams. c_i is the viscous damping constant between the ith mass and $(i + 1)$th mass. This coefficient is usually presumed proportional to the mass m_i or the soil spring constant k_i

From the equilibrium equations at all masses $m_i \sim m_n$, the following vibration equation of multi-masses-springs-dampers model can be obtained as follows;

$$
\begin{bmatrix} m_1 & & & \\ & m_2 & & 0 \\ & & \ddots & \\ & 0 & & \ddots \\ & & & & m_n \end{bmatrix}
\begin{Bmatrix} \ddot{x}_1(t) \\ \ddot{x}_2(t) \\ \cdot \\ \cdot \\ \ddot{x}_n(t) \end{Bmatrix}
+
\begin{bmatrix} c_1 & -c_1 & & \\ -c_1 & c_1+c_2 & & 0 \\ & & \ddots & \\ & 0 & & \ddots \\ & & & & c_{n-1}+c_n \end{bmatrix}
\begin{Bmatrix} \dot{x}_1(t) \\ \dot{x}_2(t) \\ \cdot \\ \cdot \\ \dot{x}_n(t) \end{Bmatrix}
$$

$$
+
\begin{bmatrix} k_1 & -k_1 & & \\ -k_1 & k_1+k_2 & & 0 \\ & & \ddots & \\ & 0 & & \ddots \\ & & & & k_{n-1}+k_n \end{bmatrix}
\begin{Bmatrix} x_1(t) \\ x_2(t) \\ \cdot \\ \cdot \\ x_n(t) \end{Bmatrix}
= -
\begin{bmatrix} m_1 & & & \\ & m_2 & & 0 \\ & & \ddots & \\ & 0 & & \ddots \\ & & & & m_n \end{bmatrix}
\begin{Bmatrix} 1 \\ 1 \\ \cdot \\ \cdot \\ 1 \end{Bmatrix} \ddot{y}(t). \tag{14.28}
$$

Using the mass matrix M, the damping matrix C, and its stiffness matrix K, the following differential equation can be obtained for the vibration of multi-masses, springs, and damper systems based on the equilibrium condition at each mass point.

$$
\boldsymbol{M} \cdot \ddot{\boldsymbol{x}} + \boldsymbol{C} \cdot \dot{\boldsymbol{x}} + \boldsymbol{K} \cdot \boldsymbol{x} = -\boldsymbol{M} \cdot \mathbf{1} \cdot \ddot{y}(t) \quad \mathbf{1} = \begin{Bmatrix} 1 \\ 1 \\ \vdots \\ 1 \end{Bmatrix}, \tag{14.29}
$$

$$
\boldsymbol{M} = \begin{bmatrix} m_1 & & & & \\ & m_2 & & 0 & \\ & & \ddots & & \\ & & & m_i & \\ & 0 & & \ddots & \\ & & & & m_n \end{bmatrix}, \quad
\boldsymbol{C} = \begin{bmatrix} \boldsymbol{c_1} & -\boldsymbol{c_1} & & & \\ -\boldsymbol{c_1} & \boldsymbol{c_1}+\boldsymbol{c_2} & & \boldsymbol{0} & \\ & & \ddots & & \\ & & & \ddots & \\ & \boldsymbol{0} & & & \ddots \\ & & & & \boldsymbol{c_{n-1}}+\boldsymbol{c_n} \end{bmatrix}
$$

$$
\boldsymbol{K} = \begin{bmatrix} k_1 & -k_1 & & & \\ -k_1 & k_1+k_2 & & 0 & \\ & & \ddots & & \\ & & & \ddots & \\ & 0 & & & \ddots \\ & & & & k_{n-1}+k_n \end{bmatrix}, \quad
\boldsymbol{x} = \begin{Bmatrix} \boldsymbol{x_1} \\ \boldsymbol{x_2} \\ \vdots \\ \vdots \\ \vdots \\ \boldsymbol{x_n} \end{Bmatrix}. \tag{14.30}
$$

where $\ddot{x}$, $\dot{x}$, and x are relative acceleration, velocity, and displacement vectors, respectively. Elements of the vector $\mathbf{1}$ are all 1.

14.2.2 Solution of Dynamic Response of Multi-masses-Springs-Dampers Model

14.2.2.1 Direct Integration in Time Domain

Under the condition in which $x(t_{j-1})$ and $x(t_j)$ are given at $t = t_{j-1}$ and $t = t_j$, $x(t_{j+1})$ can be obtained as follows (Fig. 14.9). For a constant time interval Δt, the velocity vector $\dot{x}(t_j)$ can be expressed as:

$$\dot{x}(t_j) = \{x(t_{j+1}) - x(t_j)\}/\Delta t. \tag{14.31}$$

Likewise, the acceleration vector $\ddot{x}(t_j)$ is:

$$\ddot{x}(t_j) = \{x(t_{j+1}) + x(t_{j-1}) - 2x(t_j)\}/\Delta t^2. \tag{14.32}$$

By substituting Eqs. (14.31) and (14.32) into (14.29), the following is obtained:

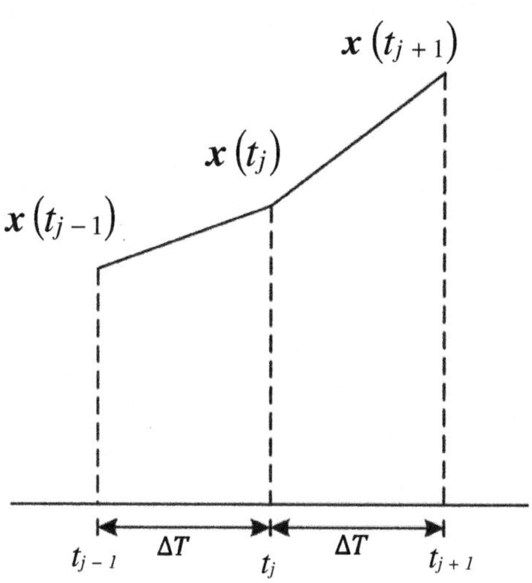

Fig. 14.9 Solution of vibration equation by direct integral method in the time domain

$$\boldsymbol{M} \cdot \{\boldsymbol{x}(t_{j+1}) + \boldsymbol{x}(t_{j-1}) - 2\boldsymbol{x}(t_j)\}/\Delta t^2 + C \cdot \{\boldsymbol{x}(t_{j+1}) - \boldsymbol{x}(t_j)\}/\Delta t + K \cdot \boldsymbol{x}(t_j) \\ = \boldsymbol{M} \cdot \boldsymbol{1} \cdot \ddot{y}(t_j) \tag{14.33}$$

$\boldsymbol{x}(t_{j-1})$, $\boldsymbol{x}(t_j)$, and input acceleration $\ddot{y}(t_j)$ at time t_j are given, $\boldsymbol{x}(t_{j+1})$ can be obtained. Repeating this procedure, the time history of $x(t)$ is acquired.

14.2.2.2 Integration in Frequency Domain by Fourier Transformation

Fourier transformation $X_i(\omega)$ of the ith element $x_i(t)$ of the displacement vector $\boldsymbol{x}(t)$ is:

$$X_i(\omega) = \int_{-\infty}^{\infty} x_i(t) \cdot \mathrm{e}^{-i\omega t} \cdot d\omega. \tag{14.34}$$

It therefore follows that Fourier transformation $\boldsymbol{X_F}(\omega)$ of the displacement vector $\boldsymbol{x}(t)$ becomes:

$$\boldsymbol{X_F}(\omega) = \begin{Bmatrix} X_1(\omega) \\ X_2(\omega) \\ \vdots \\ X_n(\omega) \end{Bmatrix}. \tag{14.35}$$

Fourier transformation of the velocity vector $\dot{\boldsymbol{X}}_{\boldsymbol{F}}(\omega)$ and acceleration vector $\ddot{\boldsymbol{X}}_{\boldsymbol{F}}(\omega)$ are expressed as:

$$-i\omega \boldsymbol{X_F}(\omega), \quad \text{and} \quad -\omega^2 \boldsymbol{X_F}(\omega). \tag{14.36}$$

Therefore, Fourier transformation of Eq. (14.29) can be carried out as follows:

$$-\omega^2 \boldsymbol{M} \cdot \boldsymbol{X_F}(\omega) - i\omega \cdot \boldsymbol{C} \cdot \boldsymbol{X_F}(\omega) + \boldsymbol{K} \cdot \boldsymbol{X_F}(\omega) = -\boldsymbol{M} \cdot \boldsymbol{1} \cdot Y(\omega), \tag{14.37}$$

where $Y(\omega)$ is Fourier transformation of the input acceleration $\ddot{y}(t)$.

From Eq. (14.38), the following can be obtained:

$$\boldsymbol{X_F}(\omega) = \left[\omega^2 \boldsymbol{M} + i\omega \cdot \boldsymbol{C} - \boldsymbol{K}\right]^{-1} \cdot \boldsymbol{M} \cdot \boldsymbol{1} \cdot \boldsymbol{Y}(\omega), \tag{14.38}$$

where $x_i(t)$ is displacement of ith element of vector $\boldsymbol{x}(t)$, which can be attained by inverse transformation $X_i(\omega)$, as shown below:

$$x_i(t) = \frac{1}{2\pi} \int_{-\infty}^{\infty} X_i(\omega) \cdot e^{i\omega t} \cdot d\omega. \tag{14.39}$$

14.2.2.3 Modal Analysis

The equation of free vibration in which both the damping term and input motion $y(t)$ are zero, can be written from Eq. (14.29), as follows:

$$\boldsymbol{M} \cdot \ddot{\boldsymbol{x}} + \boldsymbol{K} \cdot \boldsymbol{x} = \boldsymbol{0}, \tag{14.40}$$

where $\boldsymbol{0}$ is a vector, all the elements of which are zero. When the displacement vector $\boldsymbol{x}(t)$ is written as:

$$\boldsymbol{x} = \boldsymbol{X} \cdot e^{iwt}, \tag{14.41}$$

the following is obtained:

$$-\omega^2 \cdot \boldsymbol{M} \cdot \boldsymbol{X} + \boldsymbol{K} \cdot \boldsymbol{X} = \boldsymbol{0}, \tag{14.42}$$

where $\boldsymbol{X}$ expresses the vibration mode of the free vibration, and ω is circular frequency. Equation (14.42) is rewritten as:

$$\omega^2 \cdot \boldsymbol{X} = \boldsymbol{M}^{-1}\boldsymbol{K} \cdot \boldsymbol{X}. \tag{14.43}$$

Thus, this becomes the eigenvalue problem of matrix $\boldsymbol{M}^{-1}\boldsymbol{K}$. Different n values of ω satisfying Eq. (14.43) are obtained, and vector $\boldsymbol{X}$ (vibration mode) is obtained as shown below:

$$\begin{array}{lll} \text{1st vibration} & \omega_1 & X_{11}, X_{21}, \ldots, X_{n1}, \\ \vdots & & \\ i\text{th vibration} & \omega_i & X_{1i}, X_{2i}, \ldots, X_{ni}, \\ \vdots & & \\ n\text{th vibration} & \omega_n & X_{1n}, X_{2n}, \ldots, X_{nn}. \end{array} \tag{14.44}$$

In the above equation, X_{ji} expresses the vibration mode value of the jth mass point of the ith vibration. The ith vibration mode is

$$\boldsymbol{X}_i = \begin{Bmatrix} X_{1i} \\ X_{2i} \\ \vdots \\ X_{ni} \end{Bmatrix}. \tag{14.45}$$

X_i satisfies the following equation:

$$-\omega_i^2 \cdot \boldsymbol{M} \cdot \boldsymbol{X}_i + \boldsymbol{K} \cdot \boldsymbol{X}_i = \boldsymbol{0}. \tag{14.46}$$

Likewise, the *j*th vibration mode $\boldsymbol{X}_j$ satisfies the following:

$$-\omega_j^2 \cdot \boldsymbol{M} \cdot \boldsymbol{X}_j + \boldsymbol{K} \cdot \boldsymbol{X}_j = \boldsymbol{0}. \tag{14.47}$$

By multiplying Eq. (14.46) by the transpose vector $\boldsymbol{X}_j^T$ and Eq. (14.47) by transpose vector $\boldsymbol{X}_i^T$, the following are obtained:

$$-\omega_i^2 \boldsymbol{X}_j^T \cdot \boldsymbol{M} \cdot \boldsymbol{X}_i + \boldsymbol{X}_j^T \cdot \boldsymbol{K} \cdot \boldsymbol{X}_i = \boldsymbol{0}, \tag{14.48}$$

$$-\omega_j^2 \boldsymbol{X}_i^T \cdot \boldsymbol{M} \cdot \boldsymbol{X}_j + \boldsymbol{X}_i^T \cdot \boldsymbol{K} \cdot \boldsymbol{X}_j = \boldsymbol{0}. \tag{14.49}$$

where each of $\boldsymbol{X}_j^T \cdot \boldsymbol{K} \cdot \boldsymbol{X}_i$ and $\boldsymbol{X}_i^T \cdot \boldsymbol{K} \cdot \boldsymbol{X}_j$ is scalar. Therefore, the following are obtained:

$$\left(\boldsymbol{X}_j^T \cdot \boldsymbol{K} \cdot \boldsymbol{X}_i\right)^T = \boldsymbol{X}_j^T \cdot \boldsymbol{K} \cdot \boldsymbol{X}_i, \tag{14.50}$$

$$\boldsymbol{X}_i^T \cdot \boldsymbol{K}^T \cdot \boldsymbol{X}_j = \boldsymbol{X}_j^T \cdot \boldsymbol{K} \cdot \boldsymbol{X}_i. \tag{14.51}$$

Because the stiffness matrix $\boldsymbol{K}$ is symmetric:

$$\boldsymbol{X}_i^T \cdot \boldsymbol{K} \cdot \boldsymbol{X}_j = \boldsymbol{X}_j^T \cdot \boldsymbol{K} \cdot \boldsymbol{X}_i. \tag{14.52}$$

Likewise, because mass matrix M is diagonal, the following is obtained:

$$\boldsymbol{X}_i^T \cdot \boldsymbol{M} \cdot \boldsymbol{X}_j = \boldsymbol{X}_j^T \cdot \boldsymbol{M} \cdot \boldsymbol{X}_i. \tag{14.53}$$

Subtracting Eq. (14.49) from (14.48) produces:

$$(\omega_j^2 - \omega_i^2) \cdot \boldsymbol{X}_j^T \cdot \boldsymbol{M} \cdot \boldsymbol{X}_i = 0, \tag{14.54}$$

where $\omega_i \neq \omega_j$; that is, if $i \neq j$, the following is obtained:

$$\boldsymbol{X}_j^T \cdot \boldsymbol{M} \cdot \boldsymbol{X}_i = 0. \tag{14.55}$$

This is called the orthogonal condition of the vibration modes. Likewise, we obtain:

$$X_j^T \cdot K \cdot X_i = 0. \tag{14.56}$$

The modal analysis method solves a partial differentiation equation using the orthogonal condition of the vibration modes. The solution to Eq. (14.29) is:

$$x(t) = \begin{bmatrix} X_1 & X_2 & \cdots & X_i & \cdots & X_n \end{bmatrix} \cdot \begin{Bmatrix} q_1(t) \\ q_2(t) \\ \vdots \\ q_i(t) \\ \vdots \\ q_n(t) \end{Bmatrix}, \tag{14.57}$$

where X_i is the ith vibration mode, and $q_i(t)$ is a time function relating to the ith vibration mode. Suppose that damping matrix C in Eq. (14.30) is proportional for mass matrix M or stiffness matrix K. By substituting Eq. (14.57) into Eq. (14.29) and using the orthogonal condition of the vibration mode mentioned above, the following is obtained:

$$\ddot{q}_i(t) + 2\omega_i h_i q_i(t) + \omega_i^2 q_i(t) = -\frac{X_i^T \cdot M \cdot 1}{X_i^T \cdot M \cdot X_i}\ddot{y}(t). \tag{14.58}$$

In the above equation, h_i is the critical damping constant regarding the ith vibration mode. When the damping matrix C is given, the critical damping constant can be calculated. However, h_i corresponding to each vibration mode is directly determined in general cases. Using masses m_1 to m_n of the individual mass, on the right side of Eq. (14.58) is:

$$-\frac{X_i^T \cdot M \cdot 1}{X_i^T \cdot M \cdot X_i}\ddot{y}(t) = -\frac{\sum_{j=1}^{n} m_j X_{ji}}{\sum_{j=1}^{n} m_j X_{ji}^2}\ddot{y}(t). \tag{14.59}$$

The fraction on the right side is called an excitation function of the ith vibration mode. A vibration mode with a larger excitation function has greater influence on vibration of the multi-masses-springs-dampers model. Equation (14.58) is the same as that of the dynamic response of the single mass-spring-damper model shown in Eq. (14.2). When the input earthquake motion $\ddot{y}(t)$ is given, the time history of $q_i(t)$ can be obtained. Furthermore, substituting $q_i(t)$ from Eq. (14.58) into (14.57) makes it possible to calculate for vector $x(t)$ of the time history of response displacement.

The maximum values of displacement, velocity, acceleration, stress, and strain are necessary for the earthquake-resistance examination of the surface ground and structures. For an example, the approximate maximum response displacement $x_j(t)_{\max}$ at j-th mass point is obtained as follows:

$$x_j(t)_{\max} \fallingdotseq \sqrt{\sum_{i=1}^{n} \{X_{ji}(q_i(t))_{\max}\}^2}, \tag{14.60}$$

where X_{ji} is the i-th vibration mode value at j-th mass point, and $q_i(t)_{\max}$ is the maximum value of the time history $q_i(t)$, which can be evaluated from the given displacement response spectra.

Appendix 14.1: Basic Concept of Finite Element Method

The Finite Element Method (FEM) is often used to examine the seismic stability of the reactor and turbine buildings, and foundation grounds and surrounding slopes. This numerical method was developed to analyze the deformation, stress, and strain of a continuum body. The FEM analysis process is shown in Fig. 14.10.

Step (1) The structures and the ground are divided into finite number of elements. For two-dimensional analysis, triangle and square elements are used. For the three-dimensional analysis, triangular pyramid elements and cube elements are used.

Step (2) The relationship between the external forces acting at nodal points of the elements and the displacements of the nodal points is obtained based on the elastic theory. Because the stress and strain are assumed to be uniform inside the elements in most cases, division into small elements at the points of stress concentration is required.

Step (3) Equilibrium equations are obtained at the nodal point by considering the external force and inertia forces. The number of equilibrium equations is the product of the degree of freedom at the nodal points (displacements in x, y, and z directions, as well as rotation of the nodal points) with the number of the nodal points.

Step (4) Boundary conditions (for example, fixed boundary and forced displacements at the nodal points) are taken into consideration when developing the total vibration equation.

Fig. 14.10 Analysis by the finite element method

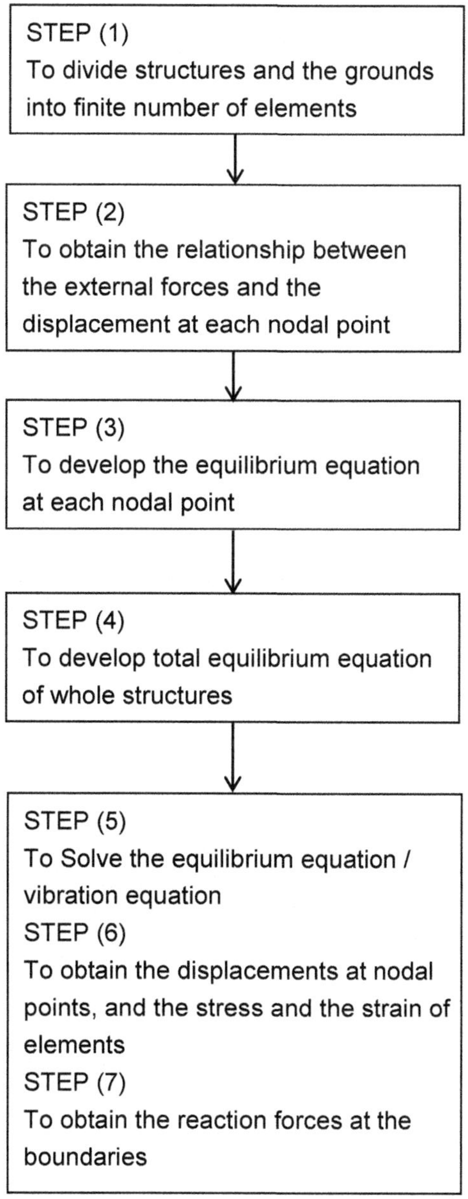

Steps (5), (6), and (7)
For a static problem, the equilibrium equation is solved, and for a dynamic problem, the vibration equation is solved by the numerical method shown in Sect. 14.2. The stress and strain in each element are calculated from the displacement of the nodal points.

Examples of the FEM model for the reinforced-concrete containment vessel, and the underground duct and ground are shown in Figs. 4.16 and 6.6, respectively.

Reference

1. "Engineering for Earthquake Disaster Mitigation", Springer, Masanori Hamada, 2014

Further Readings

1. Shuzo Okamoto "Taishin Kougaku(Earthquake Engineering)", Ohmsha, Ltd,1971
2. Toshiro Miyoshi "Yuugen Youso-hou Nyuumon (Introduction to Finite Element Method)", Baifukan Co.,Ltd.,1978

Chapter 15
Earthquake-Resistant Design

Satsuya Soda, Michiya Kuno and Masanori Hamada

Abstract Static methods and/or dynamic response analysis are applied to the earthquake-resistant design of nuclear facilities, buildings, foundation grounds, and surrounding slopes. In this chapter, the static methods, i.e., the seismic coefficient and modified seismic coefficient methods are described. For the dynamic response analysis of nuclear reactor buildings and other important structures, the methods used to prepare models for analysis, the procedure for evaluating restoring force and damping characteristics are explained. This chapter also explains the method used to examine the seismic safety of buildings and structures by dynamic response analysis as well as by the static method. Furthermore, this chapter introduces the response displacement method for underground structures of nuclear power plants, such as underground ducts for the emergency cooling water supply.

Keywords Seismic coefficient method · Modified seismic coefficient method · Regional coefficient · Time-history dynamic response analysis · Allowable stress · Ductility factor · Ultimate strength · Energy absorption

15.1 Seismic Coefficient Method [1]

In this method, static forces in both the horizontal and vertical directions are taken into consideration for the design of structures, as is the loads of the structures themselves as shown in Fig. 15.1. The horizontal force H and the vertical force V are obtained as product of the structures own weight W and the seismic coefficient K_H in the horizontal direction and K_V in the vertical direction.

$$H = K_H \cdot W \tag{15.1}$$

S. Soda (✉) · M. Hamada
Waseda University, Shinjuku, Tokyo, Japan
e-mail: satsuya.soda@waseda.jp

M. Kuno
Chubu Electric Power Company, Nagoya, Japan

M. Hamada and M. Kuno (eds.), *Earthquake Engineering for Nuclear Facilities*,
DOI 10.1007/978-981-10-2516-7_15

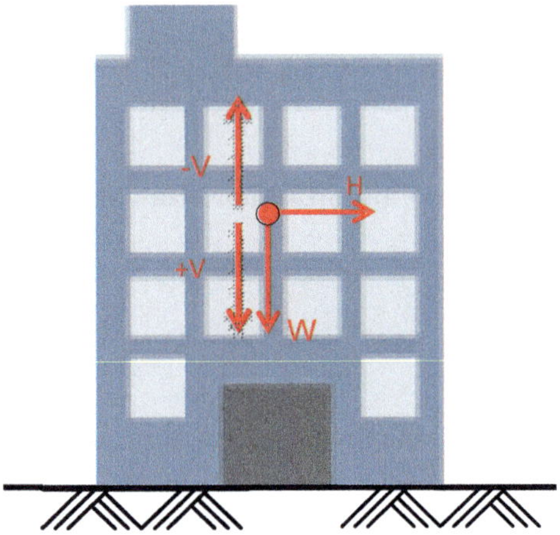

Fig. 15.1 Earthquake-resistant design by seismic coefficient method

$$V = \pm K_V \cdot W \tag{15.2}$$

For example, K_H = 0.1 means that 10 % of the structure's weight is considered in the design for the horizontal external force, and the stability of the structures, and the stress and strain of the structural members are examined.

The magnitude of the seismic coefficients is determined depending on the degree of the importance of structures and the impact of their failure on people and society. The following values are currently used:

Ordinary buildings/bridges ≒ 0.2
Oil and petrochemical products ≒ 0.3 − 0.6
Dams ≒ 0.15
Port and harbor facilities ≒ 0.15 − 0.20
Nuclear facilities ≒ 0.2 − 0.6

The horizontal force H, the inertia force caused by an earthquake that acts on structures, therefore can be expressed as

$$H = \alpha_m \cdot M \tag{15.3}$$

where α_m is the maximum horizontal acceleration on the structures by earthquake ground motion, and M is its the mass. Equation (15.3) can be rewritten as

$$H = \frac{\alpha_m}{g} \cdot Mg \tag{15.4}$$

where g is gravitational acceleration (980 cm/s^2) and Mg the weight W of the structures. From Eqs. (15.3) and (15.4), the following can be obtained:

$$K_H = \frac{\alpha_m}{g} \tag{15.5}$$

In Eq. (15.5), the seismic coefficient K_H is the ratio of the maximum horizontal acceleration of the structures to the gravitational acceleration. For example, $K_H = 0.1$ means that the maximum horizontal acceleration of the structure during an earthquake is approximately 100 cm/s^2.

The downward or upward direction of the vertical force V in Eq. (15.2) is determined by considering which direction is severer when examining the stability and safety of the structures.

In earthquake-resistant design by the seismic coefficient method, the seismic forces statically and constantly act in same direction. The inertia force by earthquake motion is dynamic, repetitively changing its direction with time. Compared with designs in which repetitive loading is considered, the design by a static external force in one direction generally has a large safety margin. Such safety margin in structures designed with static and constant seismic force varies with dynamic characteristics and the failure process of structures. Large shaking-table tests and numerical analyses have been conducted to clarify the failure process of structures, to evaluate safety margins. A three-dimensional, actual-size experimental facility (E-Defense, Part I, Chap. 5, Sect. 5.6) was constructed for this purpose, after the 1995 Kobe earthquake [2].

15.2 Modified Seismic Coefficient Method

The dynamic response of structures depends on the natural periods of structures and the dominant periods of the earthquake ground motion as the input. When natural periods of structures are close to dominant periods of earthquake ground motion, dynamic response of those structures is amplified. However, the external forces obtained by the seismic coefficient method are always constant without any relationship to the amplification of response acceleration. Therefore the modified seismic coefficient method is proposed to solve this limitation. In the modified method the seismic intensity coefficients, K_H and K_V vary depending on the natural periods of structures. Figure 15.2 depicts the K_H in Specifications for Highway Bridges and Explanation Part V, Seismic Design [2]. The horizontal axis indicates the natural period of bridges, and the modified horizontal seismic coefficient K_H is determined depending on natural periods of bridges and subsurface ground. The subsurface ground conditions are classified into three categories based on the natural period of the ground as shown in Chap. 13, Sect. 13.4. The external force in the horizontal direction H for the design can be obtained as follows:

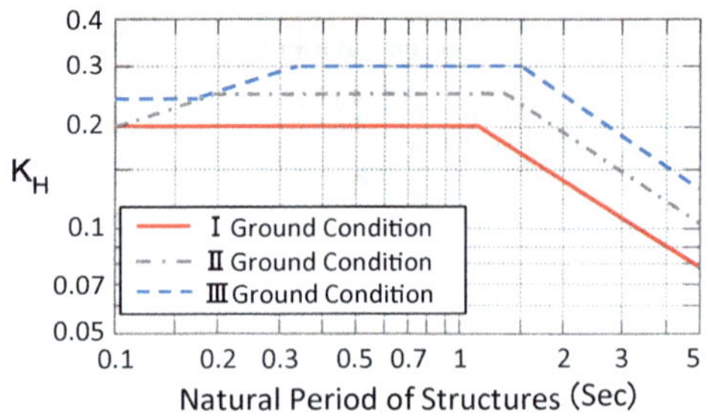

Fig. 15.2 Seismic coefficients for the modified seismic coefficient method [2]

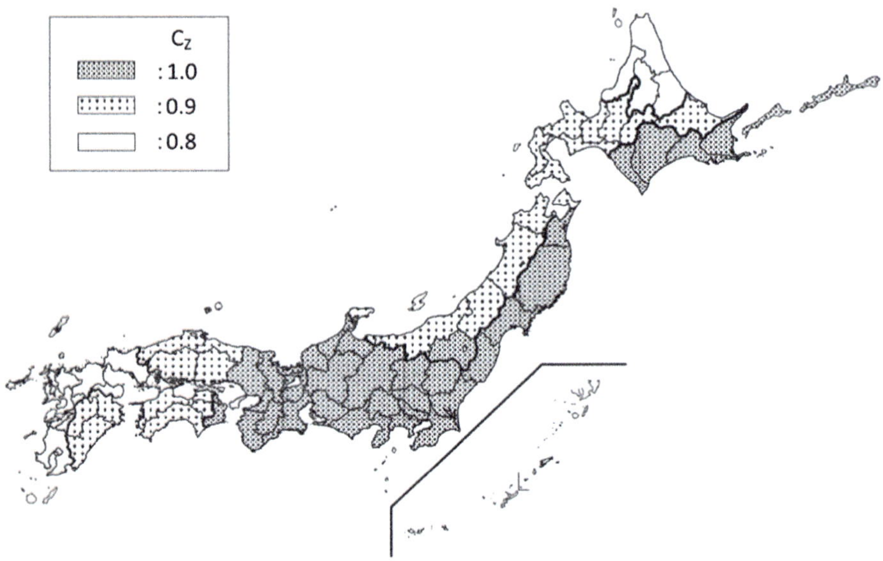

Fig. 15.3 Regional coefficients for the modification of seismic coefficients [2]

$$H = C_z \cdot K_H \cdot W, \tag{15.6}$$

where C_z is the regional coefficient depending on the seismic activity on the region as shown in Fig. 15.3; values of 0.8–1.0 are adopted. The horizontal seismic coefficient K_H is relatively large for periods between 0.3 and 1.5 s under soil condition III (soft ground). This follows from the fact that earthquake ground motions on soft ground are generally dominant in these periods.

15.3 Time-History Dynamic Response Analysis

15.3.1 Purpose of Dynamic Analysis

As described in Part II, Chap. 14, time-history dynamic response analysis is conducted by of expressing the dynamic response of the structure as an equation of motion (differential equation), and determining the response at every moment by means of integral computation. This is believed to be the most exact method for evaluating the dynamic response of structures. Time-history dynamic response analysis is not always necessary in evaluating the safety of normal buildings and structures. However, it must be performed for skyscrapers having a height of 60 m or more, for important buildings and structures of nuclear power plants.

In recent years, structures employing seismic isolation and/or damping system have become increasingly popular. However, those buildings are susceptible to a currently insufficiently characterized seismic ground motion with long-period vibration component. Therefore, it is necessary to evaluate safety by means of time-history dynamic response analysis.

Time-history dynamic analysis requires preparing a structure model and defining the input seismic motion. It is important that safety is judged based on the comprehensive evaluation of displacement, velocity, acceleration, stress, and deformation estimated by time-history dynamic response analysis.

15.3.2 Models for Dynamic Response Analysis

(1) Mass Point Configuration

Response analysis is performed using a discrete model representing the structure by a multi-masses-springs-dampers model. A building is formed of components, such as beams, columns, walls, and floors. Depending on the building design and construction plan, these components may be combined three-dimensionally in a complex geometry. Considering this, a three-dimensional model like the one in Fig. 15.4 is ideal. However, if the dynamic characteristics in two directions are mostly independent of the dynamic characteristic in the other direction, it is acceptable to use a two-dimensional planar model like the ones in Fig. 15.5 to individually address different plane frames.

In this case, the deformation of beams and floors is unlikely to have a significant influence on the overall response of the building. Therefore, it is acceptable to use one-dimensional model as shown in Fig. 15.6. If the ground is stiff enough to ignore the effect of dynamic force interaction between the foundation and ground, the dynamic response analysis may be conducted using a model of the superstructure only, like the one shown in Fig. 15.6a. If the interaction cannot be disregarded, performing dynamic analysis using a soil–structure interaction model shown in Fig. 15.6b is preferable.

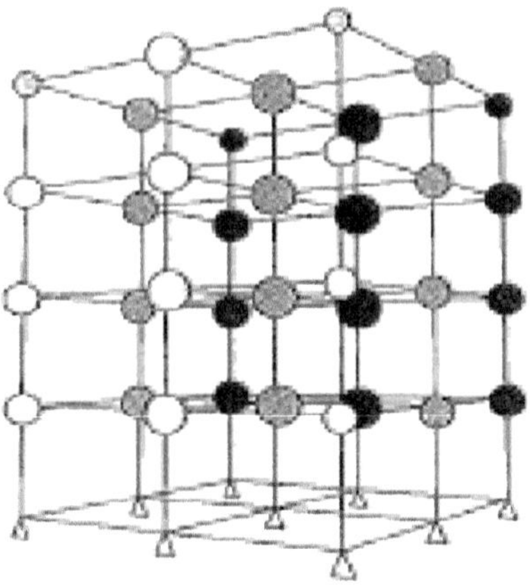

Fig. 15.4 Three-dimensional model [3]

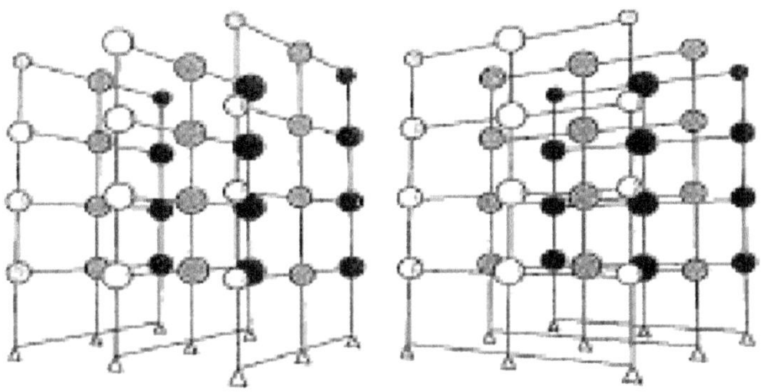

Fig. 15.5 Two-dimensional model [3]

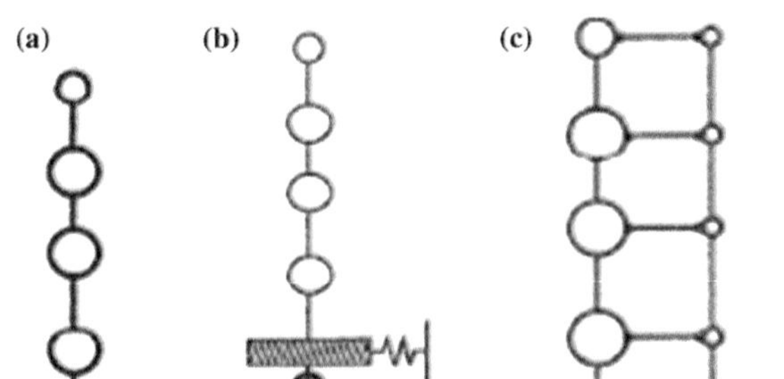

Fig. 15.6 One-dimensional model **a** Stiff foundation mode, **b** Structure–ground interaction model, **c** Bending-shear model [3]

The deformation of the entire building can be complicated because there are various mechanical characteristics involved. In many cases of bending moment frame structures, deformation of the building in the horizontal direction, resulting from the horizontal deformation of columns, is dominant. This may be represented accurately enough by springs deformable in the horizontal direction only (referred to as shear springs). A vertically continuous wall (multi-storey shear wall), like a cantilever beam, allows the wall to bend, which contributes significantly to bending of the entire building, making it impossible to accurately perform response analysis using a shear model. In this case, it is acceptable to use a bending-shear model like the one in Fig. 15.6c, which is configured by adding parallel bending model to the shear model.

In the case of a building with two- or three-dimensional irregularities, the point of application of seismic force (the center of mass) of each story may differ from the point of concentrated resistance (the center of stiffness), which can result in two-dimensional rotational deformation of the entire building. This requires attention given the fact that with such a rotational deformation, destruction may start at the external frames before spreading to the internal frames. If there is a significant risk of such destruction, it is advisable to employ a three-dimensional model.

As described above, the choice of modeling approach depends mostly on the required accuracy of the analytical results. In the designing of nuclear power facilities, analyses are performed using models that are appropriately prepared in consideration of the particularities of the given structures.

(2) Restoring Force Characteristics

Large earthquake ground motion may cause plastic deformation to different parts of structures, leading to nonlinearity in the relationship between deformation and external force. The load–deformation relationship exhibited by a structure across the elastic and plastic regions is referred to as either the load–deformation relationship or restoring force characteristics.

Differences in the type of structure and the configuration result in a wide variety of dynamic characteristics. The restoring force characteristics of each storey in a building model for dynamic response analysis can often be represented by two or three lines approximating a load–displacement relationship derived by means of push-over analysis, which determines the stress and deformation at gradually incremented load levels. The load–displacement relationship derived by a load increment analysis, however, assumes the application of static load. The load–displacement relationship actually produced by dynamic ground motion, which acts alternatingly in opposite directions, must be redefined based on the results of either experiments or detailed analyses.

Figure 15.7 shows an outline of the load–displacement relationship. The thin lines representing cyclic deformation are referred to as hysteresis loops. The bold broken line enveloping the hysteresis loops is referred to as the envelope curve. The load–displacement relationship of structures can be represented using its stiffness, strength, and deformability as parameters. Because the load–displacement

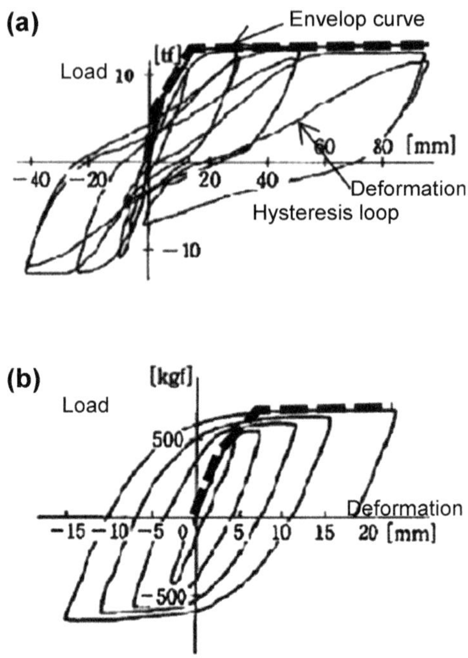

Fig. 15.7 Envelope curve and hysteresis loops **a** Frame structure of reinforced concrete **b** Steel frame structure [3]

relationship may vary greatly between the types of structures, accurate modeling is necessary.

(3) Damping

Damping has a significant influence on the dynamic response of a building. The building exhibits two types of damping effects: the damping effect of the building itself and the damping effect by installed dampers. The origin of the first kind of damping effect has not been determined, but it can be approximated assuming a proportional relationship with the relative velocity of each storey. In the context of analysis, therefore, the damping can be represented by a dash pot inserted between mass points in the structure model.

In the case of a multi-masses-springs-dampers model of buildings, the damping coefficient matrix is often approximated based on a proportional relationship with either the stiffness or the mass matrix. This should be done carefully because assuming a proportional relationship between damping and stiffness may result in overestimating the damping of higher order mode vibration.

Moreover, as plasticity increases, the damping effect tends to be overestimated and the response is underestimated. If the stiffness is expected to be low in the plastic range, allowing significant plastic deformation, the damping matrix may be defined based on a proportional relationship with the tangential stiffness, which corresponds to the inclination of a line tangent to the load–displacement curve. This approach, however, tends to underestimate the damping effect.

15.4 Seismic Safety Examination Without Dynamic Response Analysis

Because simplified methods of seismic safety examinations without dynamic response analysis are capable of accounting for dynamic effects, the current Building Standards Law of Japan and related laws and regulations allow the use of the following three methods: allowable stress calculation, ultimate strength calculation, and energy calculation. Structural calculations for normal buildings, excluding special constructions like skyscrapers, are performed by these methods. Other equally or more capable seismic safety verification methods may be added in the future subject to approval by the Ministry of Land, Infrastructure, Transport and Tourism (MLIT).

15.4.1 Allowable Stress Method

When applied to normal buildings, the allowable stress method is implemented in two stages. With any type of structure, the first stage (referred to as the primary design process) is to calculate allowable stress, while the secondary design process is to calculate deformation and the horizontal load bearing capacity. The horizontal load bearing capacity is a measure of the building's capability to withstand a horizontal load from seismic motion, and is defined as the sum of the horizontal shearing loads borne by columns, structural walls, and bracings at different story levels. This calculation method is simpler than time-history dynamic response analysis regarding the following two processes.

(1) Determination of Seismic Force

In the allowable stress calculation method, calculation formulas are provided based on engineered findings, without depending on dynamic response analysis. Specifically, as long as the assumed intensity of ground motion allows the building to remain in the elastic region (the stress appearing at each story level within the allowable stress level), the shearing loads acting on the building at each storey level $(Qud \cdot i)$ is determined by Eqs. (15.7) and (15.8)

$$Q_{ud,i} = C_i \, W_i \tag{15.7}$$

$$C_i = ZR_tA_i \, C_0 \tag{15.8}$$

where W_i represents the load arising at the story level i from the building weight that exists above the given level. This is multiplied by C_i from Eq. (15.8) to give the design basis shear force at the given story level. The coefficient C_i is referred to as the story shear force coefficient. It is calculated as the product of the zone factor (Z), the coefficient of amplification by the ground (R_t) the coefficient of seismic force distribution across the building height (A_i), and the coefficient of standard shear force from medium seismic force (C_0: 0.2 or larger).

(2) Determination of Structural Characteristics

A_i in Eq. (15.8) is given by Eq. (15.9) as a coefficient determined by the building's natural period and the distribution of mass across the building height.

$$A_i = 1 + \left(\frac{1}{\sqrt{a_i}} - a_i \right) \cdot \frac{2T}{1 + 3T}, \tag{15.9}$$

where a_i is the ratio of weight supported by the story level i to total building weight.

The natural period of a building T can be estimated as follows:

$$T = (0.02 + 0.01\alpha) \times H, \tag{15.10}$$

where H is the building height in meters and α is the ratio of the height of building stories having a main load bearing structure of either wooden or steel construction to the total building height H. In the case of a building fully employing a reinforced concrete construction, the value of α is zero. In the case of a steel structure, the value of α is 1.0.

15.4.2 Primary Design Process

In this process, i.e., the verification of safety by means of allowable stress calculations, it is necessary to account for the combination of stress from stationary loads and various temporary loads (seismic load, wind load, etc.) that can act simultaneously pursuant to the provisions in Article 82 of the Enforcement Ordinance for the Building Standards Law (EOBSL) in Japan.

The design strength of different materials, which should serve as the criteria for safety, are prescribed by a notification from the MLIT. The allowable stress level is specified in the EOBSL as a ratio, dependent upon the type of stress, to the design strength value.

15.4.3 Secondary Design Process

In the initial stage of the development of earthquake-resistant codes in Japan, seismic force was considerably smaller than the level of seismic force actually loaded when based on recent findings. However, many of those buildings have survived past strong earthquakes without collapsing. This is believed to have resulted from the practice of adding a safety margin when designing the strength of structural components, and also the fact that most structural members do not always collapse immediately after being stressed beyond the elasticity limit but still retain some capability to deform and absorb energy.

The secondary design process, prescribed by the Building Standards Law in Japan, is based on the idea that the building's capability to safely withstand strong earthquakes is dependent upon "the capability to balance out the seismic energy input by an earthquake, which is the accumulated energy absorption achieved through elastic and plastic deformation." Accordingly, the process consists of the quantitative evaluation of the strength (load bearing capacity) required for absorbing the seismic input energy through deformation. With reference to Fig. 15.8, this can be explained as follows. The vertical axis and the horizontal axis show the shear force and the deformation of a building, respectively; the area inside triangle OCD represents the seismic input energy that would respond to the maximum deformation if the building could remain elastic up to that point B. In reality, however, the building is still considered safe if it can safely allow deformation up to point E, a point plotted with a view to making the area inside trapezoid OABE equal to the area inside triangle OCD, even after the building reaches the limit of load bearing capacity at point A.

In Fig. 15.8, Q_{er} represents the response shear force calculated assuming the elasticity of the building, while Q_{ur} represents the shear capacity that the building will maintain as long as its deformability permits. μ is a dimensionless quantity expressed as $\mu = \delta_{ur}/\delta_{er}, \cdot$ *i.e.*, the ratio of the maximum displacement δ_{ur} of the

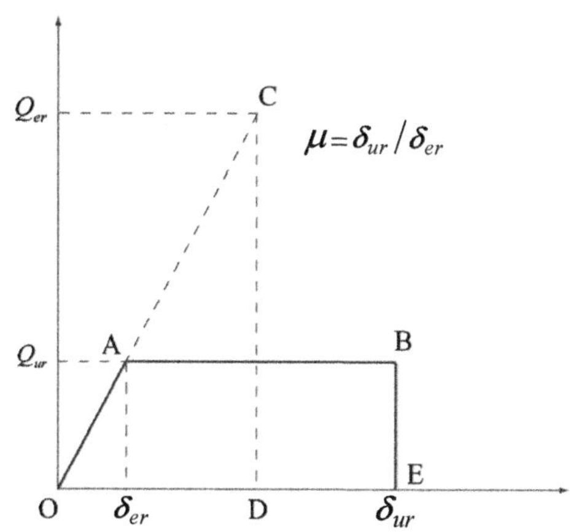

Fig. 15.8 Relationship between shear force and deformation of buildings

building to the displacement at the elasticity limit δ_{er} is referred to as the ductility factor. The ratio of Q_{ur} to Q_{er} is referred to as the coefficient of structural characteristics D_s.

Equation (15.11) equates the energy absorbed by the structure, assuming the elasticity of the structure, with the energy absorbed by the structure with plastic deformation, and relates it to the ductility factor. Because energy is also absorbed by viscous damping, a correction factor β, given by Eq. (15.12) that accounts for the damping factor h, appears in Eq. (15.11). The higher the deformability (the greater the ultimate value of μ at the collapse of the building), and the greater the value of the damping factor, the more likely it is that the building can manage with lower load bearing capacity.

$$D_s = \frac{\beta}{\sqrt{2\mu - 1}}, \tag{15.11}$$

$$\beta = \frac{1.5}{1 + 10h}, \tag{15.12}$$

The coefficient of structural characteristics is determined either by following a legally prescribed procedure for the evaluation of deformability of various structural components or with reference to the load–displacement relationship ascertained for the building based on experiments, etc. Using the coefficient of structural characteristics D_s thus determined, the following expressions are used to calculate the value of Q_{un}, the horizontal load bearing capacity (the required horizontal load bearing capacity), for each story level as a quantity dependent upon the deformability of the building. (Note that the subscript i, representing the story level, is omitted hereafter except with A_i)

$$Q_{un} = D_s F_{es} Q_{ud}, \tag{15.13}$$

$$Q_{ud} = Z R_t A_i C_0 W, \tag{15.14}$$

where C_0 is the coefficient of standard shearing force, the value of which should be ≥ 1.0. In Eq. (15.14), Z is the zone factor concerning the intensity of ground motion. R_t is the coefficient of ground motion amplification by the ground. A_i is an amplification coefficient used to allow the consideration of greater responses at higher story levels. Provided that the standard value of 1.0 is set to all these coefficients including C_0, Q_{ud} equals the building weight W.

F_{es} in Eq. (15.13) is the geometry factor, which allows the consideration that some parts of the building may become more vulnerable as result of the adverse conditions produced by the vertical and/or planar irregular distribution of story-specific stiffness. F_{es} is the product of two independent coefficients as shown below

$$F_{es} = F_e \times F_s, \tag{15.15}$$

where F_s is a measure of irregularity in the vertical distribution of story-specific stiffness, and is determined based on the stiffness ratio (R_s), which is the ratio of stiffness at the given story level to the average stiffness of the building (average of all stories). To prevent the deformation from concentrating in the story levels with extremely low stiffness, as shown in Fig. 15.9, F_s should be >1.0 in any story level having a value of $R_s < 0.6$.

F_e, however, is the coefficient used to discourage any structural design that may impair safety by a high eccentricity ratio (R_e in Fig. 15.10), a measure of torsion that may result from eccentricity in the two-dimensional distribution of either stiffness or mass in the building structure. Higher R_e causes greater deformation to a less stiff plane of the building structure. If the value of the eccentricity ratio R_e

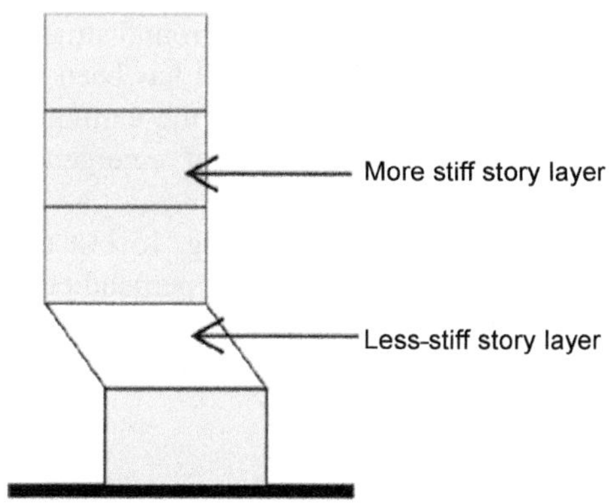

Fig. 15.9 Stiffness ratio

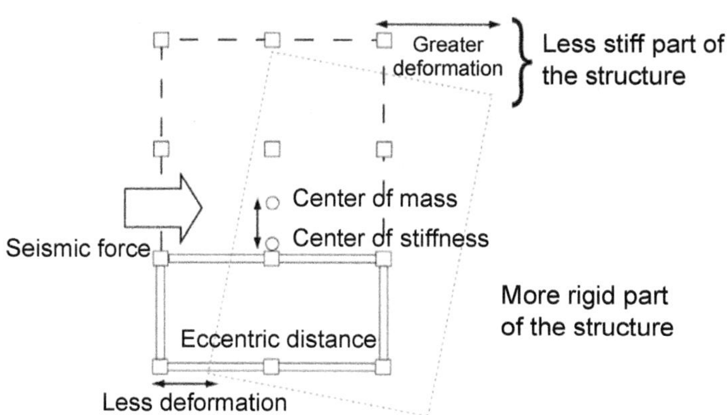

Fig. 15.10 Eccentricity ratio

is ≥ 0.15, the load bearing capacity requirement for the given story level should be increased by a factor of 1.5 at the maximum.

The horizontal bearing load maintained by the building must exceed the required level calculated above. In the case of nuclear power facilities, which are important from the viewpoint of seismic safety, the load bearing capacity should be greater than the required level by a factor of at least 1.5.

15.5 Response Displacement Method

15.5.1 Concept of Response Displacement Method

The dynamic response of underground structures during earthquakes, such as buried pipes and tunnels, is governed by the displacement of the surrounding ground, not by the inertia force as it is in aboveground structures, such as buildings and bridges. The response displacement method has been proposed based on this observational result. The method is applied to the earthquake-resistant design of underground ducts for the seawater intakes of emergency cooling systems at nuclear power plants.

The concept of this method is shown in Fig. 15.11a using the example of a buried pipe. Here, $u_g(x)$ is ground displacement perpendicular to the pipe axis at a given time, and x is a coordinate indicating the axial direction. Deformation of the pipe $u_p(x)$ when subjected to ground displacement depends on the following

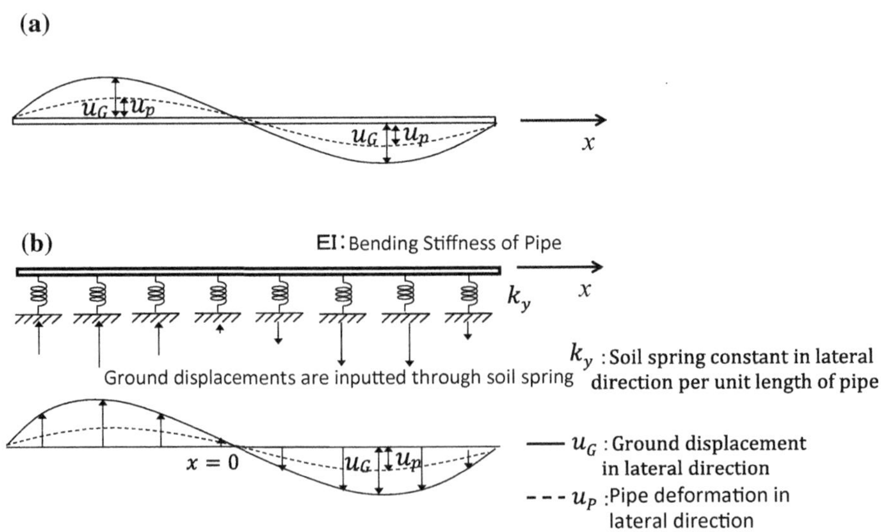

Fig. 15.11 Response displacement method and beam model on an elastic foundation. **a** Ground displacement and pipe deformation. **b** A beam model on an elastic foundation

relationships. When the buried pipe is very stiff and ground stiffness is low, the displacement of the pipes $u_p(x)$ will be small. If pipe stiffness is very low, but the ground stiffness is high: $u_p(x)$ approaches ground displacement $u_g(x)$.

15.5.2 Deformation Analysis of Buried Pipe

A model of a beam on an elastic foundation (Fig. 15.11b) can be used to appropriately display the relationship between ground displacement and buried pipe deformation. The pipe is modeled as a beam with bending stiffness and the ground as a spring. The soil spring constant is determined by the stiffness of the surrounding ground. Ground displacement $u_g(x)$ is inputted to the pipe via the soil spring, and the displacement of the pipe is numerically obtained by taking the bending stiffness of the pipe and the rigidity of the ground. Deformation $u_p(x)$ can be derived from

$$\mathrm{EI}\frac{\mathrm{d}^4 u_P}{\mathrm{d}x^4} + k_y u_P = k_y u_G, \tag{15.16}$$

where EI is the pipe bending stiffness, and k_y is the soil spring constant perpendicular to the pipe axis per unit length of the pipe.

That equation can be modified as

$$\frac{\mathrm{d}^4 u_P}{\mathrm{d}x^4} + 4\beta_y^4 u_P = 4\beta_y^4 u_G, \tag{15.17}$$

where

$$\beta_y = \sqrt{4}\frac{k_y}{4EI} \tag{15.18}$$

The general solution to Eq. (15.17) can be obtained from

$$u_P(x) = \exp(\beta_y x)(C_1 \cos\beta_y x + C_2 \sin\beta_y x) + \exp(-\beta_y x)(C_3 \cos\beta_y x + C_4 \sin\beta_y x) \tag{15.19}$$

C_1–C_4 are integral constants derived from boundary conditions.

Taking ground displacement $u_G(x)$ as a sinusoidal wave of wave length L and amplitude $\bar{u}_G$:

$$u_G(x) = \bar{u}_G \cdot \sin\frac{2\bar{u}}{L}x. \tag{15.20}$$

A special solution to $u_p(x)$ can be obtained as follows:

$$u_P(x) = \frac{4\beta_y^4}{4\beta_y^4 + \left(\frac{2\pi}{L}\right)^4}\bar{u}_G \cdot \sin\frac{2\pi}{L}x \tag{15.21}$$

Bending moment $M(x)$ generated in the buried pipes is obtained from

$$\begin{aligned} M(x) &= -\mathrm{EI}\frac{\mathrm{d}^2 u_P}{\mathrm{d}x^2} \\ &= EI\left(\frac{2\pi}{L}\right)^2 \frac{4\beta_y^4}{4\beta_y^4 + \left(\frac{2\pi}{L}\right)^4}\bar{u}_G \cdot \sin\frac{2\pi}{L}x \end{aligned} \tag{15.22}$$

Similarly, deformation $v_P(x)$ in the axial direction of the pipe is modeled by a bar on an elastic foundation, as shown in Fig. 15.12; the deformation can be obtained by the following equation:

$$\mathrm{EA}\frac{\mathrm{d}^2 v_P}{\mathrm{d}x^2} - k_x v_P = -k_x v_G. \tag{15.23}$$

where $v_G(x)$ is ground displacement in the direction of the pipe axis, EA is pipe stiffness with regard to axial deformation, and k_x is the soil spring constant in the axial direction per unit length of the pipe. Equation (15.23) can be modified as

$$\frac{\mathrm{d}^2 v_P}{\mathrm{d}x^2} - \beta_x^2 v_P = -\beta_x^2 v_G, \tag{15.24}$$

$$\beta_x = \sqrt{\frac{k_x}{\mathrm{EA}}}. \tag{15.25}$$

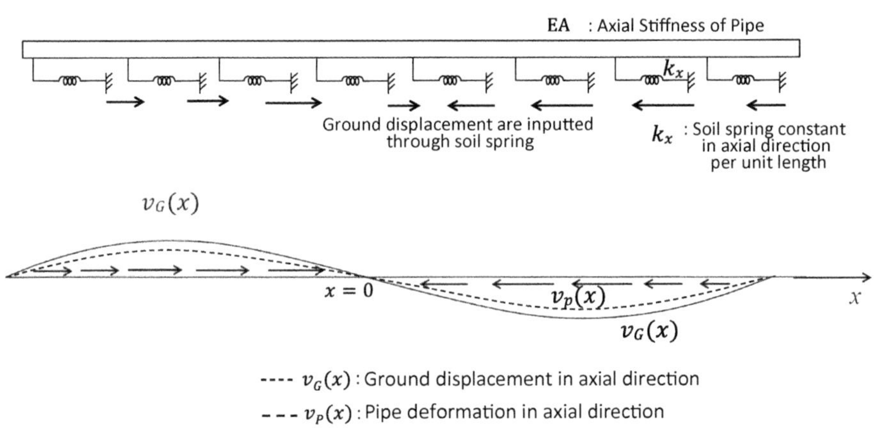

Fig. 15.12 Model to numerically analyze the deformation of a buried pipe in the axial direction

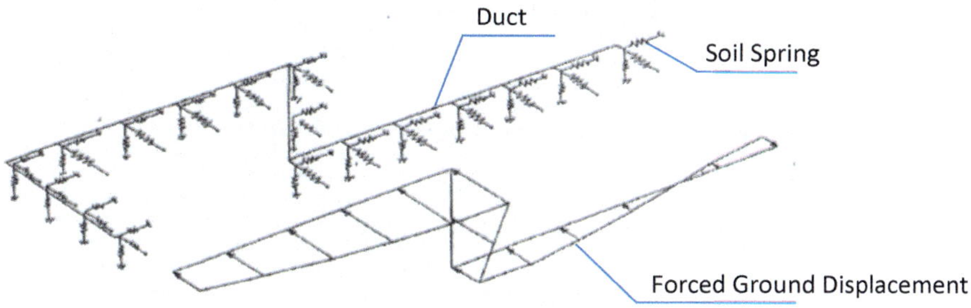

Fig. 15.13 Model of underground duct by the response displacement method

The general solution to the above equation can be obtained as follows:

$$v_P(x) = C_1 \exp(-\beta_x x) + C_2 \exp(\beta_x x) \quad (15.26)$$

Taking ground displacement $v_G(x)$ as a sinusoidal wave form of wave length L and displacement amplitude $\bar{v}_G$, a special solution to $v_P(x)$ can be determined by

$$v_P(x) = \frac{\beta_x^2}{\beta_x^2 + \left(\frac{2\pi}{L}\right)^2} \bar{v}_G \cdot \sin \frac{2\pi}{L} x \quad (15.27)$$

The axial force $N(x)$ as a result of deformation in the axial direction of the buried pipe can be expressed as

$$N(x) = EA\left(\frac{2\pi}{L}\right) \frac{\beta_x^2}{\left(\frac{2\pi}{L}\right)^2 + \beta_x^2} \bar{v}_G \cdot \cos \frac{2\pi}{L} x \quad (15.28)$$

Figure 15.13 depicts a model for underground ducts of three-dimensional distribution. The ducts are modeled as beams and bars on the elastic foundation of the ground. The input ground displacement is calculated taking either earthquake wave propagation or surface ground response into account.

References

1. "Engineering for Earthquake Disaster Mitigation", Springer, Masanori Hamada, 2014
2. "Design Specifications for Highway Bridges Part V Seismic Design", Japan Road Association, 2002
3. "Kenchiku Shindou (Dynamic Vibration of Buildings Applications)", Asakura Publishing Co, Ltd. 2008